Alpine Ecosystems in the Northwest Caucasus

Geobotany 29

Series Editor

M.J.A. WERGER

The titles published in this series are listed at the end of this volume.

Alpine Ecosystems in the Northwest Caucasus

Edited by

Vladimir G. Onipchenko
Department of Geobotany,
Moscow State University, Moscow, Russia

Kluwer Academic Publishers
Dordrecht / Boston / London

A C.I.P. Catalogue record for this book is available from the Library of Congress

ISBN 1-4020-2382-0 (HB)
ISBN 1-4020-2383-9 (e-book)

Published by Kluwer Academic Publishers,
P.O. Box 17, 3300 AA Dordrecht, The Netherlands.

Sold and distributed in North, Central and South America
by Kluwer Academic Publishers,
101 Philip Drive, Norwell, MA 02061, U.S.A.

In all other countries, sold and distributed
by Kluwer Academic Publishers,
P.O. Box 322, 3300 AH Dordrecht, The Netherlands.

Printed on acid-free paper

Printed in the Netherlands.

To the memory of Professor Tikhon A.Rabotnov

Contents

Contributors

Aleksandra A. **Aksenova** – Department of Geobotany, Biological Faculty, Moscow State University, 119992 Moscow, Russia. Tel. (+7 095) 939 43 10

Ol'ga V. **Blinkova** – Department of Geobotany, Biological Faculty, Moscow State University, 119992 Moscow, Russia.

Mikhail S. **Blinnikov** – Department of Geography, St. Cloud State University, 720 4th Ave. South, St. Cloud, MN 56301-4498, tel. (320) 255 22 63, e-mail: mblinnikov@stcloudstate.edu

Irina V. **Bulatnikova** – Department of General Soil Sciences, Faculty of Soil Sciences, Moscow State University, 119992 Moscow, Russia.

Oksana V. **Cherednichenko** – Department of Geobotany, Biological Faculty, Moscow State University, 119992 Moscow, Russia. Tel. (+7 095) 939 43 10

Alexander N. **Dobrolyubov** – Zapovednik Privolgskaja Lesostep, Ul. Fedotova 21, 440003 Penza, Russia, Tel. Private: (+7 841) 236 29 45, e-mail: zapoved@penza.com.ru

Sergei V. **Fomin** – Department of Vertebrate Zoology, Biological Faculty, Moscow State University, 119992 Moscow, Russia. Tel. (+7 095) 939 50 11

Maria A. **Gerasimova** - Department of Geobotany, Biological Faculty, Moscow State University, 119992 Moscow, Russia. Tel. (+7 095) 939 43 10

Kirill A. **Golikov** – Botanical Garden, Biological Faculty, Moscow State University, 119992 Moscow, Russia.

Margarita N. **Ivanenko** - Department of Geobotany, Biological Faculty, Moscow State University, 119992 Moscow, Russia.

Aleksander S. **Komarov** - Laboratory of Ecosystem Modelling, Institute of Physico-Chemical and Biological Problems in Soil Science of Russian Academy of Sciences, Pushchino, Moscow Region, 142290 Russia. Tel. (+7 096) 773 07 55, e-mail: komarov@issp.serpukhov.su

Tat'ayna N. **Leinsoo** – Department of Tropical and Subtropical Plants, Main Botanical Garden of Russian Academy of Sciences, Botanicheskaya ul, 4. Moscow, 127276 Russia.

Mikhail I. **Makarov** – Department of General Soil Sciences, Faculty of Soil Sciences, Moscow State University, 119992 Moscow, Russia.

Tat'yana I. **Malysheva** – Department of General Soil Sciences, Faculty of Soil Sciences, Moscow State University, 119992 Moscow, Russia.

Andrei P. **Mikhailenko** – Department of Entomology, Biological Faculty, Moscow State University, 119992 Moscow, Russia.

Aleksei V. **Mikhailov** Laboratory of Ecosystem Modelling, Institute of Physico-Chemical and Biological Problems in Soil Science of Russian Academy of Sciences, Pushchino, Moscow Region, 142290 Russia, Tel. (+7 096) 773 07 55, e-mail: mikhailov@ibbp.psn.ru

Natal'ya V. **Mikhailova –** Pushchino State University, Pushchino, Moscow Region, 142290 Russia. Tel. (+7 096) 773 07 55.

Vladimir G. **Onipchenko** – Department of Geobotany, Biological Faculty, Moscow State University, 119992 Moscow, Russia. Tel. (+7 095) 939 43 10, e-mail: vonipchenko@herba.msu.ru

Vadim N. **Pavlov** – Department of Geobotany, Biological Faculty, Moscow State University, 119992 Moscow, Russia.

Dmitrii A. **Petelin** – Department of Geobotany, Biological Faculty, Moscow State University, 119992 Moscow, Russia.

Leonid **Rasran** – Ecology research Centre, University of Kiel, Schauenburger Str. 112, D-24118 Kiel, Germany, phone: (++49+431) 8804062 or (++49+431)8804294, e-mail: leonid@ecology.uni-kiel.de

Ibragim K. **Semenov** – Karachaevo-Cherkessian Pedagogical University, Faculty of History. Karachaevsk, 369200, Russia

Galina V. **Semenova** - Department of Geobotany, Biological Faculty, Moscow State University, 119992 Moscow, Russia. e-mail: galinasemenova@hotmail.com

Igor' E. **Sizov** – Institute of Mathematical Problems in Biology of Russian Academy of Sciences, Pushchino, Moscow Region, 142290 Russia, Tel. (+7 096) 773 07 55.

Mikhail V. **Sogonov** – Department of Mycology and Algology, Biological Faculty, Moscow State University, 119992 Moscow, Russia.

Alexei L. **Stepanov** – Department of Soil Biology, Faculty of Soil Sciences, Moscow State University, 119992 Moscow, Russia. e-mail: stepanov@soil.msu.ru

Konstantin L. **Tarasov** – Department of Mycology and Algology, Biological Faculty, Moscow State University, 119992 Moscow, Russia.

Lev L. **Velikanov** – Department of Mycology and Algology, Biological Faculty, Moscow State University, 119992 Moscow, Russia.

Alexander V. **Volkov** – Department of General Soil Sciences, Faculty of Soil Sciences, Moscow State University, 119992 Moscow, Russia.

Elena V. **Volkova** – Department of Geobotany, Biological Faculty, Moscow State University, 119992 Moscow, Russia.

Aleksei A. **Zakharov** – Department of Geobotany, Biological Faculty, Moscow State University, 119992 Moscow, Russia.

Acknowledgements

During 20 years more than 250 people took part in our field study. We would like to thank all of them, especially A.Baikalova, A.Kondrashov, A.Kaverina, G.Pokarzhevskaya, A.Sennov, I.Malyshev, O.Zhakova, V.Tsyperson, O.Yurtseva, Y.Ustinova, I.Pavlova, N.Lyubeznova, O.Vertelina, K.Golikov, S.Zenyakin, A.Egorov, D.Sukhova, E.Semenova, O.Starodubtseva, M.Rabotnova, D.Orlov. The work could not have been done without permanent support and help from the administration and staff members of Teberda Reserve. We thank D.S.Salpagarov, N.N.Polivanova, F.M.Vorob'eva, L.G.Tatarnikova, A.Bok, V.Onishchenko, Y.Sarkisyan, G.Bobyr', T.Bagrova, I.Tkachenko, P.Morozov, U.Semenov, and all other staff members of the Reserve for their kind help. A lot of useful comments on the manuscript were given by Prof. A.Mark, Dr. K.Thompson, Dr. M.Blinnikov, Prof. W.Gams, Dr. G.Belyakova. Dr. I.Sizov, A.Zakharov, and A.Akhmetzhanova prepared a lot of figures for our book. We thank Prof. P.Edwards (Geobotanical Institute SFIT, Zurich), Prof. E. van der Maarel (Opulus Press), Prof. T.Herben (Folia Geobotanica), Dr. M.Janiga (Oecologia Montana), G.Pokarzhevskaya and C.Oehring (Arctic, Antarctic, and Alpine Research), B.Fenton and G.Houle (Ecoscience) for their kind permissions to reproduce some tables and figures from published papers. Dr. E.Zubchaninova permitted to reproduce several unpublished pictures of V.Smirin. We are particularly thankful to Prof. M.J.A.Werger for great editorial work on the manuscript.

Our investigations were supported by our teachers, professors of Moscow State University, T.A.Rabotnov and L.A.Grishina.

Our research was supported by many organisations. We thank the Russian Foundation for Fundamental Research (projects 93-04-6498, 96-04-

48420, 99-04-48012, 00-04-63001, 02-04-48060), Russian President Foundation for Supporting Scientific Schools (project 2125.2003.4), Program "Universities of Russia" (project 07.03.010), the John D. and Catherine T. MacArthur Foundation, the Royal Swedish Academy of Sciences, and the Netherlands Organisation for Scientific Research (NWO, N 047.02.012.96).

Preface

The Great Caucasus (Bol'shoi Kavkaz) is a huge mountain area on the border between Europe and Asia. Its rich flora and varied vegetation were (and are!) very attractive for naturalists and plant ecologists. The complicated and dramatic history of the area and the harsh alpine environment are favorable for a relatively good protection of nature. Several large national reserves on the north slopes of the Great Caucasus (North-Osetia, Kabardino-Balkaria, Teberda, and the Caucasian Reserve in the Krasnodar region) were established in the twentieth century.

The Teberda State Reserve was founded in 1936. Since 1950 it has been one of the field practice areas for hundreds of students from several Russian Universities including Moscow State University. Since 1977 a permanent expedition of the Biological Faculty and the Faculty of Soil Science has studied the alpine ecosystems of the Reserve. This collective monograph represents the main results of more than 20 years of the investigations.

Our primary objectives were to describe 1) the general structure of the plant cover in the region, 2) the detailed characteristics of widespread alpine communities which were the objects of long-term ecological researches, 3) the structure and composition of algae communities as well as the main groups of heterotrophic components of the ecosystems (fungi, grasshoppers, mammals), 4) the dynamics of the alpine communities, 5) the results of various ecological experiments which might explain mechanisms of plant coexistence and the maintenance of biological diversity. The studies were done by many scientists and were partly published previously in Russian and international journals. Kluwer Publishers invited us to undertake the present attempt to represent the whole spectrum of our research in one book.

After hot discussions within the authors group we decided to represent our data in a collective monographic style. So, the authors of separate parts (chapters, subchapters etc.) are noted in the contents and their names are not repeated in titles within the text.

V.G. Onipchenko

Chapter 1

Geography, geology, climate and the communities studied

1.1. Geographical position

The present volume deals with the alpine vegetation of the Karachaevo-Cherkessian Republic, which is one of the autonomous republics of the North Caucasus region within the Russian Federation. Our study area included the Teberda State Biosphere Reserve and neighbouring territories.

The Teberda State Biosphere Reserve ("*Teberdinskii Zapovednik*" in Russian) is located on the northern side of the Great Caucasus mountain range. The total reserve's area is about 850 sq. km. The main section of the Reserve occupies the upper reaches of the Teberda river valley between 41° 35' and 41° 55' E, and 43° 13' and 43° 28' N (Fig. 1.1).

1.2. Relief

Elevations inside the Reserve span about 2500 m: from the lowest point at the bottom of the Teberda valley (1259 m a.s.l.) to the highest (Mt. Dombai-Ulgen, 4046 m a.s.l.). Several peaks exceed an altitude of 3800 m (which is the upper climatic limit of vascular plants here), namely Bu-Ulgen (3917 m), Dzhuguturlyuchat (3896 m), Ertsog (3863 m), Belala-Kaya (3861 m), and Kyshkadzher (3822 m). About 83% of the reserve's territory lies above 2000 m a.s.l. and can be considered true alpine area.

The hydrological network is well developed. Overall, there are 130 glacial lakes and 30 rivers. All rivers belong to the Teberda river watershed. Lakes and rivers cover about 0.3% of the reserve area (Polivanova 1990).

V.G. Onipchenko (ed.), Alpine Ecosystems in the Northwest Caucasus, 1–7.

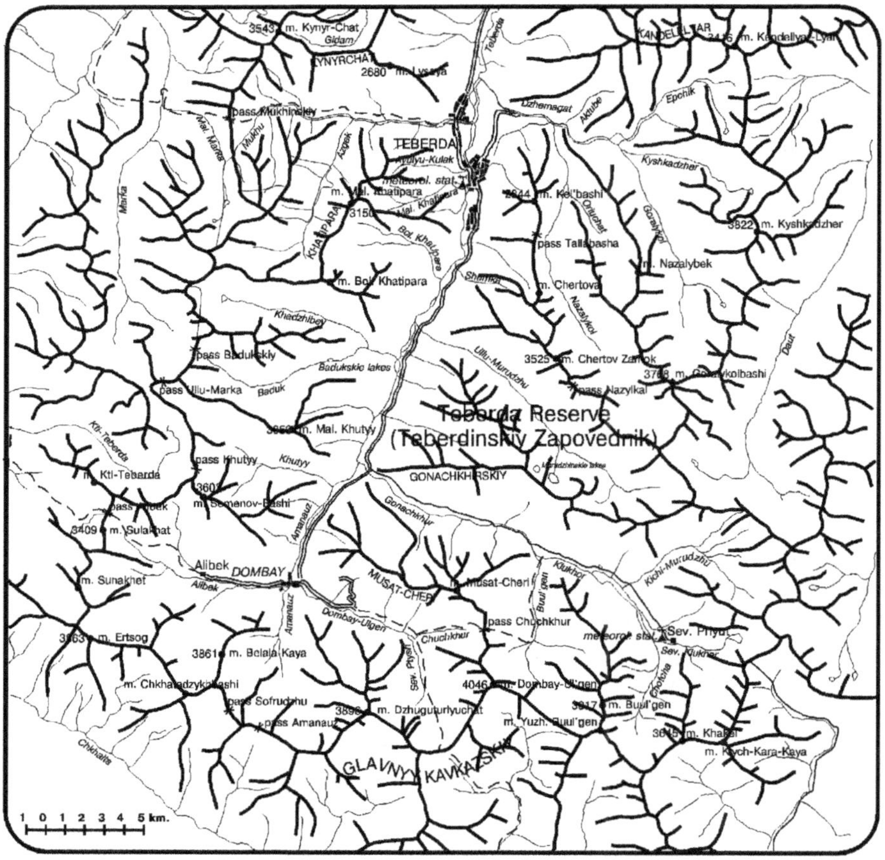

Figure 1.1. Map of the Teberda Reserve (from Onipchenko 2002 with permission of Geobotanical Institute SFIT)

Three main geological factors shaped the relief: glacial processes (ancient and modern), water erosion, and slope processes, including avalanches, rock falls and creep. There are 85 existing glaciers, occupying about 10% of the reserve area.

The late Pleistocene glaciers covered most of the mountains from their summits down to about 2400 m. The valley glaciers extended further down to 1300 m (elevation of the town of Teberda) and even lower (Shcherbakova 1973). The U-shaped ("trog") form of many valleys indicates their former glaciation. Tushinskii (1957) described four phases of the local glaciation during the Holocene. The glaciers have been steadily retreating since 1850.

The typical alpine glacial landforms are well represented in the southern part of the Reserve near the Great Caucasus Range (*Glavnyi Kavkazskii khrebet*). Pointed peaks, steep slopes and U-valleys are common in the area. Mountains in the northern part of the Reserve are smoother with more gentle slopes, especially in the upper part of the mountains, due to erosion and frost action. Former slope glaciers formed alpine cirques as depressions enclosed by frontal moraine deposits which are currently occupied by small lakes.

Rock and scree outcrops are well represented in all vertical zones, but their role is most important in the alpine and subnival zones. They cover about 26% of the total area of the Reserve.

Permafrost landforms are restricted to several high stony ridges above 3400 m. The distribution of permafrost is, therefore, very limited and permafrost has negligible influence on vegetation.

1.3. Geology

Siliceous rocks are prominent in the Teberda Reserve. Local outcrops of limestone exist two km to the north outside of the Reserve boundary. Four geological zones can be distinguished (Tushinskii 1957):

Zone 1 (northern part of the Reserve) with a predominance of paleozoic chlorite schist and some outcrops of Permian sandstones;

Zone 2 (most of the Reserve) is granitic area with absolute predominance of Precambrian gray granites with outcrops of biotite schists and gneisses;

Zone 3 (Alibek and Dombai valleys) is characterized by schist outcrops;

Zone 4 (the Great Caucasus Range) is formed by gneisses, granites and crystalline schists.

1.4. Climate

There are two permanent state meteorological stations in the Reserve: Teberda and Klukhor Pass (Fig. 1.1). Climatic diagrams of the stations are presented in Fig. 1.2(a, b). The climate of different parts of the Reserve varies considerably due to the two main factors: elevation and distance from the Great Caucasus Range, which forms the southern boundary of the Reserve. Overall, the mean annual temperature decreases by 0.5-0.6 °C per 100 m increase in altitude. Precipitation increases with altitude up to about 2200-2500 m. Moist air masses are brought mainly by south winds from the Black Sea. Thus, precipitation sharply decreases away from the Great Caucasus Range towards the north. South winds are predominant in the area, with the south slopes being windward and north slopes leeward with heavy snow accumulation ("snow beds"). They determine the pattern of the vegetation cover.

A typical climate diagram for the alpine zone in the northern part of the Reserve (Malaya Khatipara temporary meteorological station) is represented in Fig.1.2c. The climate of the alpine zone can be considered as a mountain climate of the temperate zone (type X (VI), according to Walter et al. 1975). The mean annual temperature is about –1.2 C° and the mean annual precipitation is 1,400 mm. These numbers resemble those of some areas in the Swiss Alps, for example, Weissfluhjoch (2667 m a.s.l.) near Davos (Zingg 1961). Air humidity averages about 79% during summer. The average duration of time with 100% air humidity is about 4 hours per day. Insolation at soil level is about one half of the potential because of clouds (Grishina et al. 1986). The warmest month is August with a mean monthly temperature of + 8.3° C, but frost can occur throughout summer.

1.5. Studied communities

Four plant communities were the main objects of our study. Detailed characteristics of their floristic composition, species richness, syntaxonomic position, soil properties, etc. are represented below in several chapters. Here we introduce detailed descriptions in the form of a short sketch of the communities and their respective locations, which they occupy within the study area. The following communities were investigated (their abbreviations are shown in parentheses):

1. Alpine lichen heaths (ALH)
2. *Festuca varia* dominated grasslands (FVG)
3. *Geranium gymnocaulon* - *Hedysarum caucasicum* dominated meadows (GHM)
4. Snow bed communities (SBC)

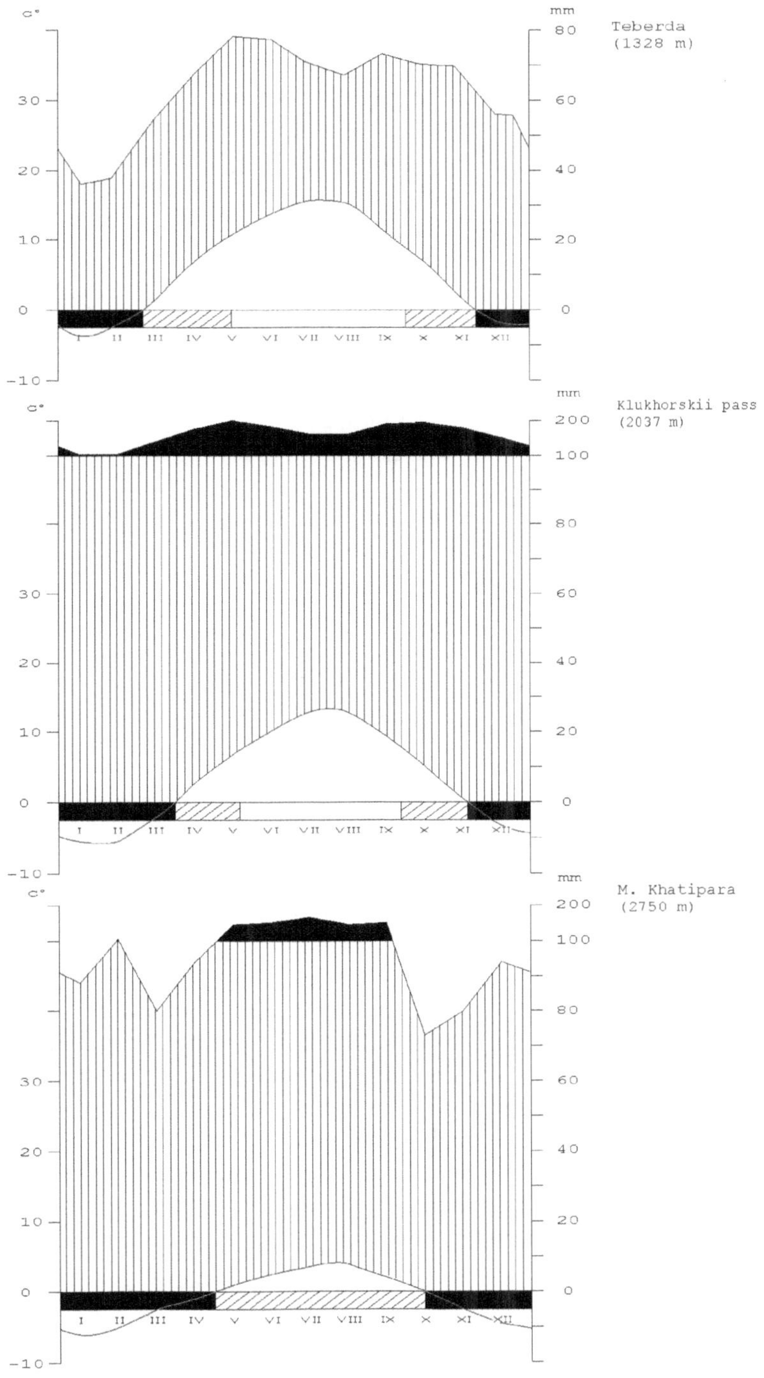

Figure 1.2. Climatic diagrams of meteorological stations in the Teberda Reserve

These communities, because of their landscape positions and resulting snow regimes, represent gradients in both soil water content and temperature. They can be therefore considered as a catena (toposequence) (Fig. 1.3, Table 1.1.). These communities are the most widespread closed alpine communities in the study area.

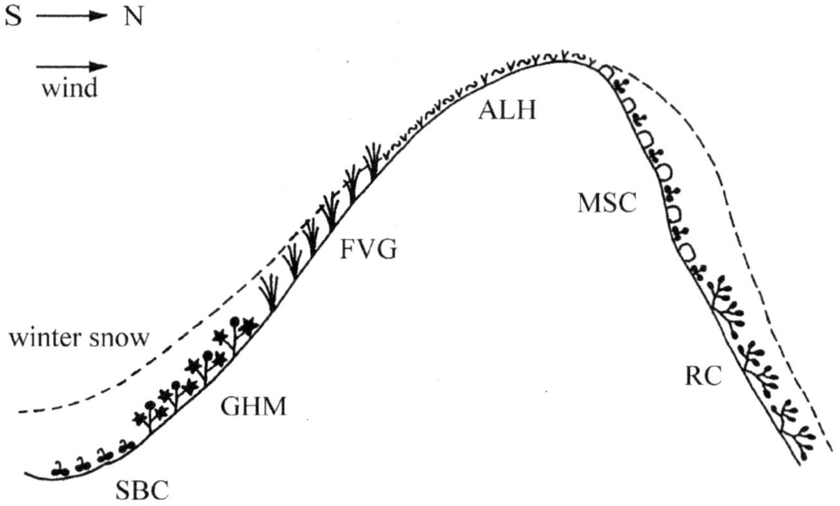

Figure 1.3. Distribution of investigated communities over mesorelief forms. ALH - Alpine lichen heaths, FVG - *Festuca varia* dominated grasslands, GHM - *Geranium gymnocaulon - Hedysarum caucasicum* dominated meadows, SBC – snow bed community, MSC – communities on moist screes, RC – *Rhododendron caucasicum* community (from Onipchenko 1994a with permission of Geobotanical Institute SFIT)

ALH occupy windward crests and slopes with little (up to 20-30 cm) or no snow accumulation in the winter. Deep freezing is typical for the soils there. The growing season lasts about 5 months (from May to September). Fruticose lichens are the main dominants (mostly *Cetraria islandica*) (see chapter 4.1. for nomenclature sources).

FVG are firm-bunch grass communities. *Festuca varia* and *Nardus stricta* are the main dominants. The communities occupy slopes with little snow accumulation (about 0.5 – 1 m). Snow cover stays until the second half of May or the first half of June, with the growing season lasting about 4 months.

Table 1.1. General characteristics of the studied communities on Mt. Malaya Khatipara (from Onipchenko 1994a with permission of Geobotanical Institute SFIT)

Community	ALH	FVG	GHM	SBC
Occupied relief forms	crests, ridges, windward slopes	Slopes (often steep)	leeward slopes, small depressions	deep depressions
Depth of snow cover at the end of winter (m)	0-0.3	0.5-1.5	1.5-3	4 and more
Duration of vegetative season, months	4.5-5.5	3.5-4.5	2.5-3.5	2-2.5
Exposition (aspect)	S	SSW	SW	SW
Steepness, (°)	15	10	5	1

GHM occupy the lower portions of slopes and small depressions with typical snow depths of 2-3 m. They become snow free by the end of June or by early July, and their growing season lasts for only 2.5 – 3.5 months. *Geranium gymnocaulon* and *Hedysarum caucasicum* are dominants and typical species of this community type. The population density of voles [*Pitymys (Microtus) majori* Thomas (Glires, Microtinae)] can reach 940 animals per hectare during a "peak-year" in these communities (Fomin et al. 1989, Chapter 10). The animals can consume approximately 15% of the annual net-biomass production and dig up to 14,300 holes per ha in such years (See chapter 10). Severe disturbances are thus typical in this community due to the burrowing activity of the voles.

SBC occupy depressions and bottoms of nival and glacial cirques with heavy winter snow accumulation (4 metres or more). They have the shortest vegetative season of about 2 – 2.5 months from the end of July until September. Short rosette and dwarf trailing plants (*Sibbaldia procumbens, Minuartia aizoides, Gnaphalium supinum, Taraxacum stevenii*) are dominants here. In some plots *Sibbaldia procumbens* is an absolute dominant. The height of most plants in SBC is a mere 2-3 cm.

Chapter 2

Soils and nutrient turnover

Mountain soils are highly variable and their properties depend on several abiotic (climate, geology, slope processes etc.) as well as biotic (vegetation, animal and microbial activity etc.) factors. In this chapter we consider soils of the studied area at several spatial scales with the increasing detail: 1) main soil types of the Teberda Reserve, 2) soils of the alpine study area, 3) soils of studied toposequences (catena) of 4 alpine plant communities (alpine lichen heath, *Festuca varia* grassland, *Geranium-Hedysarum* meadow, snowbed community) which are the main objects of experimental research; 4) features of nutrient storage and turnover in alpine communities. Soil variability within the communities is described in Chapter 5.3. In a separate chapter (5.4.) we present the influence of plant species on soil nutrients in monoculture experiment.

2.1. Main soil types of the Teberda Reserve

Soils of the Teberda Reserve were studied by several authors (Antipov-Karataev et al. 1936, Serebryakov 1957, Shalnev et al. 1977, Vladychenskii & Grishina 1987, Grishina et al. 1993, Makarov et al. 1997, Onipchenko et al. 1998b). As a whole, shallow profiles and high stone contents characterize these mountain soils. The stone content may reach 80% in the lower horizons at a depth of 40-50 cm.

V.G. Onipchenko (ed.), Alpine Ecosystems in the Northwest Caucasus, 9–23.

2.1.1. Forest soils

The main type of forest soil of the Reserve is **Cambisol** (according to the FAO Classification System), in particular Humic Cambisol. The soil profile has O, A, B, and C horizons. The organic O horizon, which can be up to 5-10 cm thick, is divided into OL, OF and OH sublayers. Such differentiation is clear in fir forests on northern slopes, but it does not occur in pine forests on southern slopes. The texture of the mineral horizons is silt loam. The B horizon is slightly enriched by clay. The upper border of the C horizon lies at the depth of 50-70 cm. Soils on northern slopes are usually darker due to higher humus content (up to 11% C). Soils on southern slopes contain less humus. They are also characterized by a thinner profile.

Humic Cambisols of the Teberda Reserve are acidic (pH 5.5-6.5). A horizon with pH about 5.5 is considered umbric, otherwise (pH > 6.0) it is mollic. The sum of exchangeable base cations is about 200 - 300 mmol(+) kg^{-1}. *Ca* is the main exchangeable cation.

2.1.2. Subalpine and alpine soils

The main type of subalpine and alpine grassland soils is **Leptosol**, most frequently Umbric Leptosols. Their soil profile consists of A, B, and C horizons. A horizon can be divided into two sublayers. The upper sublayer (about 0-10 cm) is dense root turf, where roots and rhizomes constitute 4-11% of the horizon's volume. The lower sublayer (about 10-20 cm) does not form a firm turf. The texture of Leptosols is silt loam; clay accumulation is not differentiated throughout the profile. The upper border of the C horizon lies at the depth of about 50 cm. Morphological features of Umbric Leptosols in the alpine zone are similar to those of the subalpine zone. The only difference is that the alpine soils are shallower.

Humus content of the acidic unsaturated Umbric Leptosol might be up to 20%. These soils also accumulate high amounts of total N and P. The humus contains a large proportion of slightly humified plant debris. This is primarily due to low microbial activity within the accumulated raw humus. The subalpine and alpine soils have little available nitrogen and phosphorus, but they are enriched by potassium due to potassium-rich parent rock material.

Umbric Leptosols of subalpine grasslands have the pH of 4.7-5.6 and a sum of exchangeable base cations of about 150-170 mmol(+) kg^{-1}. Soils of subalpine tall herbaceous communities, which are formed on gentle slopes disturbed by landslides and temporary streams, differ from soils of subalpine meadows. The former accumulate less organic C and N and are characterized by the relatively low concentrations of available N (NH_4^+), while accumulation of total and available phosphorus is better expressed.

Umbric Leptosols of the alpine zone are more acidic (pH 4.3-5.2), have low base saturation (14-52%) and a lower sum of exchangeable base cations (30-150 mmol kg^{-1}).

The influence of absolute altitude was reliably established for the following soil properties: contents of C, N, P, Ca and Mg, base saturation, pH, and the depth of the A horizon (Volkov 1999). As the absolute altitude increases, contents of total N and available Ca, Mg, P decrease. Carbon content increased from 1900 to 2790 m, and decreased further up the slope. The depth of the A horizon also decreased with increasing altitude (Fig. 2.1.).

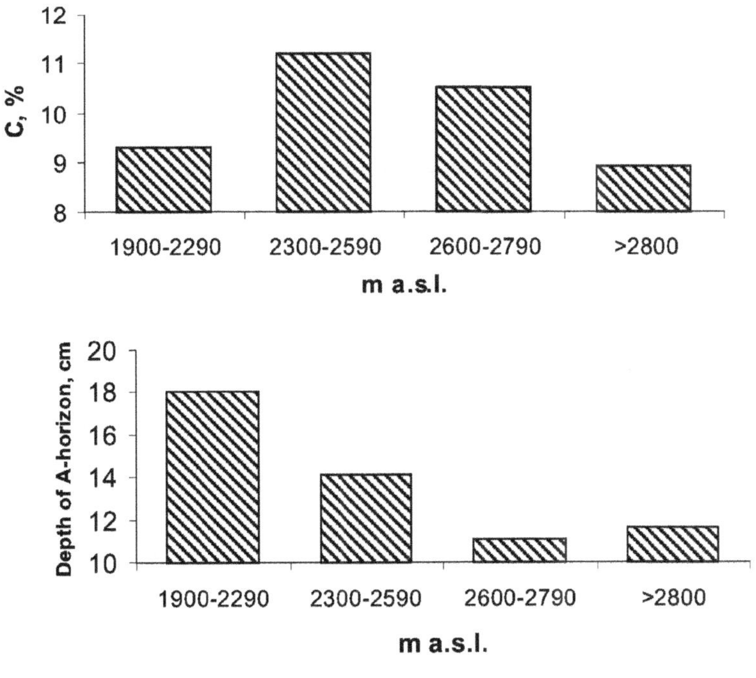

Figure 2.1. Influence of absolute altitude on the Carbon content and depth of the A horizon in soils of the Teberda Reserve

Another type of subalpine and alpine soil is **Histosol.** Terric and Fibric Histosols are formed in fen and swamp communities and Folic Histosol is the soil of *Rhododendron caucasicum* elfin shrubland. All Histosols are characterized by the high level of C and N accumulation and have high C:N and C:P ratios. Nevertheless, the Histosols of dry and moist positions strongly differ in acidity and available nutrient concentrations. Histosols of subalpine and alpine fens and swamps have high values of pH, sum of exchangeable base cations and base saturation, and also contain relatively high concentrations of available N and P. In contrast, Folic Histosols, have a low pH, low base saturation, and low available N and P concentrations.

Lithic Leptosols develop in rock cracks and on screes. They occupy large areas in all vegetation zones, but are especially widespread in the alpine and subnival zones. Primitive soils in rock cracks were studied in different altitudinal belts (forest, subalpine, alpine and subnival) and slope expositions (Bulatnikova & Makarov 2001). These soils are weakly acidic and contain low concentrations of exchangeable Ca and Mg, but they are rich in K due to parent rock. Acidity is higher and Ca and Mg concentrations are lower at higher elevations and on northern slopes. It was shown that the organic matter content and total nitrogen and phosphorus contents decreased in the sequence from forest to subnival rock soils. Soil acidity was highest at upper elevations and on north slopes (Fig. 2.2.).

2.1.3. Variability of soil properties under different subalpine and alpine communities

We determined properties of A horizons (0-10 cm) of subalpine and alpine Umbric Leptosols, which are formed under different plant communities (Table 2.1.). The study of alpine Leptosols under various plant communities, which occupy different positions along alpine toposequences in all of the Teberda Reserve, confirmed some of our earlier findings (Volkov 1999). Alpine heath soils of ridges and prominent upper reaches of slopes have the highest values of exchangeable base cations and relatively low concentrations of available N and P (Table 2.1.). The C:P ratio in these soils is high, which is a result of lower P values.

The Leptosols of the alpine meadows at the typical slope positions are characterized by some decrease of exchangeable base cations and an increase of available N and P. Low concentrations of organic carbon may be caused by more active organic mineralization in the most favourable conditions of the alpine zone. Soils of slope depressions (ass. *Hedysaro caucasicae - Geranietum gymnocauli*) are more acidic and unsaturated with base cations.

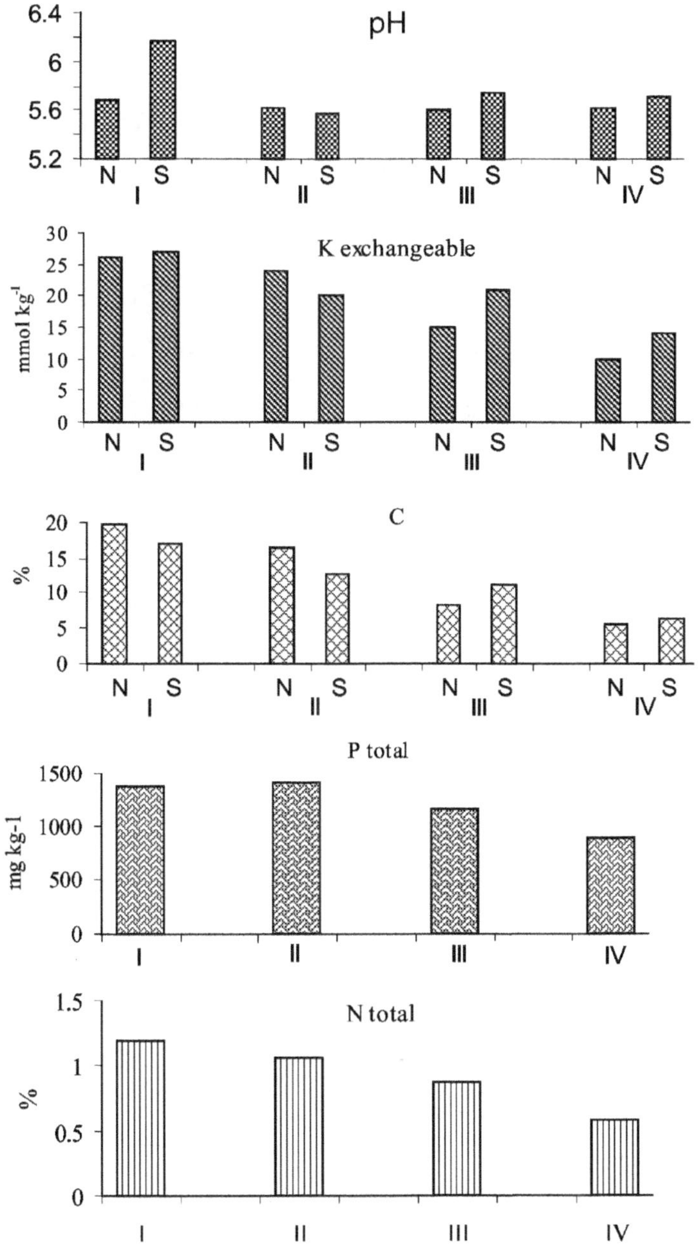

Figure 2.2. Changes of properties of Lithic Leptosols through the toposequence. I – forest , II – subalpine, III – alpine, IV – subnivale belts, N – the north slope, S – the south slope

Table 2.1. Some properties of the A horizon of subalpine and alpine soils (mean values for 11-17 soil profiles from the whole territory of the Teberda Reserve) (from Volkov 1999). HA – hydrolitic acidity, BS – rate of base saturation

Association	N, %	P, %	NH_4	P_m	pH - H_2O	HA	K	Ca	Mg	BS, %
			mg kg^{-1}			mmol(+)kg^{-1}				
Hyalopoo – Pedicularietum (SBC)	0.75	0.15	43	19	4.3	233	8	25	6	14
Caro caucasici – Caricetum nigrae	0.99	0.12	34	57	5.4	96	7	92	17	52
Ranunculetum brachylobi	0.43	0.11	32	39	4.6	131	4	20	5	21
Hedysaro – Geranietum (GHM)	0.70	0.11	46	17	5.0	181	7	46	9	26
Pediculari – Eritrichietum (ALH)	0.85	0.13	28	6	4.8	128	7	85	16	43
Polygono viviparum-Salicetum kazbekensis	0.95	0.10	31	10	4.9	178	5	87	17	38
Violo - Festucetum (FVG)	0.84	0.12	35	28	4.7	119	7	70	14	44
Rhytidio rugosi-Kobresietum capilliformis	0.88	0.14	25	13	5.2	138	7	118	23	52
Cephalario giganteae-Ligusticetum alani	0.70	0.13	34	56	5.6	99	10	134	27	63
Anthrisco sylvestris-Rumicetum alpini	0.70	0.18	31	69	4.7	109	8	112	29	54
Lerchenfeldio-Rhododenretum caucasici (RC)	1.12	0.10	24	35	3.8	161	12	157	38	58
Aconito nasuti-Juniperetum	1.14	0.13	33	14	4.5	130	7	144	24	62
Betonici macranthae-Calamagrostietum arundinaceae	0.98	0.10	45	16	5.1	133	9	121	27	54
Swertio ibericae-Caricetum nigrae	1.76	0.11	45	48	5.7	43	22	287	53	89

Most of the acidic and unsaturated soils are formed at the low end of the alpine hydrologic gradients (alpine snowbeds). However, soils of the two snowbed communities differ from one another. Development of the *Ranunculetum brachylobi* in accumulative positions near the upper boundary of the alpine zone causes relatively low concentrations of soil C, N and P, while in soils of the *Hyalopoo ponticae-Pedicularietum nordmannianae* accumulation of organic matter and associated elements is better expressed.

Soils under alpine fens and tall-forb meadows have a high amount of organic matter and the highest concentration of mobile phosphates of all high-mountain soils. Terric Histosols (peat) under subalpine fens and swamps are rich in different nutrients (Table 2.1).

Among the alpine soils, the Leptosols of *Kobresia*-dominated grasslands are characterized by the highest value of pH, exchangeable cations and base saturation, as these soils are formed in the areas with easily weathering porphyritic granite. Together with soils of the alpine lichen and dwarf shrub heaths, these soils contain low concentrations of phosphates.

It seems to us that the comparatively high ammonium contents in the soils of the alpine grasslands are caused by the disturbance regime (Tilman 1988), while the same in the soils of snow bed communities is connected with the short vegetative season that prevents full uptake of ammonium by plants.

There are no obvious signs of podzolization in the alpine soils, except some slight signs in Folic Histosols under *Rhododendron caucasicum* elfin shrublands. Only slight gleyization can be observed in the soils of snow beds, which experience water saturation during somewhat prolonged snowmelt. Thus, alpine soils of the area differ considerably from podzolic and pseudogleyic soils of the Alps (Bouma et al. 1969, Bouma & van der Plas 1971, Gracanin 1972, Posch 1977, Nestroy 1984, Muller 1987), but they are similar to the Alpine Turf soils of the Rocky mountains (Colorado) (Retzer 1956, 1974, Johnsson & Cline 1965) and to the Alpine Humus Soils in Australia (Costin et al. 1952, Costin 1955).

2.2. Soils of the four studied alpine communities

Topography influences plant and soil types formation in the alpine zone by determining precipitation gradients, particularly snowpack, and geochemical migration of elements. The depth of the snow cover is particularly important, because it regulates water supply during the growing season. We focused our attention on the four communities: alpine lichen heath (ALH), *Festuca varia* grassland (FVG), *Geranium-Hedysarum* meadow (GHM), snow bed community (SBC) (see detailed descriptions below in chapters 3 and 4, Onipchenko 2002). These communities, because of their landscape position and resulting snow regimes, represent gradients in both soil moisture content and temperature. Descriptions of the soil properties are based on 8 soil profiles established in experimental plots on Mt. Malaya Khatipara. Two profiles were studied in each community – "upper" and "lower" – according to the geomorphological position and time of snowmelt (earlier on "upper" plots). Spatial variability of soil properties within communities in connection with spatial structure of the plant community is discussed in Chapter 5.3.2.3.

Parent rock material is not uniform in the four studied communities. Two main types of rock are abundant – granite and biotite schist (Fig. 2.3.) (Vertelina et al. 1996). Schist prevailed in soils of ALH and FVG, granite dominated soils of GHM. In SBC soils the share of the two rocks was about equal. Biotite and quartz are the main minerals in the studied soil profiles. Feldspar (acid plagioclases), epidote, clinocoesite are the minor components. Several weathering indices decreased from the top to the bottom of profiles. Soils under the most productive GHM community showed the greatest degree of weathering among the studied communities.

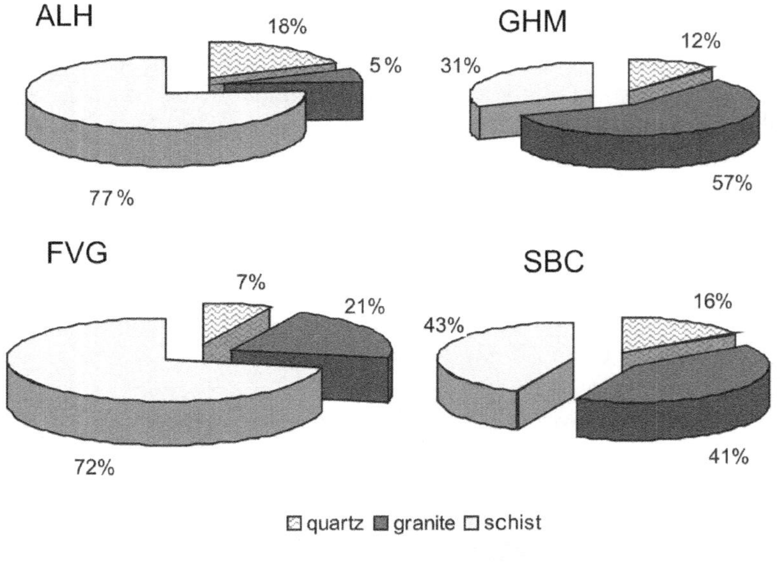

Figure 2.3. Composition of fraction of 2-10 mm in the alpine Umbric Leptosols

Stony content (minerals and rock fragments) greater than 1 mm in diameter increase in the lower horizons and decrease throughout the toposequence (from ALH to FVG to GHM to SBC) (Fig. 2.4.). It is caused by the redistribution of parent material along the topographic gradient.

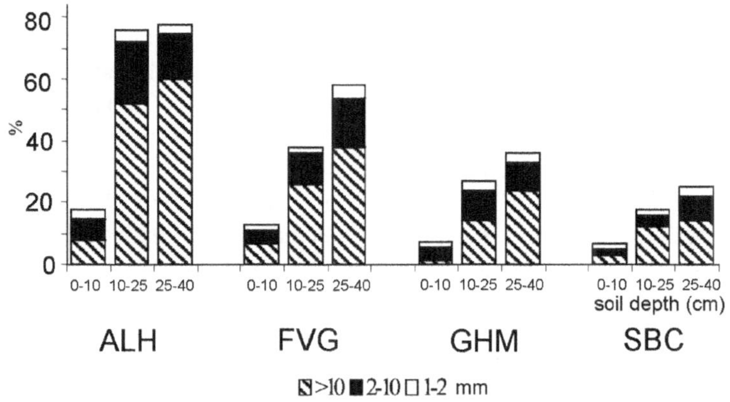

Figure 2.4. Content of stones (particle size fractions >1 mm) in the alpine Umbric Leptosols

Umbric Leptosols of the studied communities have silt loam texture. The proportion of sand decreases a little from ALH to SBC. The clay content is the same in upper and lower horizons, except in the ALH soils, where we discovered some increase in clay from the A to the B horizon (Fig. 2.5.). Micromorphology studies showed that this differentiation was not related to clay migration (Vertelina et al., 1996). More likely, it was caused by mineral transformation *in situ*.

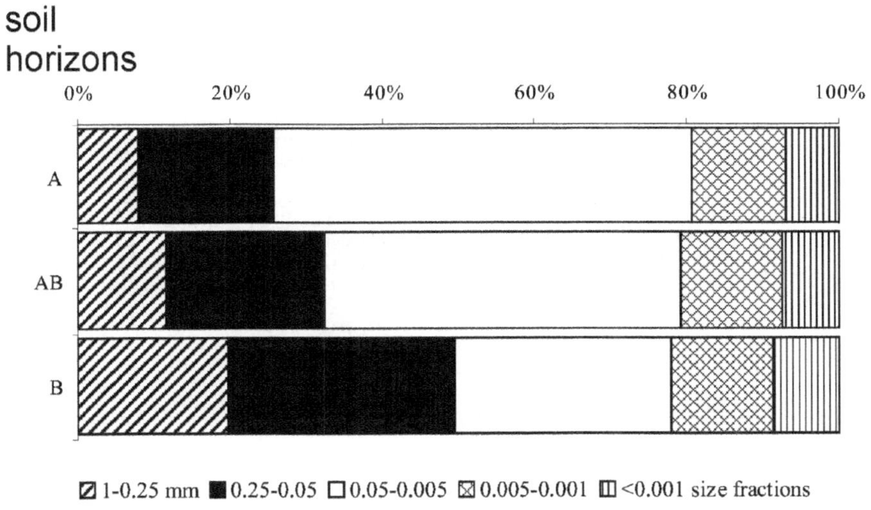

Figure 2.5. Texture of alpine Umbric Leptosols of the ALH community

The silt loam texture, high concentration of organic matter and high porosity provide high levels of water retention in the alpine soils, especially in sod layers of the A horizon. Alpine Umbric Leptosols have high water permeability with filtration coefficients of 1 to 5 mm/min. Therefore these soils are capable of retaining precipitation without much surface runoff.

Umbric Leptosols contain 52.2-63.5% SiO_2, 18.5-28.6% Al_2O_3, 6.1-13.2 % Fe_2O_3, 0.41-1.35% CaO, 1.0-1.9% MgO, and 0.21-0.61 P_2O_5 (Fig. 2.6). Concentrations of these oxides usually decrease with depth. There was no regularity in the distribution of elements along the toposequence. The maximum concentration of SiO_2 was observed in soils of the FVG and GHM communities on slopes, the minimum – in ALH and SBC located on flat land. The highest concentration of Fe_2O_3 was found in the soils of the ALH community, the highest CaO and MgO – in soils of the ALH and SBC communities.

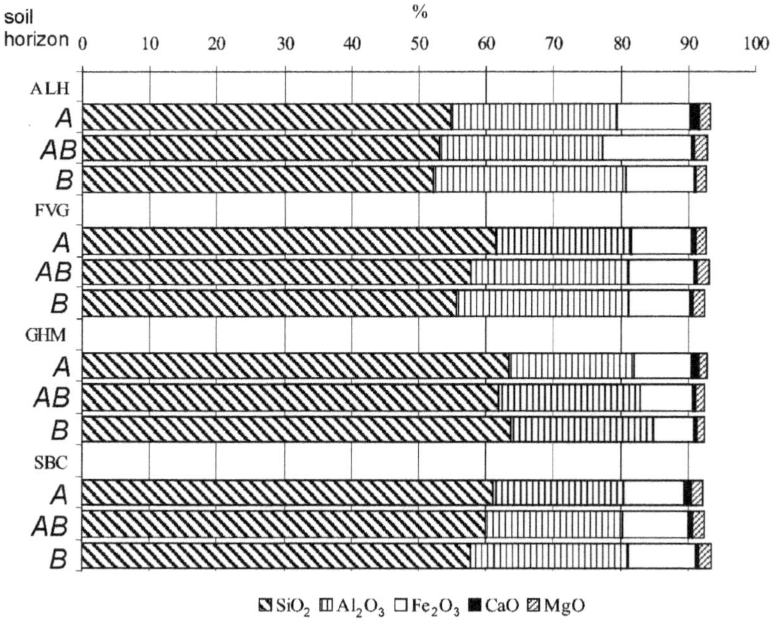

Figure 2.6. Total values (%) of mineral elements in alpine Umbric Leptosols

Alpine soils are relatively poor in available nitrogen and phosphorus, while they are rich in potassium due to the chemical composition of the bedrock. Soils formed under various plant communities differ in acidity, contents of organic C, N and P, available N and P, and exchangeable base cations (Fig. 2.7.). The Leptosols of alpine heaths had the highest pH and content of exchangeable base cations and relatively low concentrations of available N and P. The soils of *Festuca varia* grasslands at the typical slope positions were characterized by somewhat lower exchangeable base cations and higher available N and P. Lower concentrations of organic carbon in these soils may be caused by faster organic mineralization rates under the most favourable conditions of the alpine zone. Soils of slope depressions, covered by the *Geranium gymnocaulon-Hedysarum caucasicum* meadow, are more acidic and less saturated with base cations. The most acidic and unsaturated soils with the highest amount of available N and P are found under snowbeds (Grishina et al. 1993, Makarov et al. 1997).

2.3.Nutrient turnover in the alpine communities

Differences in soil properties of various alpine plant communities can be explained both by redistribution of elements as a result of geochemical migration and by specific influence of the plant community or individual species. Except for the obvious increase of nitrogen concentration under leguminous plants, little information is available about the influences of wild plants on soil properties (Wedin & Tilman 1993, Reynolds et al. 1997). Virtually nothing is known about this aspect of community function at high elevations. Therefore, we were interested to study some parameters of nutrient turnover in the alpine plant communities of the northwestern Caucasus.

2.3.1. Element accumulation in communities

Amounts of individual elements accumulated in the total plant biomass allow to estimate the factors (biological accumulation or geochemical element redistribution) that contribute to differentiation of soil properties and nutrient supply under various communities.

Low amounts of exchangeable base cations in soils of GHM and SBC imply leaching of Ca and Mg from soils of snowy habitats. The plants of these communities are adapted to the leaching of Ca and Mg from the soil and actively accumulate these elements in their biomass (Fig. 2.8). In the same way, the maximum amount of P in the SBC soil was not connected with its biological accumulation in this community.

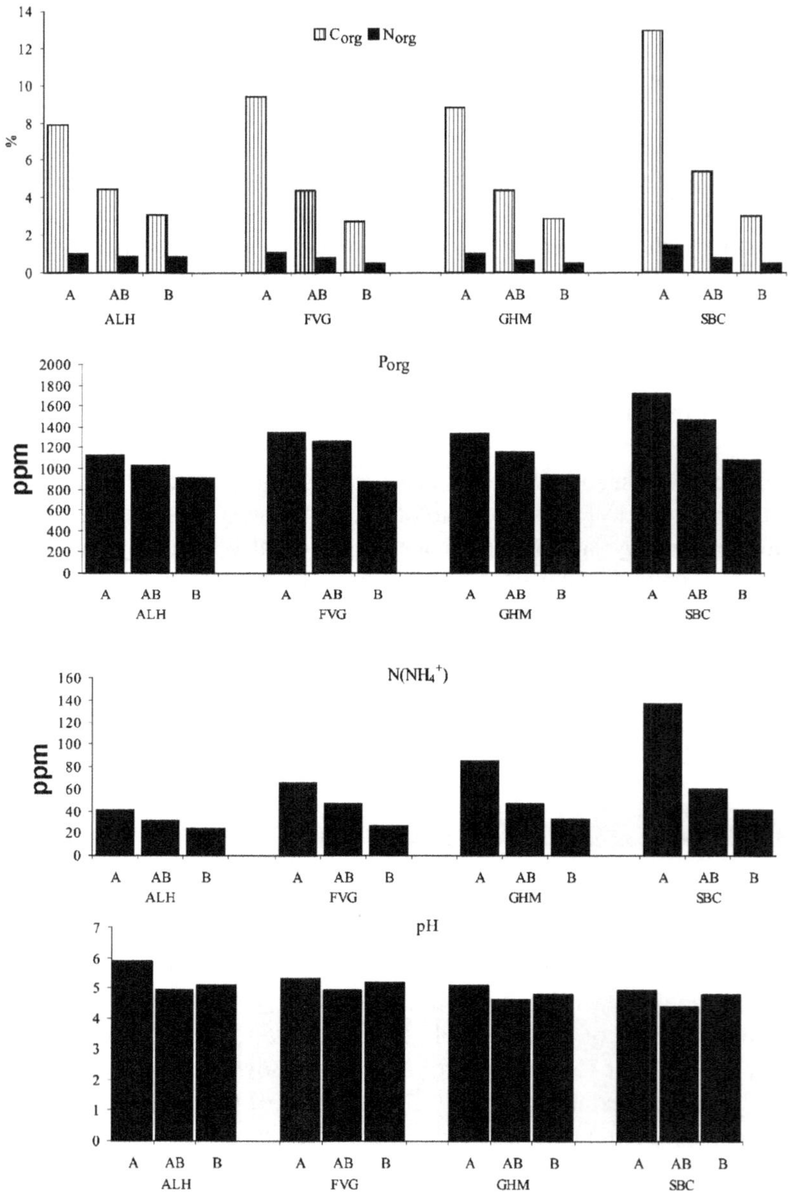

Figure 2.7. Properties of Leptosols under different plant communities

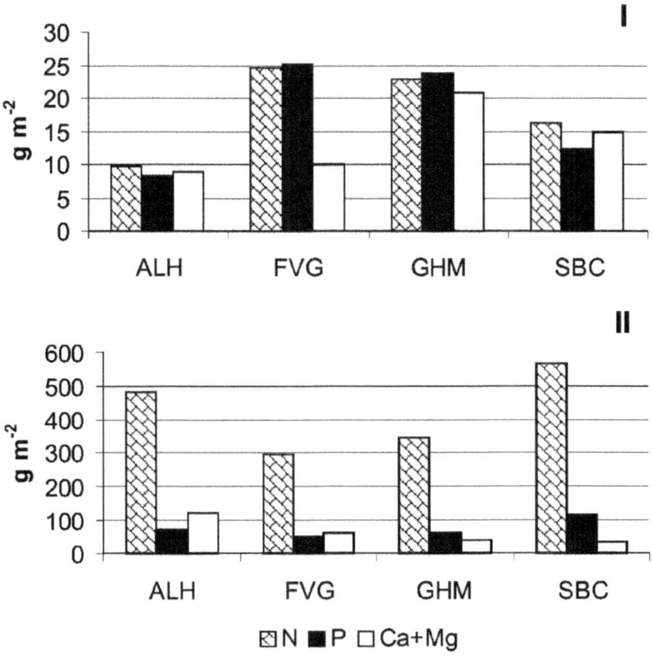

Figure 2.8. Contents of elements expressed on an area basis in total biomass (I) and in the A horizon of the soil (II) of various plant communities

2.3.2. Coefficients of biological uptake (CBU)

Plants are characterised by selective uptake of different elements. The Russian geochemist Polynov (1956) suggested to use a coefficient for assessment of uptake intensity of elements. He calculated this coefficient as the relation between the concentration of an element in the plant to it's concentration in the soil. Later, Perelman (1975) called this coefficient the "coefficient of biological uptake" (CBU). It usually varies from 0.001 to 1000. According to the value of CBU all elements can be divided into several groups – strong, middle and weakly accumulated.

Species with similar chemical makeup, but belonging to different communities, were different in their CBU (Table 2.2). For example, *Trifolium polyphyllum* (ALH) was characterized by more intensive nitrogen uptake compared with *Geranium gymnocaulon* (GHM) or *Sibbaldia procumbens* (SBC). By analogy, *Festuca varia, Leontodon hispidus* and

Scorzonera cana (FVG) absorbed nitrogen more intensively than either *Hedysarum caucasicum* or *Sibbaldia procumbens* (GHM). The highest CBU of phosphorus were typical for some of the ALH (*Anemone speciosa, Antennaria dioica, Trifolium polyphyllum)* and the FVG (*Festuca varia, Scorzonera cana*) species. The highest CBU of Ca and Mg were typical for plants of the GHM and SBC, where soils were the most acidic and devoid of exchangeable base cations.

Table 2.2. Coefficients of biological uptake (CBU) of elements by alpine plant species

Species	Community	CPU				
		N	P	K	Ca	Mg
Anemone speciosa	ALH	298	1167	55	2	3
Antennaria dioica	ALH	350	867	46	2	10
Trifolium polyphyllum	ALH	551	883	23	3	10
Carex umbrosa	FVG	275	548	17	2	7
Festuca varia	FVG	583	905	18	2	4
Leontodon hispidus	FVG	671	214	53	9	4
Scorzonera cana	FVG	638	857	37	7	9
Geranium gymnocaulon	GHM	361	787	24	14	14
Hedysarum caucasicum	GHM	414	540	18	11	13
Sibbaldia procumbens	SBC	453	625	25	15	8
Taraxacum stevenii	SBC	458	493	60	16	19

Table 2.3. Coefficients of relative concentration of elements (RCE) in alpine plant species

Species	Community	RCE				
		N	P	K	Ca	Mg
Anemone speciosa	ALH	0.8	2.0	1.6	1.0	0.5
Antennaria dioica	ALH	1.0	1.1	1.3	0.9	1.5
Trifolium polyphyllum	ALH	1.5	1.1	0.6	1.7	1.5
Carex umbrosa	ALH	0.5	0.7	0.9	0.6	2.0
Festuca varia	FVG	1.1	1.2	1.0	0.9	1.1
Leontodon hispidus	FVG	1.2	0.3	2.9	3.8	1.1
Scorzonera cana	FVG	1.2	1.1	2.0	2.8	2.4
Geranium gymnocaulon	GHM	1.0	1.2	1.0	1.2	1.1
Hedysarum caucasicum	GHM	1.1	0.8	0.8	0.9	1.1
Sibbaldia procumbens	SBC	1.2	0.9	1.0	1.3	0.6
Taraxacum stevenii	SBC	1.2	1.3	2.1	0.9	1.0

2.3.3. Relative concentration of elements (RCE)

RCE coefficients showed species in relation to their relative concentrations of individual elements in the plant community. In communities on upper slopes there were 2–3 species that concentrated some elements most intensively (Table 2.3). In ALH *Trifolium polyphyllum* concentrated N, Ca and Mg, and *Anemone speciosa* concentrated P and K. Species of FVG did not concentrate N and P, but they had the highest RCE coefficients for K, Ca and Mg. All three elements were intensively concentrated by *Scorzonera cana*, K and Ca – by *Leontodon hispidus*, Mg - by *Carex umbrosa*. There were fewer species with high RCE in the two communities of the lower slopes (GHM and SBC).

Chapter 3

Plant geographical description of the area

3.1. History of floristic study

Numerous factors determine the phytogeography of the Teberda State Nature Reserve. It is located in the northwestern part of the northern slope of the Greater Caucasus mountain range. The diversity of climate within the Reserve results from the complex mountain topography with strong altitudinal differences: the lower parts of the Reserve are at the altitude of 1,300 m, while the highest point of the Reserve is at 4,046 m above sea level (Mt. Dombai-Ulgen). Variously oriented steep slopes exacerbate differences in vegetation even within one altitudinal zone. The modern geological history of this territory includes repeated Quaternary glaciations, another major factor that has played an important role in the formation of the recent plant cover of the area.

The botanical exploration of the Caucasus dates back 300 years: in 1700–1702 a French botanist Joseph de Tournefort visited the Transcaucasus — Tbilisi, Echmiadzin and the Arax River valley. He was the first European scientist known to climb Mt. Ararat and gave a detailed description of its altitudinal zones. For a long time the main attention was paid to the peripheral parts of the Caucasus: the North Caucasus Plain and the eastern foothills of the Greater Caucasus, the coastal zone of the Caspian Sea, as well as Georgia and Armenia. The research on these territories provided the first scientific information on the unique Caucasian flora and was conducted by outstanding Russian botanists of the 18th century: academicians Johann Christian Buxbaum in 1721–1724, Anton Johann von Güldenstädt in 1768–1775, Samuel Gottlieb Gmelin (Gmelin Jr.) in 1768–1774 and Peter Simon von Pallas in 1793–1794.

V.G. Onipchenko (ed.), Alpine Ecosystems in the Northwest Caucasus, 25–54.

The intensive investigation of the flora and vegetation of the Greater Caucasus Mountain Range began only in the early 19th century, when the first floristic surveys written by Friedrich A. Marschall von Bieberstein (1808–1819), Carl A. von Meyer (1831), Rudolf F. Hohenacker (1833, 1838) and Karl H.E. Koch (1848–1851) were published. All the collected data were summarized in the famous "Flora Rossica..." compiled by Carl F. von Ledebour (1841–1853).

The earliest information on plants of the Teberda valley was apparently provided by F. Jensch (1843), who specialized in forest research.

In the second half of the 19th century the floristic investigations of the Caucasus continued. It was then that the first surveys on some regions of the Caucasus were published and the important summaries were made. Nikolay M. Albov (1896) was the first scientist to provide a description of flora and vegetation of the Colchis region, an outstanding Tertiary refugium. Gustav I. Radde (1901) outlined five main vegetation types for the Caucasus as a whole.

The first attempts to divide the Caucasus into phytogeographic regions were also made at the turn of the 20th century (Medvedev 1882, 1907, Radde 1901, Kuznetsov 1901). Nikolay I. Kuznetsov in his classical work "Principles of identifying phytogeographic provinces in the Caucasus" (1909) critically analyzed all the suggested schemes. Brief sketches of the history of the botanical exploration of the Caucasus were given by Nikolay V. Pavlov (1948) and A.A. Shcherbakova (1979).

The beginning of thorough research on flora and vegetation of the northwestern Caucasus (including the Teberda River valley) dates back to the end of the 19th century, highlighted by the works by V.I. Lipskii (1891) and Italian botanists Carlo P.S. Sommier and Emilio Levier (1900) as well as publications of Nikolay A. Busch (1897, 1898) and Elizaveta A. Busch (1909, E.A.Busch & N.A.Busch 1909). After the Teberda Reserve was established in 1935, the floristic research in it was done by Sergey V. Juzepczuk, Boris I. Fedchenko, Dmitry K. Volgunov. Later this work was continued by K.I. Marusyak in 1938–1940 and by V.N. Kononov in 1959. The next inventory of the Reserve's flora conducted by Faina M. Vorob'eva in 1976–1980 (Vorob'eva & Kononov 1991). The last checklist of its flora was published recently (Vorob'eva & Onipchenko 2001).

The main altitudinal zones of the northern slope of the Greater Caucasus Mountain Range were identified in the late 19th century (Voeykov 1871, Medvedev 1882, Keppen 1885). These researchers showed that in the western part of the North Caucasus the steppes of the lowlands and plains are replaced in the mountains by broad-leaved forests at the lower levels and by dark coniferous forests at the higher levels. The forests, in their turn, are replaced by treeless alpine ecosystems in the highest mountains.

Describing the forests of the western Caucasus, Gustav I. Radde (1901) pointed out some relic tertiary species, such as *Taxus baccata, Vaccinium arctostaphylos* and others, which came to the region from the western Colchis and did not spread to the east of the Kuban River basin.

Summarizing the results of the botanical exploration of the Caucasus in the 19th and early 20th centuries, Nikolay I. Kuznetsov (1909) identified the main features of the altitudinal zonation of the Caucasus and its vegetation cover (especially the forests) in various regions. The identification of phytogeographic regions suggested by him was based upon geographical and historical principles. He pointed out that the mountain forests of the Caucasus are to a great extent similar to those of Europe: the Alps, the Carpathians and the Balkan Mountains.

N.I.Kuznetsov (1909) placed the northwestern Caucasus with the Teberda Nature Reserve into the Kuban forests province, which he saw as a 'simplified Tertiary forest type with the admixture of western steppe species' (p. 104). Later the main arguments of Kuznetsov's work were further developed by A.A. Grossheim (1936), V.P. Maleev (1939, 1941, 1947), and others. For example, many evergreen shrubs, in spite of being the thermophilous relics of a mesophytic forest flora, are found in the mountains at very high altitudes. A.G. Dolukhanov (1966) explained this fact by pointing out that the deep snow accumulation in these habitats protects plants from freezing.

The altitudinal zonation of the northwestern Caucasus, established by N.I. Kuznetsov (steppes – broad-leaved forests – coniferous forests – alpine grasslands), was adopted as the main scheme by the successive authors (Elenevskii 1940, Shiffers 1953, Stanyukovich 1955, 1973, Kononov 1957, Lavrenko 1964, Dolukhanov 1966, Gulisashvili et al. 1975, Grebenshchikov et al. in Gerasimov et al. 1980, Polivanova 1989, Ogureeva 1999) (Fig. 3.1).

In the scientific publications of the last 2–3 decades, the altitudinal zonation of the northern slope of the western part of the Greater Caucasus is referred to as a special "Kuban" type. This was also true for the Kuznetzov's classification (Kuznetsov 1909), which distinguished this territory as a certain province of Kuban forests. The full set of this zonation type consists of 10 altitudinal belts: steppes, forest-steppes, oak-hornbeam forests, beech forests, fir-spruce forests, pine forests, subalpine meadows, alpine meadows, subnival and nival belts (Vedenin et al. in Gerasimov et al. 1980, Ogureeva 1999). The lowest altitudes of the Teberda State Nature Reserve are at 1250–1300 m above sea level, where beech, beech-fir and pine forests form the lowest altitudinal belt. The total number of vascular plant species found in the Reserve is about 1,150. Thirty-five percent of the Teberda Nature Reserve is forested, 28% is covered with meadows and more than 10% is occupied by glaciers.

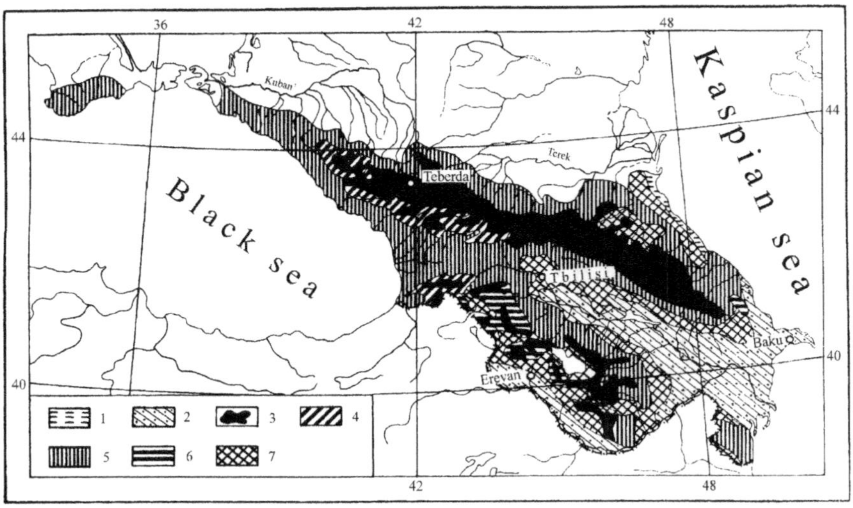

Figure 3.1. Native vegetation zones in the Caucasus: 1 – desert steppes, 2 – deserts with dwarf shrubs and dwarf semishrubs, 3 – alpine meadows, crooked-stem woodlands, glaciers, 4 – mountain fir and spruce forests, 5 – mountain broad-leaved forests, 6 – mountain meadow steppes, 7 – mountain grass steppes (based on Isachenko & Lavrenko 1980).

3.2. Steppe belt

We review the main features of the vegetation cover of the northwestern Caucasus beginning from the lowest zone, the steppe.

Steppe vegetation used to be widespread in the Caucasus, but today only fragments of steppe communities survive, primarily on the rocky slopes that can hardly be used for agriculture (Lavrenko 1940, Grossheim 1948, Dolukhanov 1966).

Valley and foothill steppe communities contain tussocks of feathergrass (*Stipa tirsa, S. capillata, S. lessingiana* Trin. et Rupr.), fescue (*Festuca valesiaca*), crested hair-grass (*Koeleria cristata*), and other species. Bluestem (*Bothriochloa ischaemum*) steppe is located in the transitional plains area between fescue-feathergrass communities and wormwood semi-deserts. Steppe vegetation is intermixed in many areas with semi-desert, arid open woodlands, and thicket communities.

Steppes of the Northern Caucasus are the natural extension of the zonal steppes of the southern Russian plains, bordering on the western Greater

Caucasus (Lavrenko 1940, Grossheim 1948, Dolukhanov 1966). At present, most of the North Caucasian steppes are cultivated, but in the past they stretched along the northern foothills of the Greater Caucasus from Krasnodar to Grozny (Grossheim 1948). The Stavropol Uplands used to be covered with steppes as well. The steppes of the plains were absolutely treeless, but in the foothills steppes had a more mesic appearance and they were alternated here with small oak woodlands, forming the subzone of forest-steppes.

The steppes of the Northern Caucasus are classified as fescue-feathergrass (Lavrenko 1940), or southern (Alekhin 1936, Agakhanyants 1986) steppes. Their typical features include the predominance of xerophilous bunch grasses (*Stipa lessingiana, S. capillata, S. zalesskii, Festuca valesiaca, Koeleria cristata*) and chernozem soils. At the same time, some characteristic xeromesophilous forbs of the northern steppes (*Vicia cracca* (incl. *V.tenuifolia*), *Filipendula hexapetala, Onobrychis arenaria, Stachys recta, Myosotis sylvatica, Thymus marschallianus*) as well as xerophilous ones of the southern steppes (*Crambe tatarica, Limonium platyphyllum, Salvia nutans, Veronica austriaca, Jurinea consanguine,* etc.) can be found here too.

Forbs do not play any significant role in the herbaceous layer, which is poor in species (12–15 species per m^2) and of comparatively low density (Lavrenko 1940). There are many ephemeral plants and spring geophytes (*Tulipa* spp., *Poa bulbosa*), epigeous lichens (*Cladonia convoluta, Parmelia vagans, P. ryssolea*), cyanobacteria (*Stratonostoc commune*) and bryophytes (*Tortula ruralis, T. caninervis*).

Typical southern steppes are replaced by more mesic meadow steppes on the Stavropol Upland and in the lower altitudinal zone in the foothills. *Stipa pennata* and *Phleum phleoides* are the dominant species in the upland, while *S. pulcherrima* C.Koch plays this role in the foothills.

Zonal plain steppes do not exist in the mountains and are usually found at the elevations lower than 200 m. The foothill steppes are found between 200 and 500 m above sea level. A.A. Grossheim (1948) and L.B. Makhatadze (in Gulisashvili et al. 1975) consider them to be "secondary communities", originating on clearcuts in the oak forests. Virgin steppes do not occupy large area in the foothills. Nevertheless, they penetrate deeply into the mountains along river valleys, where one can see them on the southern slopes at much higher elevations.

In the Teberda river basin isolated patches of fescue-feathergrass steppe with *Stipa pulcherrima, S. capillata, Festuca valesiaca*, and sometimes *Bothriochloa ischaemum* are found in the Dzhemagat River valley along the northern border of the nature Reserve as well as on river terraces, terrace slopes and in the forest openings to the north (Kononov 1957). On the rocky southern slopes, steppes tend to be more xerophytic; in the forest clearings and on the terraces there are many xeromesophilous forbs, while feather grasses sometimes are not seen at all.

The most frequently found forbs are *Veronica multifida, Galium verum, Potentilla arenaria, P. humifusa, Teucrium chamaedrys, T. polium, Anthyllis vulneraria, Medicago falcata (*incl. *M.romanica), Onobrychis cyri, Trifolium alpestre, Campanula sibirica*, etc. Spring geophytes (*Crocus reticulatus, Gagea pusilla, Merendera trigyna, Muscari muscarimi*) are abundant immediately after the snowmelt. These steppes are comparatively rich with 55–65 species per 100 m^2 and 30 species per 1 m^2. The same number of species per square m is typical for the steppes in the Teberda River valley, where *Stipa capillata* prevails over other grasses.

East of Teberda, in the upper part of the Kuban River basin, steppe patches with *Festuca valesiaca, Stipa caucasica, Koeleria cristata* and *Carex humilis*, spread along the slopes' base, are often replaced by shrublands with prickly shrubs— *Juniperus communis, J. sabina, Berberis vulgaris, Rosa pimpinellifolia, Rhamnus pallasii* and so on. On the rocky slopes thorn cushions (*Astragalus denudatus, A. aureus, Teucrium polium, T. chamaedrys, Scutellaria orientalis, Salvia canescens, Thymus marschallianus*) are common.

Mountain steppes with *Bothriochloa ischaemum* are distributed on the alluvial fans and along ancient terraces on stony soils. They are quite common on the slopes and in depressions in the lower part of the Uchkulan and Khurzuk rivers valleys, which are located in the rain shadow of the Greater Caucasus mountain range between 1300 and 2300 m elevation.

Some species of mountain xerophytic communities, such as *Salvia canescens, Astragalus denudatus, Scutellaria orientalis, Sideritis montana, Teucrium chamaedrys* and some others occur in the Teberda River valley on the southern, highly insolated, rocky slopes.

3.3. Deciduous (broad-leaved) forest belt

Broad-leaved forests form the next altitudinal zone upslope. Today the lower part of the forest zone is severely altered by human activity, especially to the north of the Teberda River mouth. The typical feature of the broad-leaved forests in the Kuban and Teberda River basins is their diverse tree composition. Oaks (*Quercus robur, Q. petraea* L. ex Liebl.), hornbeam (*Carpinus betulus*), basswood (*Tilia begoniifolia*), maples (*Acer platanoides, A. campestre*), beech (*Fagus orientalis*), ash (*Fraxinus excelsior*), crab cherry (*Cerasus avium*), appletree (*Malus sylvestris*), cherry plumtree (*Prunus divaricata*), peartree (*Pyrus communis*) and other trees are usual in these forests covering the steep slopes of the Kuban and Teberda River valleys.

Usually broad-leaved forests are left intact on steep slopes because of their inaccessibility. They are commonly found in the narrowest parts of the Kuban River valley as well as near the mouths of its tributaries.

Orientation of the slopes largely determines the species composition of tree stands. Durmast oak (*Q. petraea*) as a rule grows on dryer and better insolated southern slopes with rock outcrops; this tree is distributed mainly in the lower part of the Teberda River valley, between the Teberda and Kuban Rivers, as well as in the valley of the latter river north of the city of Karachaevsk. English oak (*Quercus robur*) is typical on wetter sites, e.g., ravine bottoms and gentle slopes with well-developed loamy soils. It is widely distributed in the Teberda and Kuban river valleys, being found both with durmast oak and other broad-leaved trees (hornbeam, basswood, beech, etc.).

At the altitude of 1100–1200 m and higher the birch (*Betula pendula*) is often found. Its presence is a peculiar feature of the mixed broad-leaved forests of the Kuban and Teberda river basins. Speckled alder (*Alnus incana*) is a very common species too; it usually grows in the wettest habitats. *Calamagrostis arundinacea* is frequently the dominant species of the herbaceous layer in the broad-leaved forests; many forb species (*Salvia glutinosa, Aruncus vulgaris, Circaea lutetiana* L., *Dryopteris filix-mas, Galium odoratum, Polygonatum verticillatum* and so on) are common as well. A shrub layer with *Corylus avellana, Euonymus europaea, Lonicera orientalis, Berberis vulgaris* and some other shrubs can be well developed in some places.

On the dry slopes, *Rhododendron luteum* often predominates among other forest shrubs, while *Prunus spinosa* is characteristic of forest edges.

It is noteworthy that mixed and pure (with predominance of *Quercus robur, Fagus orientalis, Carpinus betulus*, often with admixture of *Alnus incana* and *Betula pendula*) broad-leaved forests are found not only on the slopes of large river valleys, but also along their tributaries. To the north of the Bokovoy mountain ridge broadleaved forests are substituted by birch (*Betula pendula*) forests, which form a distinct belt at the altitudes between 1100 to 1700 m a.s.l. They are very typical of gently inclined plateaus (Dudaron, Bechasyn) that surround the Kuban river valley (Fig. 3.2-3.3.). Birch forests form a peculiar "parkland" landscape in the upper part of the forest belt with a lot of meadow grass and herb species penetrating under the light birch tree stand: *Calamagrostis arundinacea, Brachypodium sylvaticum, Bromus variegatus, Rubus saxatilis,* etc. In depressions, birch forests are often substituted by alder forest with the predominance of both black (*Alnus glutinosa*) and grey (*A. incana*) alders. Tall grasses and forbs

(*Doronicum macrophyllum, Telekia speciosa, Senecio platyphylloides, S. nemorensis, Heracleum asperum, Symphytum asperum*) are characteristic of these habitats. At comparatively lower altitudes, the herbaceous layer of wet alder forests is rich enough with *Aruncus vulgaris, Matteuccia struthiopteris, Filipendula ulmaria, Aegopodium podagraria, Petasites albus* among its dominants. Tall forbs often dominate over other herbaceous plants in forest openings and along edges, where species of *Heracleum* as well as *Angelica purpurascens, Inula helenium* L., *Cephalaria gigantea* and some other species occur.

Figure 3.2. Betula pendula belt in the Elbrus region (Foto by A.Zakharov)

The timberline of the low plateaus (Dudaron and Bechasyn) is found at lower elevations than that of the high mountain ranges: birch and alder forests there border the wet and sometimes peaty subalpine meadows (composed mainly by tall grasses) at the altitude of 1600–1800 m above sea level. The most frequent type of these meadows is blue moor grass grasslands (*Molinia coerulea*); reed grass (*Calamagrostis arundinacea*) is the most frequent second most common species. Pure reed grass meadows can be found on well-drained slopes. Meadows alternate with hypnum-sedge or, less frequently, sphagnum-sedge moors; peaty meadows with *Deschampsia caespitosa* are widespread as well. *Carex cespitosa, C. hartmanii, C. lasiocarpa, C. limosa, C. muricata, C. vesicaria* and some others are the most common sedges. Rushes are represented by *Juncus articulatus, J. effusus* and *J. filiformis*. Cotton grasses (*Eryophorum polystachyon, E. vaginatum*) are also typical for the moors.

Figure 3.3. Steep slopes with *Betula pendula* and *Pinus sylvestris* forest in Khudes valley (Foto by A.Zakharov)

In the upper part of the Teberda River basin mixed forests extend along steep slopes up to the altitude of 1300–1500 m. On the river terraces as well as at the confluences of tributaries, beech (*Fagus orientalis*) plays an important role as the main dominant of the broad-leaved forests, forming almost pure stands (Fig 3.4-3.5). The herbaceous layer in beech forests is composed of typical species of the "nemoral complex": *Aegopodium podagraria, Sanicula europaea, Aruncus vulgaris, Galium odoratum, Festuca altissima, Hordelymus europaeus, Athyrium filix-femina, Dryopteris filix-mas, D. abbreviata, Matteuccia struthiopteris, Pteridium tauricum,*

Milium effusum, Aconitum orientale, Paris incompleta, Polygonatum orientale, Impatiens noli-tangere, Asarum europaeum, Dentaria bulbifera, D. quinquefolia, Stachys sylvatica, Circaea alpina and many others. The presence of nemoral species under the canopy of broad-leaved forests of the northwestern Caucasus was first mentioned by N.I. Kuznetsov (1909), who compared them with mountain forests of the Alps and other middle-European mountains. The nemoral understorey of the Teberda beech forests was further discussed in publications of V.B. Sochava (1946, 1949), I.I. Tumadzhanov (1940, 1959, 1963), E.M. Lavrenko (1946) and other authors.

The tallest trees are found in pure beech stands, where they reach 20–25 m, while crown density is usually 90-100%. More often, there is an admixture of other broad-leaved trees: *Carpinus betulus, Acer platanoides, A. cappadocicum, Cerasus avium, Fraxinus excelsior, Ulmus glabra* as well as *Acer trautvetteri* at altitudes more than 1,500 m above sea level.

The shrub layer of the beech forests consists of *Corylus avellana, Sambucus nigra, Lonicera orientalis, L. caprifolium, Viburnum opulus, Ribes biebersteinii, Rosa canina*, etc. *Matteucia struthiopteris, Dryopteris filix-mas* and *Aconitum orientale* are the most frequent forbs. Beech forests with *Galium odoratum* and *Aegopodium podagraria* are typical for the alluvial fans of the canyons.

Figure 3.4. Pure *Fagus orientalis* forest in the Teberda valley

Figure 3.5. Gap in the beech forest, Teberda valley

3.4. Coniferous forest belt

Forests cover about 34% of Teberda Reserve. Five species of trees are the most common: Scotch Pine (*Pinus silvestris,* 34.7% of all forested area), Litwinov's birch (*Betula litwinowii,* 27.1%), Caucasian Fir (*Abies nordmanniana,* 12.6%), Caucasian spruce (*Picea orientalis,* 8.4%), and Caucasian beech *(Fagus orientalis,* 3.3%) (Vorob'eva et al. 1986). Thus, more than 50% of the forested area of the Reserve is under coniferous forest.

Fir (*Abies nordmanniana*) and spruce (*Picea orientalis*) replace beech at altitudes above 1300–1400 m, and the fir becomes the dominant tree in the forest on slopes above 1,500 m (Fig. 3.6-3.7). In the valleys of the Dombai-Ulgen and Alibek rivers, which are close to the Great Caucasus Range, fir forests completely cover the slopes. On the lateral ridges the forests are distributed according to the following pattern: north facing slopes are covered by fir forests with the admixture of spruce and south faced slopes are occupied by pine (*Pinus sylvestris*) forests, while pure spruce forests occur in the valleys of the Gonachkhir and other rivers. In the upper reaches of the Kuban river basin there are no fir or spruce forests at all, nevertheless many forbs typical for these forests (*Vaccinium vitis-idaea, Linnaea borealis, Goodyera repens*, some boreal mosses) still occur in the understorey of pine forests. This similarity between pine and dark coniferous forests was noted already by R.A. Elenevskii (1940) and V.B. Sochava (1946).

Figure 3.6. Abies nordmanniana forest on north slope in the Teberda valley

Figure 3.7. Fir forest with subalpine meadow species in the understorey

In the Teberda Reserve, the influence of wet air masses becomes less pronounced as one travels northward away from the Greater Caucasus mountain range. This explains why fir is substituted by pine in the northern part of the Reserve. Fir stands on the southern slopes along the Alibek and Dombai-Ulgen river valleys are very dense (crown density is about 100%) and tall, up to 35–40 m high. Some other tree species (*Fagus orientalis, Sorbus aucuparia, Acer trautvetteri, Populus tremula*, etc.) mix in with fir in these forests. The understorey is composed mainly of species characteristic of taiga; at the same time one can find here relic Colchic species, e.g., *Laurocerasus officinalis, Vaccinium arctostaphylos, Rhododendron ponticum*, as well as some species typical of the tall forb communities (*Telekia speciosa, Inula magnifica, Senecio platyphylloides, S. nemorensis, Campanula latifolia*).

There are several types of fir forests in the nature reserve, which differ from each other in their species' composition, density of stands, as well as other tree stand characteristics. Each type has a distinct combination of soil type, altitude, slope steepness and orientation.

North slopes are usually occupied with "fern fir forests": well-developed fir stands with *Athyrium filix-femina, Dryopteris filix-mas, Matteuccia struthiopteris* and *Petasites albus* in the understorey. Tree trunks are usually tall. Higher up some subalpine meadow species (*Angelica purpurascens, Heracleum asperum, Ligusticum alatum, Campanula lactiflora, C. latifolia, Senecio platyphylloides, Telekia speciosa*, etc.) occur under the canopy; many of them reaching a height of 2,5–3 m.

When the crown density approaches 100%, the understorey tends to disappear. In this case, the so-called "dead cover fir forests" are developed, and only few plants (e.g., boreal herbs such as *Goodyera repens* or *Linnaea borealis* or some bryophytes) are found there.

On the south slopes "fescue grass fir forests" are very common. Mountain fescue grass (*Festuca altissima*) predominates in the herbaceous layer. They reach the subalpine mountain belt, and pine (*Pinus sylvestris*) admixture is typical in such tree stands near the timberline. The shrub layer, 70–100 cm in height, consists of *Rhododendron luteum* with *Vaccinium arctostaphylos* clumps often well developed.

"Short forb fir forests" with an addition of spruce are commonly found on gentle slopes and on the river terraces. In their herbaceous layers both boreal (for example, *Oxalis acetosella*) and nemoral (such as *Galium odoratum*) species are common. *Geranium robertianum, Osmorhiza aristata, Salvia glutinosa, Calamintha grandiflora, Senecio renifolius, Paris incompleta* and some other species are typical for the understorey of these forests too.

The pine tree (*Pinus sylvestris*) is as widely distributed in the region as the fir. In the Teberda Nature Reserve pine forests are found mainly in its northern half, where they usually occupy steep stony or rocky slopes of lateral spurs of mountain ridges surrounding the valleys of the Teberda river and its tributaries — Gonachkhir, Dzhemagat, Mukhu, Bol'shaya and Malaya Khatiparas and some others (Fig. 3.8-3.9).

The comparatively large tract of pine forest covers the slopes of the Teberda river valley to the north of the Reserve. The further one gets from the Greater Caucasus mountain range, the less precipitation and consequently the more extensive is the area occupied by the pine forests. For example, the mean annual precipitation near the northern border of the Reserve is 695 mm while in Dombai it is twice as much — 1344 mm. This contrast is especially sharp in winter and is the reason for the relatively thin snow cover in the northern part of the Reserve. East of the Reserve, in the Kuban river headwaters, fir and spruce are totally replaced by pine.

Figure 3.8. Pine forest with *Calamagrostis arundinacea* in the understorey

Pine forests allow much more sunshine through the canopy in comparison with fir and spruce forest. *Rhododendron luteum* is the most common dominant of the shrub layer, while the herbaceous layer is composed of xerophilous species of grasses, sedges and forbs: *Calamagrostis arundinacea, Trisetum rigidum, Poa nemoralis, Brachypodium pinnatum, Carex humilis, Potentilla argentea, Galega orientalis, Anthyllis vulneraria, Coronilla varia, Lathyrus roseus, Geranium sanguineum, Helianthemum nummularium, Pyrola chlorantha, Moneses uniflora, Linaria genistifolia*, etc. *Sedum gracile, Campanula collina* and some other species are typical on rock outcrops.

Figure 3.9. Pinus sylvestris near the timberline in the Teberda valley

3.5. Timberline

The mean altitude of the timberline in the region near Teberda is 2100–2300 m above sea level (Elenevskii 1940, Grebenshchikov & Ozenda in Gerasimov et al. 1980), but the real amplitude of the timberline position is much wider and depends on the topographic features including the slope orientation, as well as the influence of avalanches. In winter and spring, avalanches come down from the highlands to destroy forests and dramatically change the landscape.

Figure 3.10. Crooked-stem birch (*Betula litwinowii*) forests on the avalanche fan in the Teberda valley

Northward of the Reserve, on the slopes facing the Teberda river valley the timberline can be found at 2100 m above sea level. In the upper part of the Teberda river basin it is slightly higher: fir and spruce forests on northern slopes reach 2100–2200 m, while on the slopes of the watershed ridge — even 2450 m above sea level. At these altitudes coniferous forests are restricted to the convex parts of slopes, concave portions being the main conduits for avalanches. These areas are occupied by crooked-stem birch (*Betula litwinowii*) and beech (*Fagus orientalis*) forests (Fig. 3.10). The border of such krummholz forests could lie at a comparatively low altitude (1700–1900 m), especially along the bottoms of the glacial valleys of the

Alibek, Buulgen, Dombai-Ulgen, and Amanauz rivers. In such cases the timberline becomes dentate in shape with elevational difference of 500–700 m.

Another type of forest beneath the timberline is open maple (*Acer trautvetteri*) parklands along the bottoms of the Alibek and Dombai-Ulgen river valleys, where the timberline is found at the altitude of 1800–1900 m. The concave areas on southern slopes along these river valleys, where avalanches happen once every 20–50 years, are occupied by distinct mixed forests, with the tree stands composed of straight-stemmed birch (*Betula litwinowii*) and aspen (*Populus tremula*) with a frequent admixture of maple (*Acer trautvetteri*), alder (*Alnus incana*), goat willow (*Salix caprea*) and mountain ash (*Sorbus aucuparia*).

The timberline at the same time serves as the lower border of the subalpine communities. The subalpine zone includes crooked-stem and krummholz communities; they are especially common on the periphery of the presently glaciated region on the northern slopes of the Greater Caucasus mountain range and along its spurs.

Wet air brought in by Mediterranean cyclones from the Black Sea crosses the Greater Caucasus range, mainly through the mountain passes. The annual precipitation on some slopes near the passes is often greater than 2,000 mm and is evenly distributed through the seasons. High precipitation ensures high accumulation of snow in the highlands, a low position of the snow line, and the presence of young glaciers. Crooked-stem and krummholz communities are widely distributed (50–75% of the total forest area) on the north-facing slopes and in the bottoms of glacial valleys in the upper part of the Teberda river basin with their altitudinal amplitude between 1700 and 2400 m above sea level (Tumadzhanov 1960a,b).

There are two types of crooked-stem forests in the upper part of the Teberda river: beech (*Fagus orientalis*) and birch (*Betula litwinowii*) (Tumadzhanov 1960a,b). Beech crooked-stem forests are characteristic of the lower parts of the slopes, while birch ones occupy higher positions, and sometimes are found along the wide bottoms of U-shaped glacial valleys (Fig. 3.11). Height of trees in these communities is about 2–3.5 m; often there are both shrub and herbaceous layers under their canopy.

Evergreen *Rhododendron caucasicum* is the most typical shrub of the beech crooked-stem forests; often it forms its own zone above the crooked-stem forests. Isolated *Rhododendron* shrubs may be encountered even in the alpine zone, at 2600–2900 m above sea level. Very often one can find two other ericaceous shrubs under the canopy of beech crooked-stem forests — *Vaccinium arctostaphylos* and *Rhododendron luteum*.

Figure 3.11. Crooked-stem beech forests at the wide bottom of U-shaped glacial valley (North Klukhor)

Tall forest forbs (*Inula magnifica, Telekia speciosa, Senecio platyphylloides, Ligusticum alatum, Campanula lactiflora*), that sometimes reach 2–2,5 m in height, as well as usual forest grasses and herbs (*Calamagrostis arundinacea, Galium odoratum, Calamintha grandiflora*, etc.) are usually found in the herbaceous layer of beech crooked-stem forests.

In the tree stands of birch crooked-stem forest the admixture of mountain ash (*Sorbus aucuparia*) is typical. *Rhododendron caucasicum* predominates in the shrub layer of these forests and birch in the tree layer (Leskov 1932, Tumadzhanov 1960a,b). A lot of typical forest forbs and shrublets compose the herbaceous layer of birch crooked-stem forests on the gentle slopes: *Calamagrostis arundinacea, Vaccinium myrtillus, Pyrola media, Gymnocarpium dryopteris, Goodyera repens, Vaccinium vitis-idaea, Oxalis acetosella*, etc. The moss layer that consists of typical forest species (*Hylocomium splendens, Pleurozium schreberi, Polytrichum commune, Dicranum scoparium,* etc.) is usually well developed too. Open birch crooked-stem woodlands with reed grass (*Calamagrostis arundinacea*), absolutely prevailing in the herbaceous layer, are found on the convex slopes.

Under the birch canopy, the same subalpine tall forbs are found as in the beech crooked-stem forests; in addition, *Heracleum asperum, H. freynianum, Valeriana alliariifolia*, and *Dryopteris filix-mas* are found here.

Very peculiar birch crooked-stem forests occur on the steep concave slopes along the usual avalanche tracks. Often such tracks occur not only in the crooked-stem forest stripe, but also in the typical forest zone and are marked by massive alluvial fans in the river valleys. Quite often there is no developed soil cover, but only rocky outcrops at the bottoms of these concave areas. When it is so, pure crooked-stem forests are developed with prostrate and crawling birches and some herbs among them, which could be either forest species or alpine and subalpine ones (*Anemone caucasica, Fritillaria latifolia* and *F. collina, Gagea* spp., *Ornithogalum oligophyllum*), brought in by avalanches.

Further to the north, because of the decreasing precipitation the appearance and the composition of subalpine crooked-stem forests change dramatically. *Fagus orientalis, Acer trautvetteri, Vaccinium arctostaphylos* and some other species disappear, while *Betula litwinowii, Sorbus aucuparia, Salix pantosericea* become dominant.

Both on the north-facing and south-facing slopes, as well as at the bottoms of glacial valleys, crooked-stem and krummholz communities are combined with open and picturesque subalpine forests. A lot of typical forest forb and shrublet species (*Vaccinium myrtillus, V. vitis-idaea, Goodyera repens, Gymnocarpium dryopteris, Neottia nidus-avis, Oxalis acetosella, Pyrola media*, etc.) penetrate into the highlands under the canopy of birch or *Rhododendron caucasicum*.

3.6. Subalpine belt

On the north-facing slopes of some valleys (Dombai-Ulgen and Alibek) there is a rich combination of various alpine, subalpine, rock and scree elements. On the south-facing slopes of these valleys subalpine meadows cover more area and determine (together with crooked-stem and krummholz communities) the landscape character.

Most of the authors distinguish three types of Caucasian subalpine meadows: grass, grass-forb and forb (Makhatadze in Gulisashvili et al. 1975). All of them are represented in the northwestern Caucasus and can be found in the upper part of the Teberda river basin.

In this region, *Calamagrostis arundinacea, Bromus variegatus* (especially in the northern part of the Reserve), *Festuca varia* (mainly on southern slopes), *Helictotrichon versicolor,* and *Agrostis vinealis* are the most widespread dominants of the subalpine grasslands. Many other grass species (*Festuca djimilensis, F. ovina, F. rubra, Anthoxanthum odoratum,*

Phleum alpinum, P. phleoides, Nardus stricta, Koeleria eriostachya, Briza marcowiczii, Trisetum flavescens, Brachypodium pinnatum) can be found in subalpine grasslands as an admixture. Very often the grasslands of this type are dominated by two grass species; the most typical combinations are *Bromus variegatus - Agrostis vinealis, Bromus variegatus - Calamagrostis arundinacea, Calamagrostis arundinacea - Festuca djimilensis, Bromus variegatus - Brachypodium pinnatum, Bromus variegatus – Deschampsia flexuosa.* The grass diversity in subalpine meadows can reach 8–10 species per 100 m^2 (Shiffers 1953). The altitudinal range of *Festuca varia* is more than 1 km — from 1800 to 3100 m above sea level. One can find this species dominating grasslands at various elevations mainly on southern slopes.

Figure 3.12. Betonica macranta in subalpine meadow (Foto by A. Zakharov)

In some other cases the percentage of grasses and forbs is close to being equal — these are the grass-forb subalpine meadows. The pure forb meadows are found in the upper reaches of river valleys, where one can find *Anthemis marschalliana, Anemone fasciculata, Astrantia maxima, Betonica macrantha* (Fig. 3.12), *Campanula collina, Centaurea cheiranthifolia, C.*

montana, Erigeron caucasicus, Hedysarum caucasicum, Inula orientalis, Linum hypericifolium, Pedicularis condensata, P. atropurpurea, Polygonum alpinum, P. bistorta (incl. *P. carneum*), *Pulsatilla aurea, Ranunculus caucasicus, Scabiosa caucasica, Trifolium canescens, Trollius ranunculinus, Veratrum album* (incl. *V. lobelianum*), *Veronica gentianoides* and other forbs. The number of forbs sometimes reaches 50 species per 100 m^2 with the typical height of the forb layer at 60–90 cm.

Among the numerous dominants of the subalpine forb meadows the most frequent are *Anemone fasciculata, Polygonum alpinum* and *P. bistorta* (Fig. 3.13), *Inula orientalis* (Fig. 3.14) and *Geranium renardii.* The landscape quality varies during the year depending of the main blooming forbs: in the middle of April on dry slopes it is *Pulsatilla albana*, in May it is *Primula ruprechtii*, typical of wetter habitats, from the end of May up to mid-June it is *Anemone fasciculata*, in mid to late June it is *Veronica gentianoides* and *Myosotis alpestris*, while in July the chief blooming species are *Polygonum bistorta* and *P. alpinum*, at the end of July *Betonica macrantha,* and at the beginning of August *Scabiosa caucasica.*

Subalpine tall forb vegetation is composed of *Senecio platyphylloides, S. macrophyllus, Cephalaria gigantea, Campanula lactiflora, Heracleum asperum, Ligusticum alatum, Valeriana alliariifolia,* and some other species. These herbs, 1.5–2.5 m high, prevent smaller species from growing in the deep shade of the giant herb canopy. This vegetation is usually found among crooked-stem forests and *Rhododendron caucasicum* thickets along the timberline.

Subalpine and alpine zones are treated differently by Western European and Russian botanists. Traditionally the area above the timberline is considered as the alpine zone, in the broad sense, in the western European mountains (Ellenberg 1986, Walter & Breckle 1994). Most Russian botanists, however, (Grossheim 1948, Shiffers 1953, Kononov 1957, Gulisashvili et al. 1975, Voskanyan 1977) do not hold this point of view. Traditionally, they divided the area into three zones, or "belts": subalpine, alpine (*sensu stricta*) and subnival. The subalpine zone lies near and above the actual timberline upward from the climatic timberline. We can easily distinguish this belt from the alpine belt by the type of plant communities developing in snowbed depressions (Onipchenko 2002). Communities of the *Mulgedio-Aconitetea* are typical for such places in the subalpine belt. In the alpine belt (*sensu stricta*) communities belonging to another class - *Salicetea herbaceae* (alpine "carpets") - occupy snowbed depressions. A more favourable temperature regime in the subalpine belt leads to intensive spring snowmelt, which prolongs the vegetative season in snowbeds as compared with the alpine belt.

Figure 3.13. Polygonum bistorta dominates in subalpine meadows

Figure 3.14. Inula orientalis in a subalpine tall forb community

3.7. Alpine belt

The subalpine zone gives way to the alpine zone at the altitude of 2300–2800 m above sea level. The position of the ecotone between the two highland belts varies greatly and depends on many factors: in the southern half of the Reserve this transition occurs at the altitude of 2600–2700 m on south-facing slopes and at the altitude of 2250–2300 m on the north-facing ones.

The alpine belt occupies extensive areas in the upper parts of lateral spurs (in the range from 2500 to 3000 m above sea level) northwards of the main divide, which makes up about 35% of the total Teberda Nature Reserve area (Utyakov 1960).

One can observe a great diversity of plant communities in the alpine belt (see Chapter 4). Most authors distinguish the following physiognomic types of the alpine vegetation: alpine meadows, alpine "carpets" (snowbeds), heaths, krummholz communities, alpine fens, vegetation of screes and rocks (Bush 1935, Grossheim 1948, Schiffers 1953, Kononov 1957, Tumadzhanov 1963, 1980, Gulisashvili 1964, Dolukhanov 1966, Gulisashvili et al. 1975, Vorob'eva 1977, Rabotnov 1987, etc.). A few hundred vascular plant species grow here. Both α-diversity and β-diversity of the alpine belt reflect the diversity of habitats and ecological conditions within the zone. F. M. Vorob'eva (1977) lists the following elements of the alpine landscape: a) valley bottoms (mainly glacial), filled with moraine and fluvioglacial deposits; b) moraine ridges; c) avalanche debris fans; d) bottoms of former lakes and lake-like glacial reservoirs, filled with fluvioglacial till; e) steep slopes, often notched by avalanches; f) gentle slopes and peneplain surfaces; g) mountain cirques with vast stone fields, screes, taluses, rock streams, snow patches and glaciers; h) rocks.

The opportunities for snow accumulation as well as the duration and thickness of the snow cover on slopes vary greatly and depend on various factors, with the orientation and steepness of slopes being the most important (Schröter 1926, Walter & Breckle 1994, Franz 1979, Reisigl & Keller 1987, Körner 1999, Onipchenko 2002).

Depending on the thickness of the snow cover in winter, communities of various types provide the following sequence: alpine "carpets" (snow thickness is equal to 4–8 m), alpine meadows (1–3 m), heaths and heath meadows, communities on rocks and screes, which are habitats free of snow or with insignificant snow cover in winter (Rabotnov 1987).

The development of plants in alpine snowbeds occurs under an extremely short growing period — about two months. Alpine carpets are formed by a dense cover (1.5 to 4 cm height) of low-lying perennial plants: *Campanula tridentata* (Fig. 3.15), *Sibbaldia procumbens* (Fig. 3.16), *Trifolium polyphyllum, Anthemis cretica, Taraxacum stevenii, Anemone speciosa,*

Pedicularis nordmanniana, Gnaphalium supinum, Carum caucasicum, Ranunculus brachylobus, Hyalopoa pontica, etc. They cover the terrain mainly between 2,900 and 3,200 m, but in the upper reaches of glacial valleys they can come down to even 2,200 m above sea level. Although plant productivity is low in the alpine carpets, outside the Reserve they are used intensively as summer pastures for grazing livestock.

Figure 3.15. Campanula tridentata – dominated alpine snowbed (Foto by A.Zakharov)

Various alpine meadows are the most common highland ecosystems of the Teberda Nature Reserve and in the northwestern Caucasus in general. These ecosystems are the richest in biodiversity providing some of the most picturesque landscapes. Alpine meadows are usually dominated by few forb species; besides, there are also grass and sedge alpine meadows. The most frequent species met in the alpine belt are *Alchemilla caucasica, Anthemis marschalliana, Anthyllus vulneraria, Aster alpinus, Astragalus oreades, Bromus variegatus, Campanula collina, Erigeron caucasicus, Galium verum, Gentiana pyrenaica, G. septemfida, G. verna, Geranium gymnocaulon, Hedysarum caucasicum, Matricaria caucasica, Myosotis alpestris, Pedicularis condensata, Potentilla crantzii, Polygonum bistorta,*

Primula algida, P. ruprechtii, Ranunculus oreophilus, Sempervivum caucasicum, Luzula multiflora, Senecio taraxacifolius, S. aurantiacus, Solidago virgaurea, Veronica gentianoides, Viola altaica and others. *Festuca varia* is the most typical dominant of the alpine grasslands found on south-facing slopes (Fig. 3.17).

Figure 3.16. Sibbaldia procumbens dominates in heavily grazed alpine snowbeds

As in the case of the subalpine belt, the dominants of the alpine meadows and snowbed communities demonstrate picturesque aspects created by various flowering forbs, which replace each other through the growth season.

The most chionophobic communities of the alpine belt are the alpine heaths, which occupy windward stony slopes and crests between 2700 and 3000 m above sea level. Heaths have common features both with tundra and meadows. Epigeal lichens (*Cetraria islandica, C. nivalis, Cladina mitis, Thamnolia vermicularis*) play an important role in these communities. Several grasses and sedges, reflective of poor soils, are among them: *Festuca ovina, Carex sempervirens, Kobresia capillifolia*. A few forb species (*Anemone speciosa, Campanula tridentata, Gentiana pyrenaica, Trifolium polyphyllum, Antennaria dioica*) contribute to the picturesque aspect.

Along the glaciers' edges, in intermoraine depressions on bottoms of valleys and near springs, one can find small alpine fens with *Sphagnum* mosses. *Carex cespitosa, Primula auriculata, Saxifraga hirculus* and *Caltha palustris* (incl. *C. polypetala*) are the most common species of these wetlands (see Chapter 4).

Figure 3.17. Festuca varia – dominated grassland

Considerable areas in the alpine belt are covered by rocks and screes. Hygrophilous, cryophilous and psychrophilous plants inhabit them. There are peculiar cryophilous cushion plants (*Draba rigida, Potentilla divina, Minuartia circassica, Gypsophyla tenuifolia, Saxifraga juniperifolia*) and chasmophytes (*Saxifraga paniculata, S. caucasica, S. exarata, Lloydia serotina, Viola biflora, Murbeckiella huetii, Chamaesciadium acaule, Valeriana alpestris, Campanula bellidifolia*) (Ovchinnikov 1941).

Delphinium caucasicum, Dentaria bipinnata, Draba scabra, Saxifraga flagellaris, Cerastium kasbek, Nepeta supina – Fig. 3.18, *Lamium tomentosum* – Fig. 3.19, *Veronica minuta* – Fig. 3.20, *Senecio karjaginii,* and *Jurinella moschus* are the species typical of screes, especially of moving ones. When a scree is overgrown with vegetation, it is covered by either meadow or heath, depending on the slope aspect and other factors.

3.8. Subnival belt

The upper reaches of the alpine belt (both in the Teberda region and all of the northwestern Caucasus) usually coincide with the snow line, the mean altitude of which is 2900–3000 m (Tushinskii 1957). It does not make sense to distinguish subnival and nival belts, since there really is no difference between the two both in terms of plant species or communities (Grossheim 1948). Very special conditions are found above the snow line, where only crevices in collapsing rocks and cliffs provide any opportunity for plants to grow (Fig. 3.21). The spatial distribution of plants is determined here mainly by availability of such favourable sites for plant establishment and survival.

Figure 3.18. Nepeta supina on a dry scree in the Teberda Reserve

The snow line in the region runs at about 3,000 m. This elevation can be considered the borderline dividing the alpine and subnival zones. Sparse plant communities on rocks and screes alternating with glaciers and snowfields are commonly found in the subnival zone. According to Onipchenko (2002), the upper limit of vascular plant distribution in Teberda is approximately 3,750 m. Several species (*Saxifraga exarata, S. sibirica, Draba rigida, D. siliquosa, Minuartia imbricata* – Fig. 3.22, *Primula amoena, Senecio karjaginii, Potentilla gelida, Carum caucasicum, Hyalopoa pontica*) reach this elevation on the southern slope of Mt. Kyshkadzher.

Figure 3.19. Lamium tomentosum is a typical plant for alpine and subnival screes

Figure 3.20. Veronica minuta grows on moist screes

Figure 3.21. Subnival belt on Mt. M. Khatipara

Figure 3.22. Minuartia imbricata on subnival rocks

In summary, the following altitudinal vegetation zones can be distinguished in the Teberda region (from the bottom up): 1) steppes; 2) broad-leaved forests; 3) dark coniferous forests; 4) subalpine meadows and crooked-stem forests; 5) alpine meadows. Such a sequence is typical for the North Caucasian subprovince of the Caucasian Euxine forest province, which, in turn, is included in the European broad-leaved forest region of the Holarctic kingdom of vegetation (Isachenko & Lavrenko 1980).

Chapter 4

Alpine plant communities: syntaxonomy, floristic composition and richness

4.1. Floristic sources

The flora of Teberda Reserve is rather rich, given its small territory. A recently revised checklist included 1,133 vascular plant species (Vorob'eva & Onipchenko 2001). The flora of bryophytes used to contain 73 liverworts and 227 leafy mosses (Ignatova et al. 1990), but 67 moss species have been added since (Onipchenko & Ignatova 1996). Similar floristic richness is known for other mountain regions in Europe, for example, Davos (Schibler 1937).

The botanical nomenclature of the Caucasus plants is not well developed. Most of the regional floras (Grossheim 1949, Kosenko 1970, Galushko 1978-80, Vorob'eva & Kononov 1991) as well as a general vascular plant checklist for Russia (Czerepanov 1995) are using "micro-species" (splitting approach). We had to study the nomenclature of each species separately in different publications, including a current review of some groups and floras of neighbouring countries (Davis 1965-1980). Full names and main synonyms of the vascular plants according to the recent checklist (Vorob'eva & Onipchenko 2001) are represented in the species index. We use Wirth (1980) as the main source of macrolichen names. Bryophyte nomenclature is based on checklists for the former Soviet Union (Ignatov & Afonina 1992, Konstantinova et al. 1992).

Main life forms of the vascular plants are herbaceous polycarpic perennials (75%), annuals (11%), trees and shrubs (8%), and monocarpic perennials (4%) (Vorob'eva & Kononov 1991). There are 23 threatened

V.G. Onipchenko (ed.), Alpine Ecosystems in the Northwest Caucasus, 55–75.

species listed in the "Red Data Book of Russia" found in the Reserve.

Four main vegetation zones are well represented in the Reserve (see Chapter 3), namely the forest, subalpine, alpine and subnival zones. The total number of plant species decreases from the forest through the subalpine and the alpine to the subnival zone: 900, 373, 206, and 122 species respectively.

4.2. Syntaxonomy of high-mountain communities

A detailed description and floristic classification of the alpine communities of Teberda Reserve are represented in our special publication (Onipchenko 2002). Here we present a list of syntaxa (*prodromus*) with brief characteristics and diagnostic species of associations.

4.2.1. Screes and pebbles

***Thlaspietea rotundifolii* Br.-Bl. 1948**

This class comprises open plant communities on unstable stony substrates (screes, rock streams, gravel bars along rivers, etc.). They can be relatively stable, but their stability depends upon the permanent disturbance regime and/or extremely hard climatic conditions of the subnival zone.

***Chaerophylletalia humilis* Onipchenko 2002**
***Chaerophyllion humilis* Onipchenko 2002**
***Veronico minutae-Chaerophylletum humilis* Onipchenko 2002**

(Chaerophyllum humile, Delphinium caucasicum, Corydalis alpestris, Dentaria bipinnata, Eunomia rotundifolia, Noccaea pumila, Draba scabra, Draba siliquosa, Alopecurus dasyanthus, Potentilla gelida, Cruciata valentinae)

***Androsacetalia alpinae* Br.-Bl. in Br.-Bl. et Jenny 1926**
***Murbeckiellion huetii* Onipchenko 2002**
***Hyalopoo ponticae-Oxyrietum digynae* Onipchenko 2002**

(Carex atrata, Gnaphalium supinum, Hyalopoa pontica, Primula amoena, Cladonia pyxidata, Senecio taraxacifolius, Carex pyrenaica)

***Scrophulario variegatae-Epilobietum dodonaei* Onipchenko 2002**

(Scrophularia variegata, Trifolium spadiceum, Pohlia filum, Betula litwinowii, Poa nemoralis)

***Allosuro-Athyrion alpestris* Nordhagen 1936**
***Dicranoweissio crispulae-Rubetum idaei* Onipchenko 2002**

(Dryopteris filix-mas, Rubus idaeus, Calamagrostis arundinacea, Dicranoweisia crispula, Dicranum scoparium, Lescuraea saxicola, Juniperus communis, Hypnum cupressiforme, Sempervivum caucasicum)

Epilobietalia fleischeri **Moor 1958**
Salicion incanae **Aichinger 1933**
Silene compactae-Salicetum purpureae **Onipchenko 2002**
(Salix purpurea, Rumex acetosella, Silene compacta, Taraxacum officinale aggr., *Trifolium repens, Agrostis stolonifera, Anthyllis vulneraria, Cicerbita racemosa, Lotus corniculatus, Alnus incana, Cerastium fontanum, Gnaphalium sylvaticum)*

4.2.2. Rocks

Asplenietea trichomanis **(Br.-Bl. in Meier et Br.-Bl. 1934) Oberd. 1977**

Plant communities of this class occupy rock outcrops in all vegetation zones. Rock vegetation is sparse, so the influence of the surrounding closed vegetation on the rock communities is considerable. It seems that elevation is a major factor controlling the floristic composition of the rock communities (Onipchenko 2002).

Androsacetalia multiflorae **Br.-Bl. in Meier et Br.-Bl. 1934**
Gypsophilion tenuifoliae **Onipchenko et Gorbachevskaya 2002**
Potentilletum divinae **Onipchenko et Gorbachevskaya 2002**
(Potentilla divina, Campanula tridentata, Carex sempervirens, Luzula spicata, Carum caucasicum, Trifolium polyphyllum, Empetrum nigrum, Eritrichium caucasicum, Cetraria islandica, Anemone speciosa, Helictotrichon versicolor)
Astragaletum levieri **Onipchenko et Gorbachevskaya 2002**
(Ranunculus oreophilus, Alopecurus glacialis, Festuca brunnescens, Astragalus levieri, Encalypta rhaptocarpa, Thymus nummularius, Jurinea coronopifolia)

Thalictro foetidi-Asplenion **Onipchenko et Gorbachevskaya 2002**
Galio valantioides-Polypodietum vulgaris **Onipchenko et Gorbachevskaya 2002**
(Polypodium vulgare, Homalothecium sericeum, Galium valantioides, Grimmia elatior, Homalia besseri, Homalothecium philippeanum, Frullania dilatata, Senecio renifolius, Draba supranivalis)

Thymo-Seselietum petraei **Onipchenko et Gorbachevskaya 2002**
(Thymus marschallianus, Seseli petraeum, Centaurea dealbata, Stachys recta, Grimmia laevigata, Hedwigia ciliata, Melica ciliata, Grimmia ovalis, Sedum telephium)

4.2.3. Fens

Scheuchzerio-Caricetea fuscae **R.Tx. 1937**
Communities of *Scheuchzerio-Caricetea fuscae* are common along small rivers and springs and near lakes on the flat bottoms of U-shaped valleys. They occupy few areas, but they are typical components of the vegetation. Among diagnostic species of the class (Dierssen 1982, Steiner 1993) are *Carex nigra, Aulacomnium palustre* and *Bryum pseudotriquetrum.*
Scheuchzerietalia palustris **Nordhagen 1937**
Caricion lasiocarpae **Vanden Berghen in Lebrun et al. 1949**
Caricetum rostratae **Osvald 1923 em Dierssen 1982**
(*Carex limosa, Carex rostrata*)

Caricetalia fuscae **Koch 1926 em Br.-Bl. 1949**
Caricion fuscae **Koch 1926 em Klika 1934**
Caro caucasici-Caricetum nigrae **Onipchenko 2002**
(Carum caucasicum, Polygonum viviparum, Cerastium cerastoides, Sibbaldia procumbens)
Swertio ibericae-Caricetum nigrae **Onipchenko 2002**
(Potentilla erecta, Swertia iberica, Crepis glabra, Campylium stellatum, Dactylorhiza urvilleana, Pinguicula vulgaris, Veratrum album, Geranium sylvaticum, Eleocharis quinqueflora, Scapania irrigua)

4.2.4. Cold water springs

Montio-Cardaminetea **Br.-Bl. et R.Tx. ex Klika et Hadac 1944 em Zechmeister 1993**
These moss-rich communities have a well-represented set of diagnostic species of higher syntaxa (*Montio-Cardaminetea, Montio-Cardaminetalia, Cardamino-Montion,* see Zechmeister 1993). *Philonotis fontana, Palustriella commutata, Bryum pseudotriquetrum, B. schleicheri* and *Brachythecium rivulare* are the most common.

Regional specificity of this community type depends on the vascular plants. All diagnostic species of the association are typical species of cold and moist habitats.

Montio-Cardaminetalia **Pawlowski 1928 em Zechmeister 1993**
Cardamino-Montion **Bl.-Bl. 1926 em Zechmeister 1993**
Cerastio cerastoidis-Cardaminetum uliginosi **Onipchenko 2002**
(Cardamine uliginosa, Cerastium cerastoides, Epilobium algidum, Hyalopoa pontica, Saxifraga sibirica) – Fig. 4.1.

Figure 4.1. Cold water spring with dense moss mat of *Cerastio cerastoidis-Cardaminetum uliginosi*

4.2.5. *Kobresia* grasslands

Carici rupestris-Kobresietea bellardii **Ohba 1974**

This is a circumpolar class of cold steppe-type grasslands, developing on windward crests and slopes (Oberdorfer 1978, Grabherr 1993a). Species of *Kobresia* are common dominants of the communities. The class is not well represented in the northwestern Caucasus due to mild winters and high amounts of precipitation. We suggest two associations within the alliance *Oxytropido-Elynion*, order *Oxytropido-Kobresietalia*. *Lloydia serotina*,

Polygonum viviparum, Potentilla crantzii, P. gelida, Aster alpinus and *Oxytropis kubanensis* (vicarious species of *O. campestris*) represent the diagnostic species set of the syntaxa. The communities have a considerable floristic similarity with the communities of *Anemonion speciosae* (*Juncetea trifidi*, see below), but we consider the *Kobresia* communities in a separate class according to the established European tradition (Grabherr 1993a).

***Oxytropido-Kobresietalia* Oberdorfer ex Albrecht 1969**

***Oxytropido-Elynion* Br.-Bl. 1949**

***Drabo scabri-Kobresietum schoenoidis* Onipchenko 2002**

(Kobresia schoenoides, Draba scabra, Polygonum bistorta, Pedicularis caucasica)

***Alchemillo-Kobresietum capilliformis* Tsepkova 1987.**

(Kobresia capillifolia, Vaccinium vitis-idaea, Rhytidium rugosum, Carex caryophyllea, Antennaria dioica, Hypnum cupressiforme, Bromus variegatus, Primula ruprechtii)

4.2.6. Alpine heaths

***Juncetea trifidi* Hadac 1946**

In Teberda Reserve this class comprises alpine lichen heaths (alpine tundra) and fellfields on windward crests and slopes on acid soils (Onipchenko 2002). Diagnostic species of *Juncetea trifidi, Caricetalia curvulae* are represented by *Helictotrichon versicolor, Luzula spicata, Minuartia recurva, Trifolium polyphyllum* (geographical vicarious species of *Trifolium alpinum*, Fig. 4.2*), Euphrasia ossica (*geographical vicarious species of *Euphrasia minima)* (Oberdorfer 1978; Grabherr 1993b). Due to the severe ecological regime (deep winter freezing, frequent summer drought, thin soil profile, great day-to-night temperature fluctuations), the communities have low production rates, high species richness and a significant lichen cover.

***Caricetalia curvulae* Br.-Bl. in Br.-Bl. et Jenny 1926**

***Anemonion speciosae* Minaeva ex Onipchenko 2002** – Fig. 4.3.

***Campanulo ciliatae-Chamaesciadietum acaulis* Minaeva et Onipchenko 2002**

(Anthyllis vulneraria, Dactylina madreporiformis, Chamaesciadium acaule, Primula algida, Pulsatilla albana, Aetheopappus caucasicus, Campanula ciliata, Gentiana aquatica, Thymus nummularius, Jurinella moschus, Campanula bellidifolia)

***Pediculari comosae-Eritrichietum caucasici* Minaeva et Onipchenko 2002**

(Gentiana pyrenaica, Carum caucasicum, Vaccinium vitis-idaea, Rhytidium rugosum)

Figure 4.2. Trifolium polyphyllum in the alpine belt of Mt. M.Khatipara

Figure 4.3. Anemone speciosa in an alpine lichen heath

4.2.7. Alpine grasslands

***Calluno-Ulicetea* Br.-Bl. et R.Tx. ex Klika et Hadac 1944**

This class and order include mountain and lowland grasslands on poor acidic soils. The position of the alpine grasslands within the order and the class has been a matter of extensive discussion by different authors (Grabherr 1993b, Ellmauer 1993, Julve 1993, Krahulec & Moravec 1995, Pott 1995). Due to the high frequency of the diagnostic species of the order and considerable ecological and floristic dissimilarity from the *Anemonion*, we treat the alpine grasslands of the northwestern Caucasus within *Nardetalia*. *Anthoxanthum odoratum, Nardus stricta, Luzula multiflora, Antennaria dioica, Deschampsia flexuosa, Phleum alpinum* are all well represented as diagnostic species in the grasslands of Teberda. The communities have been exposed to different intensities of grazing. They occupy the most favourable slopes within the alpine zone, where winter snow accumulation is moderate (1-3 m). Therefore, production of the grasslands is relatively high.

Figure 4.4. Festuca varia grassland with *Galium verum* in flower

***Nardetalia* Oberd. ex Preising 1949**

***Violo altaicae-Festucion variae* Onipchenko 2002**

***Violo altaicae-Festucetum variae* Rabotnova et Onipchenko 2002**

(Festuca varia, Scorzonera cana, Ajuga orientalis, Galium verum, Viola altaica, Carex umbrosa, Alchemilla caucasica, Helictotrichon versicolor, Minuartia circassica, Silene saxatilis, Calamagrostis arundinacea) – Fig. 4.4.

***Hedysaro caucasicae-Geranion gymnocauli* Onipchenko 2002**

***Hedysaro caucasicae-Geranietum gymnocauli* Rabotnova et Onipchenko 2002**

(Geranium gymnocaulon, Pulsatilla aurea, Hedysarum caucasicum, Carum meifolium, Pedicularis condensata) – Fig. 4.5-4.6.

Figure 4.5. Geranium gymnocaulon creates a dense canopy

Figure 4.6. Hedysarum caucasicum in an alpine meadow

4.2.8. Snow bed communities

***Salicetea herbaceae* Br.-Bl. 1948**

The order *Hyalopoetalia ponticae* combines alpine communities developing on siliceous rock on acid soils with considerable snow accumulation (snowbeds) and a short vegetation season. Some species of *Salicetea herbaceae* are typically found here (*Gnaphalium supinum, Sibbaldia procumbens, Cerastium cerastoides, Polytrichum sexangulare,* and *Kiaeria starkei*) (Onipchenko 2002). The class is a well-studied syntaxon in Europe (Braun-Blanquet 1975, Dierssen 1984, Sandra et al. 1988, Englisch 1993). The diagnostic set of our order includes such Caucasian species as *Taraxacum stevenii, Sedum tenellum, Hyalopoa pontica.* The order can be considered as a geographical vicarious syntaxon of *Salicetalia herbaceae* Br.-Bl. 1926. We distinguish two alliances in the order, namely *Hyalopoion ponticae* (alpine closed "carpet"-like communities) and *Saxifragion sibiricae* (open rock and scree communities).

***Hyalopoetalia ponticae* Onipchenko 2002**

***Hyalopoion ponticae* Rabotnova et Onipchenko 2002**

Ranunculetum brachylobi **Pokarzhevskaya et Onipchenko 2002**

(Ranunculus brachylobus, Corydalis conorhiza, Geranium gymnocaulon)

Hyalopoo ponticae-Pedicularietum nordmannianae **Rabotnova et Onipchenko 2002**

(Catabrosella variegata, Nardus stricta, Stereocaulon alpinum, Desmatodon latifolius, Carex oreophila) – Fig. 4.7.

Figure 4.7. Alpine snowbed with *Pedicularis nordmanniana* in flower

Saxifragion sibiricae **Onipchenko 2002**

Saxifragetum sibiricae **Onipchenko et Lubeznova 2002**

(Saxifraga sibirica, Minuartia imbricata, Matricaria caucasica, Veronica minuta, Murbeckiella huetii, Draba scabra)

4.2.9. Subalpine meadows

Mulgedio-Aconitetea **Hadac et Klika in Klika et Hadac 1944**

Subalpine meadows and tall herbaceous communities are well represented in the Teberda Reserve. Such diagnostic species of the class as

Millium effusum, Geranium sylvaticum, Campanula latifolia, Rumex alpestris are common in the Caucasian communities (Buslik 1990, Karner & Mucina 1993). We consider the communities within two orders - *Calamagrostietalia villosae* and *Rumicetalia alpini.*

***Calamagrostietalia villosae* Pawlowski et al. 1928**

***Calamagrostion arundinaceae* Oberd. 1950**

***Betonici macranthae-Calamagrostietum arundinaceae* Onipchenko 2002**

(Betonica macrantha, Bupleurum falcatum, Leontodon hispidus, Myosotis alpestris, Campanula collina, Anthemis melanoloma, Hedysarum caucasicum, Festuca varia, Trifolium canescens, Geranium renardii, Thesium alpinum, Tragopogon reticulatus, Polygala alpicola, Rhinanthus minor, Primula veris, Festuca djimilensis, Gentiana septemfida)

Figure 4.8. Ruderal tall herbaceous community with *Rumex alpinus* as the main dominant

***Rumicetalia alpini* Mucina in Karner et Mucina 1993**

***Rumicion alpini* Rubel ex Klika in Klika et Hadac 1944**

***Anthrisco sylvestris-Rumicetum alpini* Onipchenko 2002**

(Urtica dioica, Anthriscus sylvestris, Galeopsis bifida, Symphytum asperum, Senecio platyphylloides) – Fig. 4.8.

***Cephalario giganteae-Ligusticetum alani* Onipchenko 2002**

(Ligusticum alatum, Aconitum orientale, Vicia cracca, Dactylis glomerata, Cephalaria gigantea, Lapsana communis)

***Poetum longifoliae* Onipchenko 2002**

(Alopecurus pratensis, Poa longifolia, Senecio integrifolius)

4.2.10. Elfin dwarf schrub

***Loiseleurio-Vaccinietea* Eggler 1952**

The class and order combine Arctic and alpine dwarf scrub communities on poor acidic soils (Grabherr 1993c). Physiognomic features (i.e., dwarf shrub dominance) are important for the diagnosis of this community, since the diagnostic species set is represented by a few species with wide ecological ranges. There are two such species in the study area, *Vaccinium vitis-idaea* and *Empetrum nigrum*.

***Rhododendro-Vaccinietalia* Br.-Bl. in Br.-Bl. et Jenny 1926**

***Rhododendrion caucasici* Onipchenko 2002**

***Lerchenfeldio-Rhododenretum caucasici* Onipchenko et Sennov 1992**

(Rhododendron caucasicum, Vaccinium myrtillus, Solidago virgaurea, Senecio taraxacifolius, Chamenerion angustifolium, Dicranum scoparium, Deschampsia flexuosa, Anthoxanthum odoratum) – Fig. 4.9.

Figure 4.9. Rhododendron caucasicum in flower

***Aconito nasuti-Juniperion* Onipchenko 2002**
***Aconito nasuti-Juniperetum* Onipchenko 2002**
(Seseli alpinum, Aconitum nasutum, Juniperus communis, Abietinella abietina, Betonica macrantha, Bromus variegatus, Tortula ruralis, Cotoneaster integerrimus, Festuca varia, Rhytidium rugosum, Senecio aurantiacus, Campanula collina)

***Salici kazbekensis-Empetrion nigrae* Onipchenko 2002**
***Polygono viviparum-Salicetum kazbekensis* Onipchenko 2002**
(Polygonum viviparum, Salix kazbekensis, Lloydia serotina, Helictotrichon versicolor, Carex umbrosa, Carex sempervirens, Cornicularia muricata, Eritrichium caucasicum, Tortella tortuosa, Anemone speciosa, Campanula tridentata, Carum caucasicum, Cetraria cucullata, Thamnolia vermicularis, Cetraria nivalis, Cladonia gracilis, Luzula spicata)

4.2.11. Elfin birch woodland

***Vaccinio-Piceetea* Br.-Bl. in Br.-Bl. et al. 1939**
Elfin birch forests with dense cover of *Rhododendron caucasicum* are widespread in the Caucasus (Grossgeim 1949, Tumadzhanov 1960a,b, Gulisashvili et al. 1975). We suggested a new alliance *Rhododendro caucasici-Betulion litwinowii* for such communities (Onipchenko 2002). Due to high frequency of several diagnostic species of *Vaccinio-Piceetea* and *Piceetalia excelsae* (*Dicranum scoparium, Hylocomium splendens, Vaccinium myrtillus, V.vitis-idaea)* we distinguish the following communities within these syntaxa (Mirkin et al. 1989, Oberdorfer 1992, Wallnoffer 1993).
***Piceetalia excelsae* Pawlowski in Pawlowski et al. 1928**
***Rhododendro caucasici-Betulion litwinowii* Onipchenko 2002**
***Rhododendro caucasici-Betuletum litwinowii* Onipchenko 2002**
(Rhododendron caucasicum, Oxalis acetosella, Senecio renifolius, Juniperus communis, Deschampsia flexuosa, Cetraria pinastri, Hypnum pallescens, Cetraria islandica, Gymnocarpium dryopteris, Barbilophozia barbata)

Derived association

***Senecioni nemorensis-Betuletum litwinowii* Onipchenko 2001 [*Rhododendro caucasici-Betulion litwinowii / Calamagrostion arundinaceae* Oberd. 1950]**
(Senecio nemorensis, Heracleum asperum, Poa nemoralis, Geranium sylvaticum, Pulmonaria mollis, Chaerophyllum aureum, Cruciata laevipes, Astrantia maxima, Acer trautvetteri, Silene vulgaris, Cephalaria gigantea,

Lapsana communis, Dryopteris filix-mas, Campanula latifolia, Betonica macrantha, Aconitum nasutum)

4.3. Comparison of the floristic richness of alpine communities in the Northwest Caucasus and the Central Alps

4.3.1. Introduction and object description

All factors determining the floristic richness of a plant community can be divided into two groups. One group includes recent ecological factors such as soil fertility, disturbance, microclimate, etc. The other one may be called the "historical" one. It includes such parameters as general floristic richness of a territory, florogenesis, migrations, etc. It is very difficult to distinguish the relative input of these two groups on real communities. It seems that only comparative studies of communities growing under similar ecological conditions, but in different geographical regions, can help us resolve this issue. Alpine communities are convenient objects for such studies. Vegetation in high altitudes of temperate regions is under strong control of mesorelief factors (Billings 1973). We decided to compare the floristic richness for some alpine communities of the Central Alps with the northwestern Caucasus, because both developed under similar climatic conditions and on similar parent rock material, but are separated geographically (Onipchenko & Semenova 1995).

We chose to compare the floristic richness of the Teberda alpine communities with those of the Davos area in the Central Alps (Grison, Switzerland) (Onipchenko & Semenova 1995). Investigated plots were situated on mountains of the Jacobshorn, Rinerhorn, Strela and proximate ridges between 1900 to 2500 m a.s.l.

Climatic observations on the Weissfluhjoch (2667 m; Zingg 1961) demonstrated the presence of the same type of climate as in the Teberda Reserve. Mesoclimatic conditions are very variable on slopes of different exposition and steepness. There are some differences between the two regions in terms of prevailing wind direction, however. In Davos, the north-facing slopes are windward and the south-facing slopes are leeward, thus the snow-melt period is shorter and more intensive than in the Teberda Reserve (where the north-facing slopes are leeward). All of studied Davos communities (except *Seslerietum*) were studied on soils developed on siliceous bedrock.

According to Schibler (1937), the flora of Davos includes about 1,300 species in the area of about 80,000 ha. Thus, Davos and Teberda are quite similar in their total floristic richness.

4.3.2. Compared communities

Fifteen communities with different positions in mesorelief were studied. There were 6 pairs of analogous communities from the Caucasus and the Alps with similar floristic composition and position in the mesorelief (Table 4.1, Onipchenko & Semenova 1995). Three open communities were investigated, because of their high species richness and for comparison with closed communities. They included communities on dry (*Campanulo ciliatae-Chamaesciadietum acaulis* hereafter, simply *Chamaesciadietum)* and wet (*Saxifragetum sibiricae*, hereafter, simply *Saxifragetum)* screes in the Caucasus, and *Seslerietum* in the Alps.

4.3.3. Methods and results

Arrays of sample areas were selected for study of species richness. An array consisted of square plots with sides of 5, 10, 25, 50, 100 (4 replicates of each), 200, 500 and 1000 cm (one replicate of each). We counted the number of vascular plant species only, because it was difficult to identify cryptogams in the field. At least five arrays were used for each community type. Thus we had 5 replicates for larger (4, 25 and 100 m^2) areas and 20 replicates for smaller areas. The largest plot size for the Caucasian SBC and SH of the Alps (see Table 4.1. for abbreviation) was 25 m^2 because they can only be found in small depressions.

Mean floristic richness (number of species per plot) and standard error were calculated for all communities and plot sizes. A t-test was used for comparisons between analogous communities of the Caucasus and the Alps.

Two equations were fitted by regressions:

$$S=c \cdot A^z \qquad (1)$$

$$S=a + b \cdot lnA \qquad (2)$$

where S: number of species, A: area, a,b,c,z: regression parameters.

Equation (1) is from Arrhenius (1921) and Equation (2) is from Gleason (1922). We compared floristic similarity (Czekanowski's index, CI) between the Alps and Caucasian communities on different slopes. The CI was greatest between the Caucasian and Alpic communities occurring in analogous habitats (diagonal values in Table 4.2). There are no species in common between the Caucasian snow-free Lichen Heaths (ALH) and the Alpic *Salix* Snow Beds (SH).

Table 4.1. List of compared communities in Teberda and Davos. SD – snow depth in late winter, in m, VS – duration of vegetative season, in months (based on Onipchenko & Semenova 1995, with permission of Opulus Press)

Teberda				Davos			
Short name	Syntaxonomy	SD	VS	Short name	Syntaxonomy	SD	VS
ALH	*Pediculari comosae-Eritrichietum caucasici*	<0.2	5	CC	*Caricetum curvulae cetrarietosum*	<0.5	5
FVG	*Violo altaicae-Festucetum variae*	0.5-1	4	CCN	*Carici curvulae-Nardetum*	0.5-1	4.5-5
GHM	*Hedysaro caucasicae-Geranietum gymnocauli*	2-3	2.5-3	HCC	*Hygrocaricetum curvulae*	2-3	3-4
SBC	*Hyalopoo ponticae-Pedicularietum nordmannianae*	>4	2-2.5	SH	*Salicetum herbaceae*	3-5	2-3
RC	*Lerchenfeldio-Rhododendretum caucasici*	3-4	3-4	RF	*Rhododendretum ferruginei*	?	4-5
FEN	*Caro caucasici-Caricetum nigrae*	3-4	3-4	AFEN	*Caricetum goodenowii*	?	3.5-5
Chama-esciadi-etum	*Campanulo ciliatae-Chamaesciadi-etum acaulis*	<0.2	5				
Saxif-ragetum	*Saxifragetum sibiricae*	2-3	2-3				
				Sesle-rietum	*Seslerio-Caricetum sempervirentis*	?	5

Table 4.2. Similarity between some Caucasian and Alpine communities (Czekanowski's Index for frequency values of 20 common species, %) See Table 4.1. for abbreviations. (from Onipchenko & Semenova 1995, with permission of Opulus Press)

Communities Caucasian / Alpic	**CC**	**CCN**	**HCC**	**SH**
ALH	4.69	2.90	0.78	0.00
FVG	3.29	10.88	5.09	1.54
GHM	1.33	5.03	7.13	2.91
SBC	0.00	1.13	2.75	4.70

Species richness varied considerably among the studied communities (Table 4.3, Onipchenko & Semenova 1995). *Festuca varia* grasslands (FVG) have the highest value (100 m^2 plots) among the Caucasian communities, and *Seslerietum* among the ones in the Alps. Alpine fens (both of the Caucasus and the Alps), and *Sibbaldia* SBC had the lowest species richness on 25 m^2 plots.

The number of species per small plots (0.0025 m^2) was lowest for the *Rhododendron caucasicum* shrubs (RC) and *Carex nigra* Fens (AFEN) of Davos (1.9, and 2.2 species respectively).

Table 4.3. Floristic richness at different plot sizes (average number of vascular plant species). ***Bold italic*** marks values, which are significantly different between analogous communities according to *t*-test, $p<0.05$. Regions: C – Caucasus (Teberda), A – Alps (Davos). See Table 4.1 for abbreviations (from Onipchenko & Semenova 1995, with permission of Opulus Press)

Re-gion	Community	length of plot side (cm)							
		5	10	25	50	100	200	500	1000
C	ALH	2.6	6.2	***14.8***	***20.2***	***25.0***	***30.6***	***36.0***	39.8
A	CC	3.2	6.6	***10.8***	***13.7***	***16.6***	***21.2***	***25.8***	31.8
C	FVG	***2.5***	***4.6***	11.1	17.8	24.4	40.4	48.6	***56.8***
A	CCN	***3.5***	***8.4***	13.3	17.3	21.7	30.8	36.8	***42.8***
C	GHM	***2.3***	***5.3***	11.4	15.8	19.5	26.4	30.6	35.0
A	HCC	***4.5***	***8.4***	11.9	14.6	18.0	23.2	29.2	35.6
C	SBC	***3.3***	***5.1***	***7.9***	***9.4***	***11.3***	14.8	***18.0***	
A	SH	***4.4***	***7.5***	***10.7***	***13.4***	***16.1***	20.5	***25.5***	
C	RC	***1.9***	***3.5***	***6.1***	***8.8***	***12.5***	20.2	29.0	40.4
A	RH	***2.7***	***5.9***	***9.7***	***13.9***	***18.5***	27.0	34.4	39.2
C	FEN	***2.9***	4.1	5.9	8.0	9.9	14.0	18.8	23.0
A	AFEN	***2.2***	3.8	6.1	8.3	11.1	15.4	18.2	23.2
C	*Chamaesciadietum*	1.9	4.5	11.2	16.8	22.1	33.6	40.2	46.0
C	*Saxifragetum*	2.8	4.6	8.0	10.9	14.8	21.6	25.2	30.6
A	*Seslerietum*	3.9	8.9	15.9	20.9	26.8	35.0	42.6	50.6

A comparison of analogous communities from different regions is of great interest (Table 4.3.). The fens appear very similar. No significant differences were found for any but the smallest plots. The alpine grasslands and meadows are similar in their floristic richness for plots larger than 25 x 25 cm, but the Caucasian communities have fewer species on smaller plots. The alpine lichen heaths (ALH) of Teberda are richer than the *Carex-Cetraria* grasslands (CC) for all plot sizes, except for the two smallest ones. On the other hand, the plots of the *Salix* snow beds (SH) are richer in species than the Caucasian *Sibbaldia* snow beds (SBC) for all plot sizes (Onipchenko & Semenova 1995).

The *Rhododendron* shrubs of both regions are very similar in floristic richness on larger plots (4 – 100 m^2). However, there are more species found on small plots in the *Rhododendron ferrugineum* shrub (RF) than in its Caucasian counterpart (RC). In general, most of the Alps' communities are richer in species on small plots than the analogous communities from the Caucasus.

Regression coefficients "*c*" and "*a*" of equations (1) and (2) show the estimated number of species per m². The largest values of the coefficients were obtained for the *Festuca varia* grasslands (FVG); the smallest values are found in both fens and *Sibbaldia* snow beds (SBC) (Table 4.4.). Coefficients "*z*" and "*b*" show a rate of increase in species richness. There is a good correspondence between the coefficients of the two equations (Table 4.4).

Table 4.4. Coefficients (*c, z, a, b*) of regression and correlation coefficients (r^2) for two equations on species-area relations. See Table 4.1 for abbreviations (from Onipchenko & Semenova 1995, with permission of Opulus Press)

Community	Equation 1			Equation 2		
	c	z	r^2	a	b	r^2
ALH	19.56	0.233	0.925	24.44	3.61	0.997
CC	15.09	0.198	0.980	18.01	2.60	0.955
FVG	20.58	0.298	0.948	29.55	5.45	0.982
CCN	19.79	0.218	0.985	24.40	3.73	0.989
GHM	16.14	0.241	0.942	20.49	3.18	0.998
HCC	17.21	0.180	0.980	20.12	2.83	0.993
SBC	11.27	0.179	0.988	12.26	1.58	0.993
SH	15.82	0.183	0.988	17.24	2.24	0.994
RC	12.24	0.282	0.996	17.71	3.48	0.950
RF	16.29	0.249	0.967	21.31	3.60	0.988
FEN	10.00	0.196	0.996	12.14	1.89	0.974
AFEN	9.95	0.217	0.983	12.37	1.97	0.986
Chamaesciadietum	18.00	0.290	0.940	25.08	4.40	0.991
Saxifragetum	13.27	0.224	0.975	16.67	2.69	0.987
Seslerietum	23.05	0.224	0.968	28.60	4.39	0.997

It is interesting that the coefficients of the logarithmic equation are similar for *Seslerietum* and communities on dry screes in Teberda *(Chamaesciadietum).* The similarity may be connected with the open nature and significant species richness of the communities.

All but two values of residual variance are lower for Equation 2 than for Equation 1. The opposite was found for several alpine communities in the Krkonose Mountains (Leps & Stursa 1989).

The values of floristic similarity among the studied communities *within* the regions are greater for the communities of the Alps than that for the Caucasus' ones (Onipchenko & Semenova 1995). Thus, floristic differentiation between the Caucasus communities is stronger than between the ones of the Alps.

4.3.4. Discussion

Our data show that communities with similar syntaxonomic position have quite similar floristic richness. The highest similarity was found for the fens, where the differences for all plot sizes do not exceed 2 species.

In many cases, the species richness on small plots was greater for the Alps' communities than those of the Caucasus. This may be explained by the recent grazing impact. We suggest that the moderate rate of grazing in Davos results in an increase in floristic richness on small plots due to both creation of gaps and decrease in plant size (Onipchenko & Semenova 1995). The grazing in most of the Teberda Reserve ceased in 1943. The history of grazing of the NW Caucasus is different from that of the Central Alps (see Chapter 11). The rate of grazing has fluctuated considerably in the Caucasus in the last two millennia. Periods of ravaging wars coupled with negligible grazing alternated with periods of peace, when the number of cattle increased (Tsepkova 1986, Dinesman 1992). On the other hand, grazing in the Swiss Alps was apparently less severe and more constant. A small plant size might have become more adaptive under such conditions.

The snow dynamics in the studied areas differ considerably. Snow accumulation is more typical for the northern slopes in Teberda. In Davos, however, it is restricted predominantly to the southern slopes due to prevailing winter wind patterns. We suggest that the environmental stress caused by snow accumulation is weaker in Davos. It can account for higher species richness of the *Salix herbacea* snowbeds than of the *Sibbaldia* snowbeds (SBC) of the Caucasus. It seems that less differentiation in snow regime between the Alps' communities is responsible for the more closely intraregional floristic similarity between the communities there (Onipchenko & Semenova 1995).

The lichen heaths (ALH) and the *Festuca* Grasslands (FVG) of the Caucasus have significantly greater number of species on larger plots than the analogous communities of the Alps. Apparently, there are several causes to account for these differences. The FVG is one of the most widespread types in the high-mountain Caucasus. A floristic similarity exists between these communities and the species-rich meadow steppes (common species: *Galium verum, Leontodon hispidus, Anthoxanthum odoratum*). Species with Caucasian and Asia Minor distributions play a significant role in the FVG composition. On the other hand, arctic-alpine species are more frequent in the Alps than in the Caucasus, as a result of the history of flora development. It seems, therefore, that regional flora composition plays a greater role in determining floristic richness in this case than do the contemporary ecological conditions.

A few papers discuss the issue of floristic richness of alpine communities. Leps & Stursa (1989) investigated floristic richness of five high mountain communities in the Krkonose Mountains (Czech Republic). They arrived at average values of 5.5 species per 16 m^2 for *Carici fyllae-Nardetum*, 10.0 species for the *Empetro-Trichophoretum austriaci* and 9.2 species for *Cetrario-Festucetum supinae*. These communities are much poorer than the analogous communities in Teberda and Davos. We suggest that island effects (MacArthur & Wilson 1967) play an important role because the total alpine area in the Krkonose Mountains is less than 1% of the area of the Caucasus or of the Alps. The floristic richness in snow-bed communities of Japan (Kudo & Ito 1992) is also lower than in similar communities in both regions studied by us.

The alpine lichen tundra of large Asiatic mountain systems such as the Altai (Sedelnikov 1979) and Priokhot'e in the Russian Far East (Shlothauer 1990) have considerably poorer floras than the ALH of Teberda or *Carex-Cetraria* grasslands of Davos. The same is true for the *Rhododendron* shrublands, grasslands, and fens. Only some snow-bed communities of the Asiatic Mountains have similar floristic richness. It may be related to the generally lower total number of species in these mountain systems. The high-mountain flora of the Kuznetskii Alatau comprises 332 vascular plant species (Sedelnikov 1979). The Priokhot'e mountains with a total area about 100,000 km^2 have only about 730 species (Shlothauer 1990). We conclude that the species richness of the Alps' and the Caucasian alpine communities is relatively high.

The results of this investigation suggest that both recent ecological conditions and the history of flora development have a great influence on the local floristic diversity of alpine communities and can lead to considerable similarity among the communities of different regions.

Chapter 5

Structure and dynamics of closed alpine communities

5.1. Plant life forms

5.1.1. Morphological analysis of alpine communities

5.1.1.1. Introduction

The morphological analysis of plant species is necessary for a detailed study of important features of plant communities, such as plant adaptations, plant interactions and community structure (Raunkiaer 1934, Zaugol'nova et al. 1991, Pokarzhevskaya & Onipchenko 1995).

In the opinion Zaugol'nova et al. (1991) the analysis of "architectural" models and life forms can be of large value for studies of plant population behaviour and its role in the formation of community structure.

Different authors carried out morphological analyses of plants of high-mountainous communities. Spectra of life forms were made for communities of subnival and subalpine belts of the Central Caucasus (Nakhutsrishvili 1981, Nakhutsrishvili & Gamtsemlidze 1984), the Pamir (Pavlov 1986) and Tien Shan (Lebedeva 1986).

In Teberda Nature Reserve the investigation of morphological spectra was conducted (Pokarzhevskaya & Onipchenko 1995, Pokarzhevskaya 1995). For this aim different life form systems were used: Raunkiaer's system (Raunkiaer 1934), Serebryakov's system (Serebryakov 1962, Pokarzhevskaya 1995), architectural models (or models of shoot formation) (Serebryakova 1977, 1981) and the concept of vegetative mobility (clonality) (Pokarzhevskaya 1995). The extended system was applied for

V.G. Onipchenko (ed.), Alpine Ecosystems in the Northwest Caucasus, 77–179.

analysis of four alpine communities, which differ in structure and ecology and which can easily be compared ecologically. Each plant species was investigated from the point of view of the different morphological parameters.

In total 82 species were chosen for a detailed analysis: one dwarf shrub, *Vaccinium vitis-idaea*, one dwarf semishrub, *Sibbaldia procumbens*, 3 annuals and 77 herbaceous perennial plants. This list includes all species occurring with at least frequency class II (21 – 40%) in the releves of the communities, or with a high frequency in the experimental plots situated in the area (Onipchenko 1994a). Data on the aboveground biomass are presented in Chapter 5.2.

5.1.1.2. Degrees of vegetative mobility

The following types were defined for characterization: three for clonal plants and one for non-clonal plants (Pokarzhevskaya & Onipchenko 1995).

1. High mobility. Plants possess an annual growth of about 7 cm and are able to form stolons and achieve vegetative propagation. Only three species belong to this category – *Agrostis vinealis, Sempervivum caucasicum, Vaccinium vitis-idaea.*
2. Medium mobility – the annual growth ranges from 2 to 5 cm; examples are *Minuartia aizoides* or *Anthoxanthum odoratum.*
3. Low mobility. This includes plants with an annual horizontal growth of less than 2 cm; examples are *Carex* spp., *Festuca* spp. or *Geranium gymnocaulon.*
4. No vegetative mobility (non-clonal plants): *Taraxacum stevenii, Campanula tridentata.*

Results of the investigation have shown that clonal species with low vegetative mobility prevail in all community types (Fig. 5.1). Non-clonal species are also abundant in the more stressed conditions, especially in SBC and LH. In SBC they prevail on a frequency basis. The occurrence of species with high and medium mobility is quite noticeable in SBC.

On average clonal species make up ca. 70% of all the species, against 30% non-clonal (Pokarzhevskaya 1995). Nakhutsrishvili & Gamtsemlidze (1984) have reported similar results for subnival communities in the Central Caucasus.

5.1.1.3. Types as to the presence of rosettes

A rosette is a shoot with short internodes (spur) possessing more then 3 green leaves. The rosette structure of shoots is considered usually as an adaptation to different stress environmental conditions: low temperature, water stress, heavy illumination, etc (Zhmylev et al. 1993). Pokarzhevskaya & Onipchenko (1995) distinguish 3 types of plants according to rosette development.

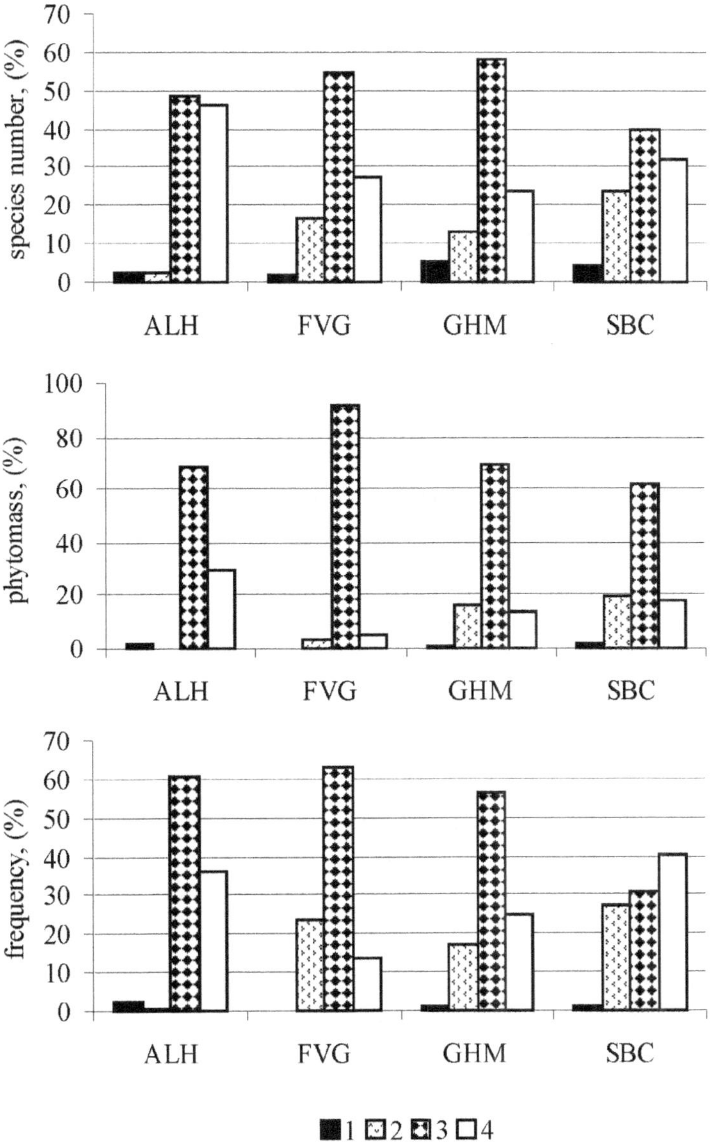

Figure 5.1. Percentage distribution of vegetative mobility in four alpine communities (ALH – alpine lichen heath, FVG – *Festuca varia* grassland, GHM – *Geranium-Hedysarum* meadow, SBC – snow bed community). 1: high mobility, 2: medium, 3: low. 4: no mobility (from Pokarzhevskaya 1995 with permission of *Folia Geobotanica*)

1. Non-rosette plants: *Gentiana septemfida, Hedysarum caucasicum, Thymus nummularius*. Their aboveground shoots have only long internodes.
2. Semi-rosette plants – with vegetative shoots with rosettes and reproductive shoots with leaves. For example: *Scorzonera cana, Anemone speciosa, Geranium gymnocaulon.*
3. Rosette plants with only rosette shoots. Their generative shoot is a scape without leaves. For example: *Plantago atrata, Taraxacum stevenii, Trifolium polyphyllum.*

The semi-rosette species are the most common in the four communities (Fig. 5.2.). Semi-rosette and rosette plants together usually make up about 90% (Pokarzhevskaya 1995). Dominance of semi-rosette plants was observed in the subnival communities of the Central Caucasus as well (Nakhutsrishvili & Gamtsemlidze 1984). The greatest participation of rosette plants is observed under the more severe conditions of the alpine lichen heath and snow bed communities.

5.1.1.4. Models of shoot formation

The following models of shoot formation (Serebryakova 1981) are distinguished for herbaceous perennial polycarpics (Fig. 5.3):

1. Monopodial long-shoot – plants have perennial vegetative shoots with long internodes and a long functioning apical meristem. Their annual reproductive shoots are lateral. For example: *Cruciata laevipes, Gentiana septemfida.*
2. Monopodial rosette – as against the first group the plants have short internodes, but also have monopodial growth: *Anemone speciosa, Campanula tridentata, Eritrichium caucasicum.*
3. Sympodial long-shoot – plants with long internodes and a sympodial growth type. The apical meristem of a shoot dies (as a rule, as a result of flowering) and it is replaced by new shoot that laterally grows out. Thus the sympodial shoot is a system of several modules. Examples are *Aconitum nasutum, Hedysarum caucasicum, Hypericum linarioides.*
4. Sympodial semi-rosette plants have vegetative shoots with rosettes and reproductive shoots with long internodes. Examples are *Leontodon hispidus, Luzula multiflora, Pulsatilla aurea.*

Species with a sympodial semi-rosette model prevail in all investigated communities (Fig. 5.4). Species with a monopodial rosette model are frequent in ALH.

5.1.1.5. Raunkiaer's life forms

Four of the Raunkiaer life form types (Raunkiaer 1934) were observed for the communities: chamaephytes, hemicryptophytes, geophytes and therophytes. Hemicryptophytes prevail in the four communities (Table 5.1).

The distribution of chamaephytes, geophytes and therophytes is negligible, except for SBC where chamaephytes account for 38% of the total aboveground phytomass (Pokarzhevskaya 1995) due to abundance of *Sibbaldia procumbens*.

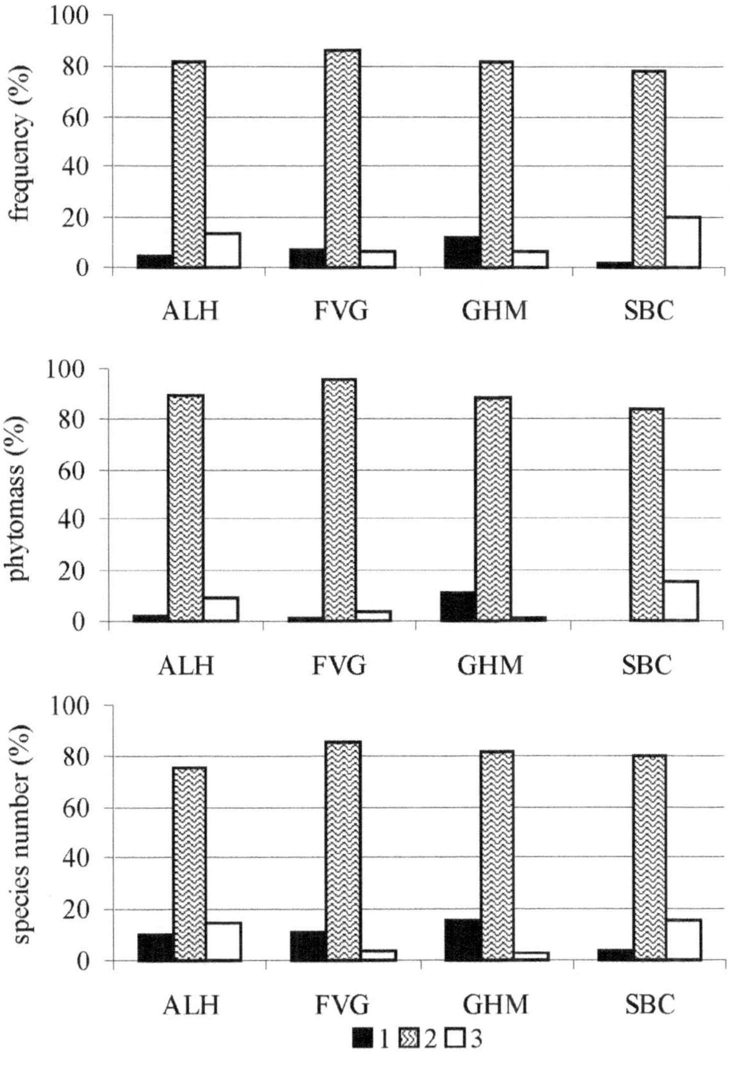

Figure 5.2. Percentage distribution of rosette types in four alpine communities (for community abbreviations see Fig. 5.1). 1: without rosettes, 2: semi-rosette form, 3: rosette form (from Pokarzhevskaya 1995 with permission of *Folia Geobotanica*)

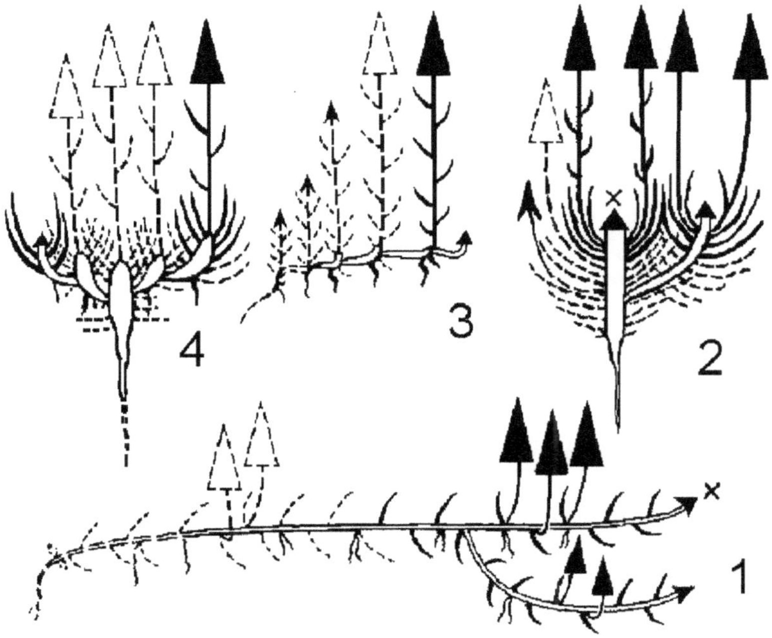

Figure 5.3 Models of shoot formation 1 – monopodial long-shoot plants, 2 – monopodial rosette plants, 3 – sympodial long-shoot plants, 4 – sympodial semi-rosette plants (from Serebryakova 1981)

Table 5.1. Percentage distribution of Raunkiaer's life forms in four alpine community types (from Pokarzhevskaya 1995 with permission of *Folia Geobotanica*)
ALH – alpine lichen heath, FVG – *Festuca varia* grassland, GHM – *Geranium-Hedysarum* meadow, SBC – snow bed community

Life form	Number of species				Above-ground biomass				Species frequency			
	ALH	FVG	GHM	SBC	ALH	FVG	GHM	SBC	ALH	FVG	GHM	SBC
Chamaephytes	7	5	5	8	4	1	<1	38	8	4	1	8
Hemicryptophytes	76	80	79	72	81	97	97	61	78	91	86	69
Geophytes	10	13	13	20	15	2	3	1	13	4	12	23
Therophytes	7	2	3	0	<1	0	0	0	1	1	1	0

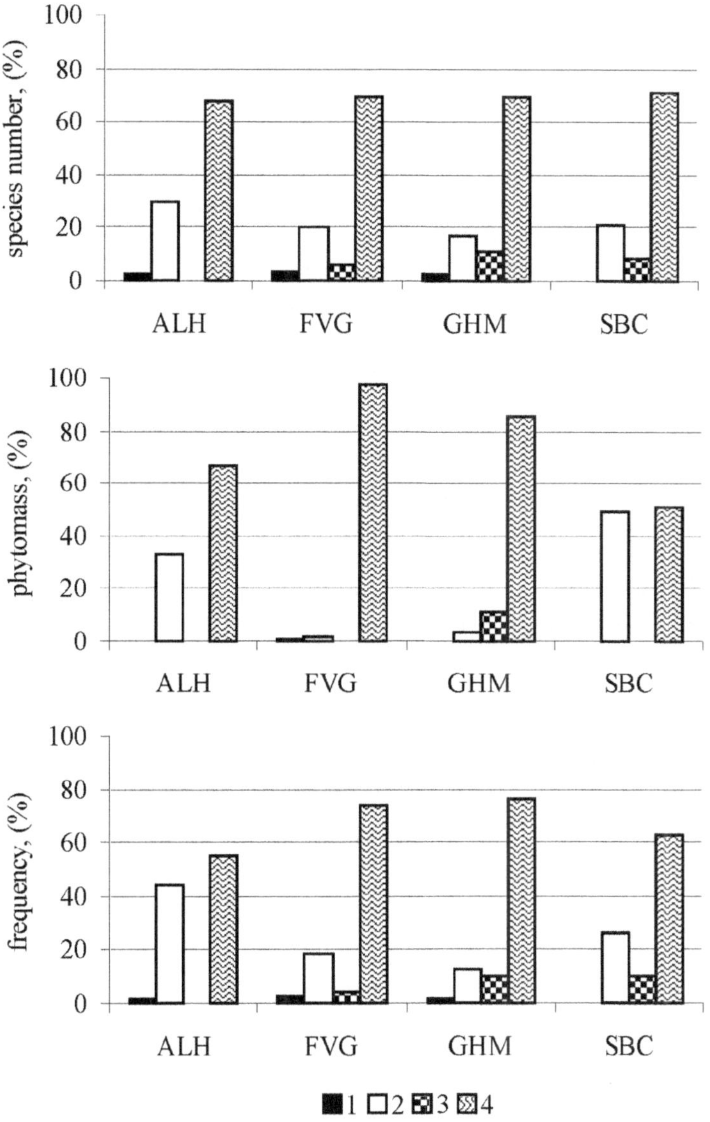

Figure 5.4 Percentage distribution of shoot formation types in four alpine communities (for community abbreviations see Fig. 5.1). 1: monopodial long-shoot, 2: monopodial rosette, 3: sympodial long-shoot. 4: sympodial semi-rosette (from Pokarzhevskaya 1995 with permission of *Folia Geobotanica*)

5.1.1.6. Life forms according to Serebryakov

Serebryakov (1962) made the main division of his classification according to the longevity of main plant axes. Pokarzhevskaya & Onipchenko (1995) have used some additions to the classification. The group of taproot-rhizomatous plants (*Minuartia aizoides, Campanula collina* etc.) was included in the spectrum of plant life forms. Plants of this group have a deep taproot as well as horizontal rhizomes or stolons providing vegetative mobility. This group is not described by Serebryakov, but mentioned in studies of other authors (Nakhutsrishvili 1981, Polozova 1981). Taproot-rhizomatous stoloniferous polycarpics were included in the stoloniferous group, while Serebryakov strictly divides tap-rooted and mobile plants. Nevertheless, Pokarzhevskaya (1995) has found it possible to classify the above mentioned species in their own group, taking into account their large vegetative mobility at the expense of stolons formation. Vegetative propagation of these plans after artificial disintegration is discussed in Chapter 5.1.2.

Species of the communities could be assigned to 14 life-form types. The three spectra obtained on the basis of species number (unweighted), aboveground phytomass and frequency (weighted) differ clearly (Table 5.2) (Pokarzhevskaya & Onipchenko 1995, Pokarzhevskaya 1995).

Table 5.2. Percentage distribution of Serebryakov's life forms in four alpine communities (from Pokarzhevskaya 1995 with permission of *Folia Geobotanica*). See Table 5.1. for community abbreviations

Life form	Number of species				Above-ground biomass				Species frequency			
	ALH	FVG	GHM	SBC	ALH	FVG	GHM	SBC	ALH	FVG	GHM	SBC
Brush-root plants	15	7	3	4	12	1	0	3	10	2	1	8
Short/rhizome plants	22	29	26	12	10	5	49	9	23	19	30	9
Tap-rooted plants	22	13	13	12	19	4	12	14	27	9	16	15
Rhizom./Tap-rooted plants	15	13	11	8	10	<1	1	11	16	12	7	8
Dense-tussock plants	7	9	11	12	41	82	17	12	12	30	13	9
Loose-tussock plants	5	14	21	24	6	7	16	9	7	15	22	21
Long-rhizome grasses	0	0	3	12	<1	0	1	4	0	0	1	7
Stoloniferous plants	2	7	3	0	<1	1	0	0	1	10	0	0
Tuberiferous plants	0	0	0	4	0	0	0	<1	0	0	0	8
Bulbiferous plants	2	2	3	4	<1	<1	<1	<1	<1	<1	<1	2
Monocarpic perennials	0	2	3	4	0	<1	1	<1	0	<1	6	4
Annuals	7	2	3	0	1	<1	<1	0	1	1	1	0
Dwarf shrubs	2	0	0	0	1	0	0	0	2	0	0	0
Dwarf semishrubs	0	2	3	4	0	<1	2	38	0	2	3	9

There are clear differences in Serebryakov's life form spectra between the four community types. Tap-rooted, short-rhizome and dense-tussock plants are most typical of ALH, short-rhizome and loose-tussock plants of GHM, dense-tussock plants of FVG, semi-shrubs and loose-tussock plants of SBC. Tuberiferous, bulbiferous and monocarpic plants are not important in any of the communities (Pokarzhevskaya 1995).

5.1.1.7. General features of the studied communities

The analysis carried out has revealed distinctions in spectra of the plants' life forms of the alpine communities of the Northwestern Caucasus.

The ecological distinctions between the communities are determined by duration of snow cover. Alpine lichen heath occurs under the severe environmental conditions of snow-free habitats: high insolation, high probability of drought and also low winter temperature and wind desiccation (Stanton et al. 1994). Perennial herbaceous polycarpics prevail here (37 species), besides that three annuals and one dwarf shrub occur too. The species with low vegetative mobility are also specific for these communities (Pokarzhevskaya & Onipchenko 1995). Tap-rooted plants have a wide distribution here: they are probably well-adapted to these conditions (Pokarzhevskaya 1995). Adaptation to the severe conditions of alpine habitats is often more important than competitive ability (Johnson & Billings 1962).

The highly productive *Festuca varia* grasslands experience a milder climate than the snow-free communities. Here 53 perennial herbaceous polycarpic, 1 perennial monocarpic and 1 annual were noted. The species with low vegetative mobility prevail here (Pokarzhevskaya and Onipchenko 1995). Wide distribution of tussock species is observed. FVGs experience relatively early snow melting, a long vegetative season and have well drained soils. These conditions could lead to strong competitive interactions (Stanton et al. 1994), with tussock plants having competitive advantage in this community type, especially the dense tussock plants, which form 82% of the phytomass in FVG (Chapter 5.2.). They produce large amounts of litter and many tillers simultaneously. They can prevent invasion of other species to this community type. Experiments with reciprocal transplantations made in the same communities (Chapter 7.1.) confirm the above hypotheses. It is interesting that tussock plants make up only 9% of the species number of FVG. But they form the larger part of the phytomass and 30% of frequency (Table 5.2.). This means that many other species, especially the short rhizome life form, can exist in this community type, although they are not very frequent (Pokarzhevskaya 1995). So FVG can be considered as a community with strong tussock competitors and a lot of subordinate species. Such a structure is very close to Tilman's model community (Tilman 1994), where one long-lived competitor coexists with several less long-lived species that are perfectly able to colonise.

GHMs are highly productive wet meadows. They have a shorter vegetative season than FVGs. In the community composition 36 perennial herbaceous polycarpics, 1 perennial monocarpic and 1 annual occur. Species with low vegetative mobility prevail, but those with medium and high vegetative mobility also have considerable presence in the species composition (Pokarzhevskaya & Onipchenko 1995). Short rhizome plants prevail here. Loose-tussock species are also abundant. The main factor influencing GHM is burrowing activity of rodents (Fomin et al. 1989). This activity, causing the creation of gaps, can provide good conditions for the development of rhizomatous species. It could also decrease the role of dominants in GHM (Pokarzhevskaya 1995).

The greatest snow cover we can see in SBC. This community type has the smallest number of species among the observed alpine communities: 23 perennial herbaceous polycarpics, 1 dwarf semishrub and 1 perennial monocarpic (Pokarzhevskaya & Onipchenko 1995). Nevertheless, the diversity of life form groups is equal to the other community types. This could be caused by variation in snow melting time for different parts of this community, with different patches providing possibilities for several life forms. Species with low vegetative mobility and immobility prevail here (Pokarzhevskaya 1995). The dwarf semishrub *Sibbaldia procumbens* dominates in the phytomass of the community. This species can exist in different communities along the mesotopographic gradient (Stanton et al. 1994), but in the late snowbeds it can have a competitive advantage (Pokarzhevskaya 1995).

It is necessary to emphasize the similarity in morphological structure of the plants of the four types of the alpine communities. The majority of species have low vegetative mobility, a semi-rosette form of growth, and a sympodial semi-rosette model of shoot formation. Hemicryptophytes prevail in the species composition. Plants with the loose-tussock growth form are rather common in these communities (Pokarzhevskaya & Onipchenko 1995).

5.1.2. Experimental study of the potential ability of taproot-rhizomatous plants to vegetative propagation

Taproot-rhizomatous plants are widespread in the high mountain region of the Caucasus and play an important role in the high-mountain communities (Nakhutsrishvili 1981, Nakhutsrishvili & Gamtsemlidze 1984). As we have seen (Table 5.2.) participation of this life form in the communities of Teberda Nature Reserve is also considerable.

The idea of this investigation was started in connection with experiments on reciprocal transplantations (Chapter 7.1.). During the transplantations taproot-rhizomatous plants often were disintegrated due to cutting the sods. Thus the question has emerged: does the disturbance of taproot-rhizomatous plants influence their survival after transplantations. What is the reason of mortality of the plants on experimental plots: change of conditions in the community or impossibility of damaged individuals to survive? On the other hand morphological features of taproot-rhizomatous plants attract attention. Are they capable of vegetative propagation? What is the role of a main taproot in the nutrition of the plants?

The study of the potential ability of taproot-rhizomatous plants to vegetative propagation was performed from 1992 to 1996 (Cherednichenko & Yurtseva 1999). Five species were included in the experiment, namely: *Arenaria lychnidea, Minuartia aizoides, Minuartia circassica, Trifolium polyphyllum* and *Galium verum*. Degree of taproot and adventitious root development of the species is different. *Minuartia circassica* and *Arenaria lychnidea* have rather underdeveloped adventitious roots, which arise on rather old shoots. Those of the other species develop on younger shoots, and the most powerful adventitious roots can be indistinctive from the taproot.

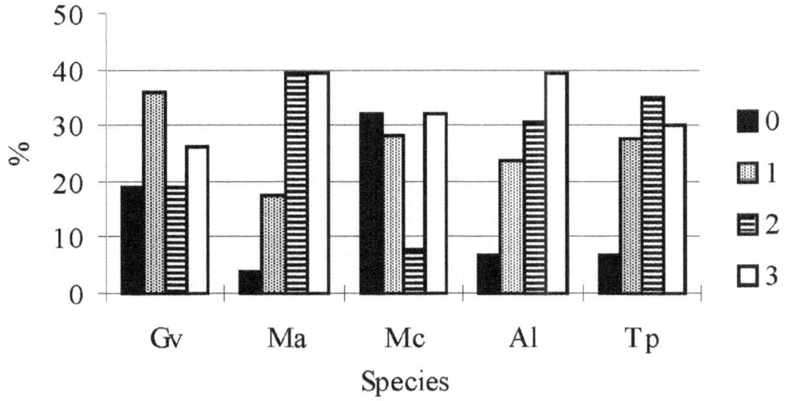

Figure 5.5. Percentage distribution of ramets in the first year after cutting. 0 – died off ramets; 1 – ramets with numerous dead shoots; 2 – oppressed ramets (vitality decreased, but a large number of shoots persisted); 3 – ramets with unchanged vitality. Plant species: Gv – *Galium verum*, Ma – *Minuartia aizoides*, Mc – *Minuartia circassica*, Al – *Arenaria lychnidea*, Tp – *Trifolium polyphyllum*

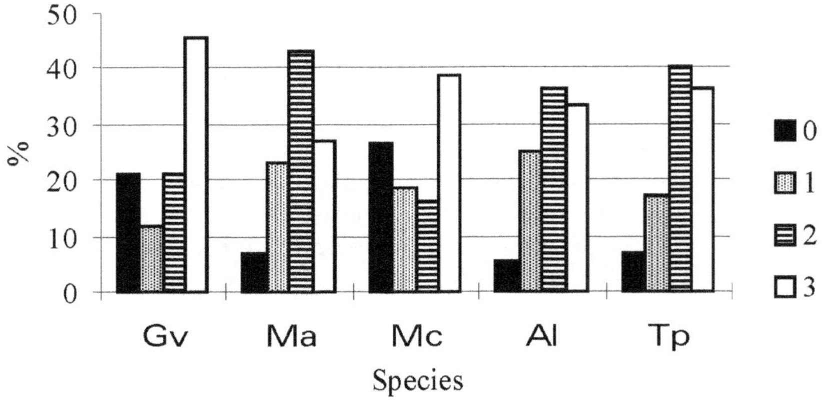

Figure 5.6 Average ramet vitality distribution for 3 years (% of the total number of ramets): 0 – died off ramets; 1 – ramets with numerous dead shoots; 2 – oppressed ramets (vitality decreased, but a large number of shoots persisted); 3 – ramets with unchanged vitality. Plant species: Gv – *Galium verum*, Ma – *Minuartia aizoides*, Mc – *Minuartia circassica*, Al – *Arenaria lychnidea*, Tp – *Trifolium polyphyllum*

In this experiment individuals of the plant species were divided *in situ* into several parts by knife. Observation of which individuals formed new shoots was conducted during three years. Their vitality was estimated on a scale (Fig. 5.5, 5.6). Percentage distribution of ramets among vitality categories on years was counted up for each species (Cherednichenko & Yurtseva 1999). The total number of ramets was set at 100%. In the last year of observations the investigators dug out some experimental plants to study the influence of the disturbance on the shoot-systems of the species (Fig. 5.7 – 5.9).

The results obtained show that the disturbance of the shoot system has strongly affects on *Minuartia circassica* due to restricted abilities for vegetative propagation of this plant. Adventitious roots develop on old parts of the rhizomes (segments at least five years old). In other species development of the roots was marked at rhizome segments of 2 – 3 years old. Probably this largely explains the high mortality of *Galium verum* ramets. This species rarely has adventitious roots on young segments. However, the number of ramets which died off, even at the case of *Minuartia circassica*, on the average, did not exceeded 30% (Fig. 5.5.). In the other investigated species the number of dead ramets was low: it was less than 10%.

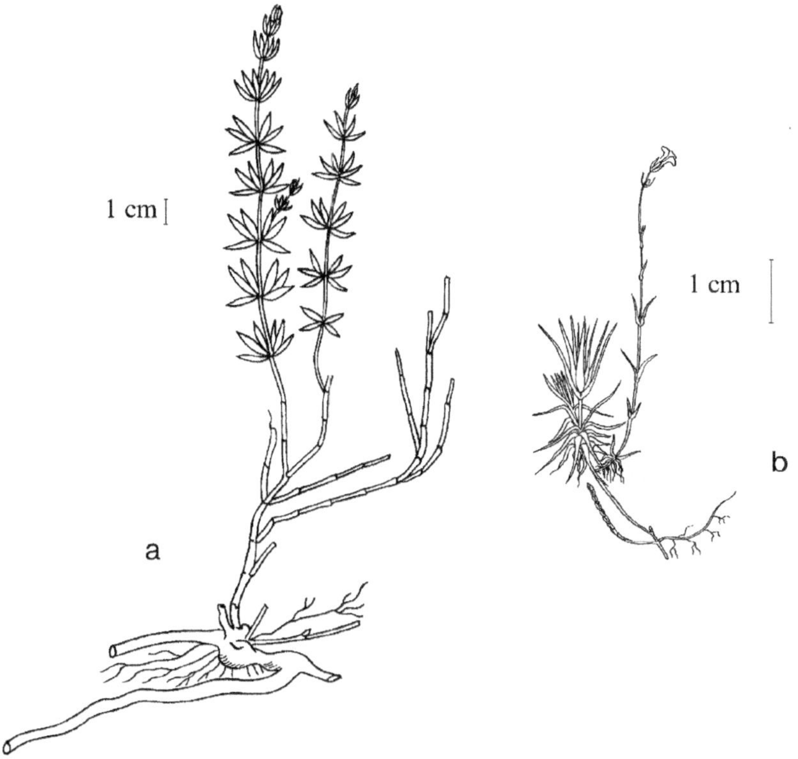

Figure 5.7 The rhizome parts of *Galium verum* (a) and *Minuartia circassica* (b) with shoots in the 4th year after cutting

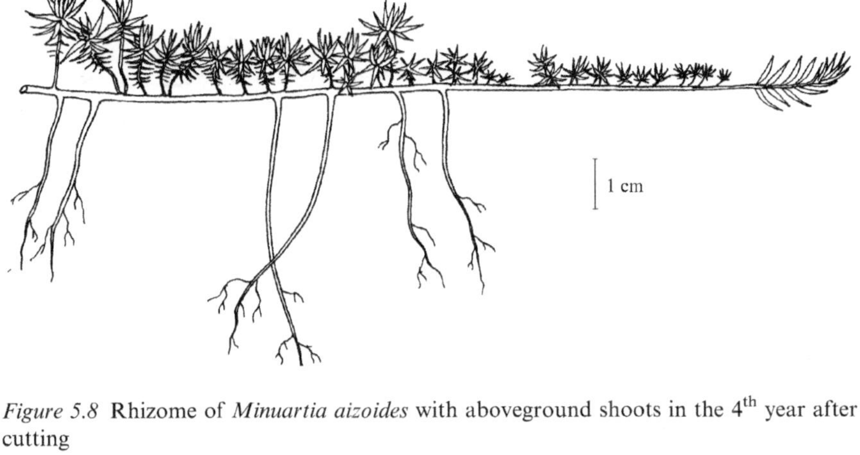

Figure 5.8 Rhizome of *Minuartia aizoides* with aboveground shoots in the 4th year after cutting

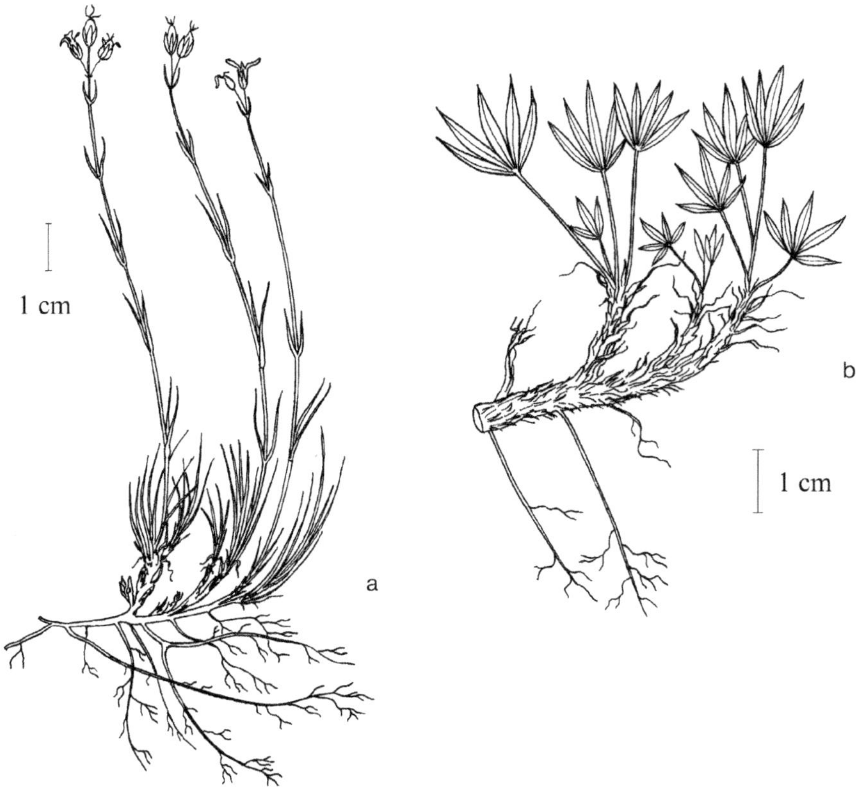

Figure 5.9 The rhizome parts of *Arenaria lychnidea* (a) and *Trifolium polyphyllum* (b) with shoots in the 4th year after cutting

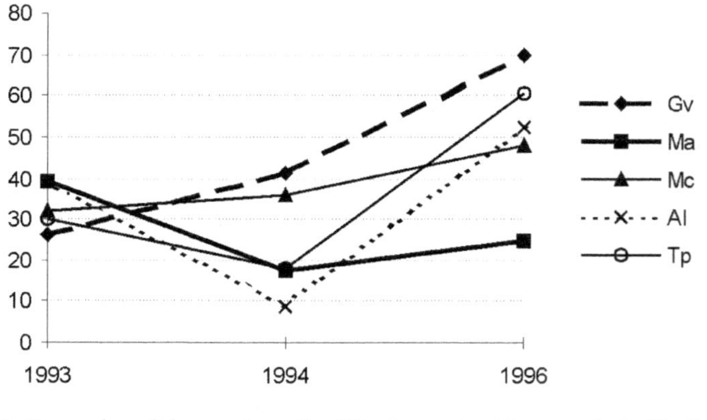

Figure 5.10 Dynamics of the number of artificial ramets with normal vitality for 3 years. Gv—*Galium verum*; Ma—*Minuartia aizoides*; Mc—*Minuartia circassica*; Al—*Arenaria lychnidea*; Tp—*Trifolium polyphyllum*

Individuals with reduced vitality are most common. However, the number of individuals with constant vitality is high in comparison with other groups (Fig. 5.5. – 5.6.). It is clear that individuals of taproot-rhizomatous plants stay viable after rhizome cutting for at least 4 years (time of the experiment), provided that main roots or adventitious roots are present. Moreover, as shown in Fig. 5.10, there was no tendency of decrease in number of ramets with normal vitality. This fact indicates that the regeneration and continued normal existence of ramets are possible.

Decreasing vitality of the ramets after cutting never has resulted in the destruction of whole individuals. The decline in vitality might be explained by the insufficient quantity of adventitious roots to provide enough nutrition for the shoot system. The number of ramets with a normal vitality in the majority of the investigated species increased during the period of observation (Fig. 5.10.). It is possible to explain this tendency by the continued development of adventitious roots in the oppressed shoot systems. Thus the damaged individuals carrying main or adventitious roots regenerate and become capable of independent existence. Excavations of the shoot systems have shown that the old adventitious roots thicken and, evidently, functionally replace the lost main root.

Thus, the artificial disintegration does not result in the complete destruction of individuals of taproot-rhizomatous plants. On the contrary, in this case vegetative propagation of the plants took place. Simultaneously the destruction of ramets with undeveloped adventitious roots is observed. Hence, the opportunities of vegetative propagation of these species are limited by the extent of root forming on the rhizome segments.

The results allow us to make the following conclusions:

1. All investigated taproot-rhizomatous plants have a potential ability for vegetative propagation. The disintegration of individuals does not result in their mortality.
2. The decrease of vitality of a significant number of ramets is observed during the first years after disturbance (since 1993 at 1994). However, in the next years there is a tendency of increased vitality.
3. The survival of ramets depends on their age and rate of adventitious root development. Among the investigated species the greatest negative influence of separation from the main taproot was shown in *Minuartia circassica*, and least in *Minuartia aizoides* and *Arenaria lychnidea*.

5.2. Biomass and production

5.2.1. Introduction

Biomass and production are very important ecosystem functional parameters. The parameters depend on efficiency in use of solar radiation as well as on decomposition rate. On the other hand, the structure of herbivore communities, grazing rates and the stability of ecosystems are dependent to a great extent on plant biomass accumulation and productivity. Biomass of different species is commonly used as an index of their abundance or competitive success in the plant community. Allocation patterns and shoot:root ratios are often used for estimation of the relative role of aboveground (light) versus belowground (water and soil nutrients) competition (Tilman 1988). So we studied basic biomass parameters for four main alpine communities (Onipchenko 1985, 1990).

From this point on we use the term "biomass" for living parts of plants, "mortmass" for dead parts of plants and litter, and "phytomass" as a sum of biomass and mortmass.

5.2.2. Methods

Aboveground biomass was estimated in the second half of summer (August) over three years. Plants were clipped at ground level in 25 × 25 cm plots (20 plots every year, total replication was 60). Each sample was sorted by species, dried (105°C for 8 hours) and weighed. Present year litter was included to approximate total values of aboveground production. Mortmass was studied on the same samples by hand separation. Aboveground production was estimated as a sum of aboveground herbaceous plant biomass and current year litter with the correction for vole's herbivory (Fomin et al. 1989).

Belowground biomass was studied in soil core samples ($100\,cm^3$, at least 10 replications for each 4 cm level). Roots were separated from soil by flotation and water sieving methods (see Böhm 1979) after 2 – 3 hours soaking in 0,27% sodium pyrophosphate solution. Dry weight of different root fractions was measured after drying (105°C for 8 hours). Dead belowground parts of plants were separated using flotation and microscopic features of roots (colour, consistence, root hairs presence etc.) (Demin 1970). Potential belowground production was assessed by annual rate of new root growth to artificial rootfree blocks (size 10 × 5 × 10 cm).

5.2.3. Aboveground biomass

Lichen biomass reached more than 400 g/m^2 (up to 600 g/m^2 on separate places) in alpine lichen heaths, but the role of lichens in other communities was negligible (Table 5.3, Fig. 5.11). *Cetraria islandica* was an absolute dominant among fruticose lichens (Fig. 5.12). Other arcto-alpine lichen species (*Cetraria nivalis, Cetraria cucullata, Thamnolia vermicularis*) were common in ALH as well.

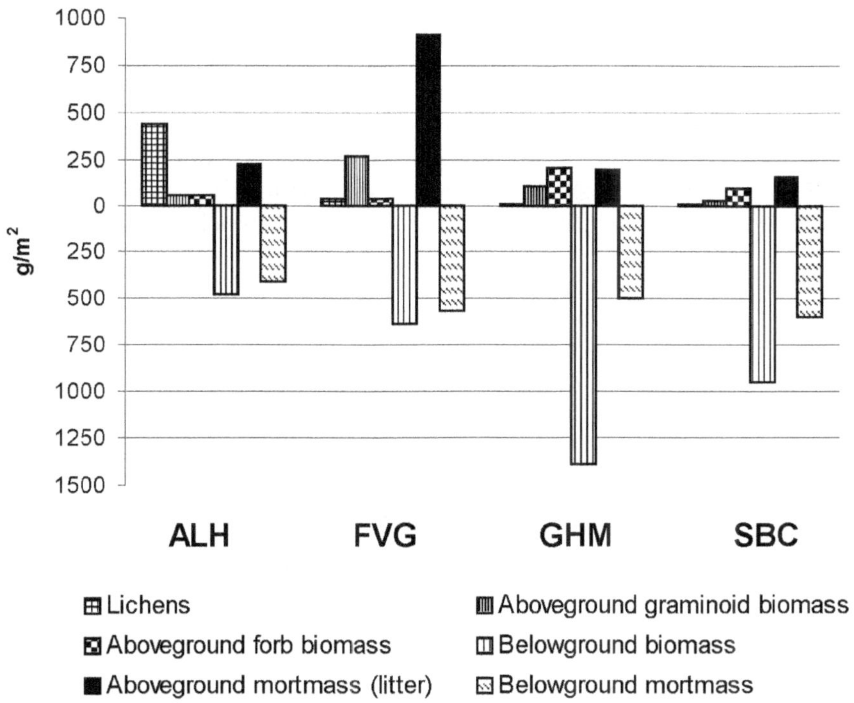

Figure 5.11. Phytomass structure of 4 alpine communities

The FVG and the GHM had the highest vascular plant biomass (on average 306 and 318 g/m^2 respectively) among studied alpine communities. The communities develop on slopes with moderate snow cover that is favourable for most alpine plants.

We may consider the following species as dominants (species contributing more than 5% of aboveground biomass) of the alpine communities:

ALH – *Anemone speciosa, Antennaria dioica, Campanula tridentata, Carex* spp. (mainly *C. umbrosa, C. sempervirens*), *Festuca ovina;*

FVG – *Festuca varia*(more than 50% of aboveground biomass), *Nardus stricta;*

GHM – Geranium gymnocaulon, Hedysarum caucasicum, Nardus stricta, Phleum alpinum;

SBC – *Gnaphalium supinum, Minuartia aizoides, Sibbaldia procumbens, Taraxacum stevenii.*

Firm-bunch grasses constitute about 80% of the FVG aboveground biomass. Forbs are the main components in communities with deeper snow (GHM and SBC). Proportions of forbs and graminoids are equal among vascular plants in ALH (Fig. 5.11). The role of bryophytes was negligible in all studied communities.

The aboveground mortmass was the highest in FVG (about 900 g/m^2). This is due to a high production of the community, and to a low rate of decomposition of dead leaves of firm-bunch grasses (Leinsoo *et al.* 1991).

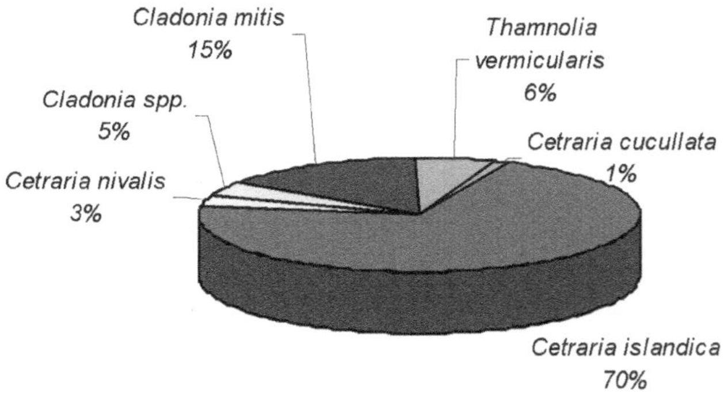

Figure 5.12. Lichen biomass species composition in the alpine lichen heath

We compared our data with aboveground biomass data of different types of tundra and alpine communities. As a rule, our communities had lower biomass than lowland meadows (Matvejeva 1980), secondary grasslands of the forest belt in the Caucasus (Bazilevich et al. 1987), subalpine meadows in Switzerland (Dietl 1986), and alpine grasslands in New Zealand (Meurk 1978). Similar values were obtained for tundras (Wielgolaski 1980, Andreyashkina 1981). The biomass of studied Caucasian grasslands (FVG,

Table 5.3. Aboveground biomass of plant species in 4 alpine communities, g/m^2, mean (m) $\pm$ st.err. (se), n=60. *Carex* spp. = *C. sempervirens* + *C. umbrosa* + *C. caryophyllea*

Species, fractions	ALH		FVG		GHM		SBC	
	m	se	m	se	m	se	m	se
Agrostis vinealis					2.2	0.5	1.8	0.4
Alchemilla caucasica	0.9	0.49						
Anemone speciosa	12.1	1.59						
Antennaria dioica	5.9	1.25	2.6	1.4				
Anthemis cretica			0.6	0.3	1	0.4		
Anthoxanthum odoratum			7.6	1.4	15.7	2.2		
Arenaria lychnidea	1.55	1.7						
Campanula tridentata	12.1	1.93	1.3	0.8	1.1	0.5		
Carex atrata				5.3	2	3.5	0.8	
Carex oreophila							2.1	0.6
Carex spp.	23.8	3.07	2.2	0.7				
Carum caucasicum	4.45	0.85						
Carum meifolium					4.1	0.7		
Catabrosella variegata					1.5	0.7	6.3	0.7
Deschampsia flexuosa			9.2	2.2	10	1.8		
Erigeron caucasicum					0.7	0.5		
Eritrichium caucasicus	0.65	0.25						
Festuca brunnescens			4.6	1.7	15.7	2.7	0.8	0.5
Festuca ovina	23.7	2.32	2.9	0.9	1.2	0.6		
Festuca varia			172	21	4	3.5		
Gentiana pyrenaica	2.6	0.6	2.2	0.5	1.1	0.5		
Gentiana septemfida			1.5	1.5				
Geranium gymnocaulon					137	14		
Gnaphalium supinum							11	1.9
Hedysarum caucasicum					33.5	5.8		
Helictotrichon versicolor	5.35	1.1						
Hieracium spp.						1.2	1.2	
Hyalopoa pontica							0.6	0.2
Leontodon hispidus			5.6	3.2	3.2	1.2		
Luzula spicata	1.45	0.35						
Matricaria caucasica					10.8	1.5		
Minuartia aizoides					3.5	1.1	13.4	1.6
Minuartia circassica	1.6	1.15						
Minuartia recurva			0.5	0.2				
Nardus stricta			67.6	9.3	34.9	4.3	15.1	2.8
Pedicularis comosa	0.75	0.41						
Pedicularis nordmanniana							3.8	0.8
Phleum alpinum					19.9	2.7	4.3	1.1
Potentilla gelida	1.45	0.58	0.5	0.5				
Rumex alpestris						0.7	0.6	
Scorzonera cana				4.4	1	4.1	0.5	
Sibbaldia procumbens					5.6	1.7	49.4	2.8
Taraxacum confusum								
Taraxacum stevenii	1.8	0.43					17.8	1.5
Trifolium polyphyllum	8.1	3.04						
Vaccinium vitis-idaea	1.55	0.7						
Veronica gentianoides	0.85	0.29	1.7	0.6	2.4	0.5		
Viola altaica			2.9	0.4				
other vascular plants	2.15		4.5	0.8	2	0.6	2.1	0.5
Total vascular plants	113	7.1	306	17	318	20	129	5
Bryophytes	3.4	3.05	8.8	2.3	1.7	1.6	2.7	0.8
Lichens	468	29.8	36.1	2.2	5.7	2.9	1.5	0.5

GHM) exceeded the values for mountain tundras of Khibiny (Politova et al. 1989), alpine meadow in Armenia (Ziroyan 1981, Barsegyan et al. 1987), and subnival plant communities in the Central Caucasus (Nakhutsrishvili et al. 1980) and the Austrian Alps (Brzoska 1973). ALH biomass is similar to the values for *Curvuletum* in the Alps (Pumpel 1977, Grabherr et al. 1978, Grabherr 1987a). Caucasian alpine grasslands (FVG, GHM) biomass is close to the values for *Seslerietum* and *Luzuletum alpino-pilosae* (Mentler & Brauer 1986), subalpine meadows in the Caucasus (Dzybov 1979, Nakhutsrishvili et al. 1980) and in the Altai (Sedel'nikov 1985), as well as *Kobresia* meadows in China (Yang Fu-tun et al. 1987).

5.2.4. Belowground biomass

Belowground phytomass increases with increasing snow depth from ALH to GHM, slightly decreasing in SBC (Table 5.4). The highest value was obtained for GHM (about 1900 g/m^2). The belowground biomass was three times as much as mortmass in this community. In the other investigated communities belowground biomass only slightly exceeded the mortmass.

Thin roots (<0.5 mm in diameter) are the main components of the belowground biomass. Their role increased from essentially snow-free ALH (275 g/m^2) to SBC (616 g/m^2). Körner & Renhardt (1987) found that allocation to thin roots is about 30% for alpine plants and 15% for lowland plants. Our data even exceeded their value for alpine plants: 47, 44, 30 and 57% for ALH, FVG, GHM and SBC, respectively.

Thick roots (>2 mm in diameter) were common only in ALH and GHM where taproot forbs play a significant role (Pokarzhevskaya 1995). Rhizome biomass was highest in GHM due to thick rhizomes of the main dominant – *Geranium gymnocaulon*.

Similar values to our belowground biomass values were obtained in alpine plant communities in Armenia (Voskanyan & Ziroyan 1979, Barsegyan et al. 1987), New Zealand alpine meadows (Evans 1980), and subalpine meadows in the Central Caucasus (Nakhutsrishvili et al. 1980). Lower biomass was noted in the subnival belt in the Caucasus (Nakhutsrishvili et al. 1980) and in the Alps (Brzoska 1973), and in the Venezuelan paramo (Smith & Klinger 1985). Cryophytic meadows in the Polar Ural were similar to SBC in terms of their biomass structure (Igosheva 1984, 1986).

The bulk of the roots was concentrated in the upper (0 – 5 cm) soil layer (Fig. 5.13). Roots of alpine plants can achieve a significant depth (50 – 100 cm and more, Kutschera et al. 1992), but their mass decreases rapidly with depth. A faster decrease was typical for GHM and SBC. On the other hand, the gradual decline in the root biomass with depth was noted in FVG (Fig. 5.13). It may serve as an adaptive feature to provide water supply during dry and hot periods.

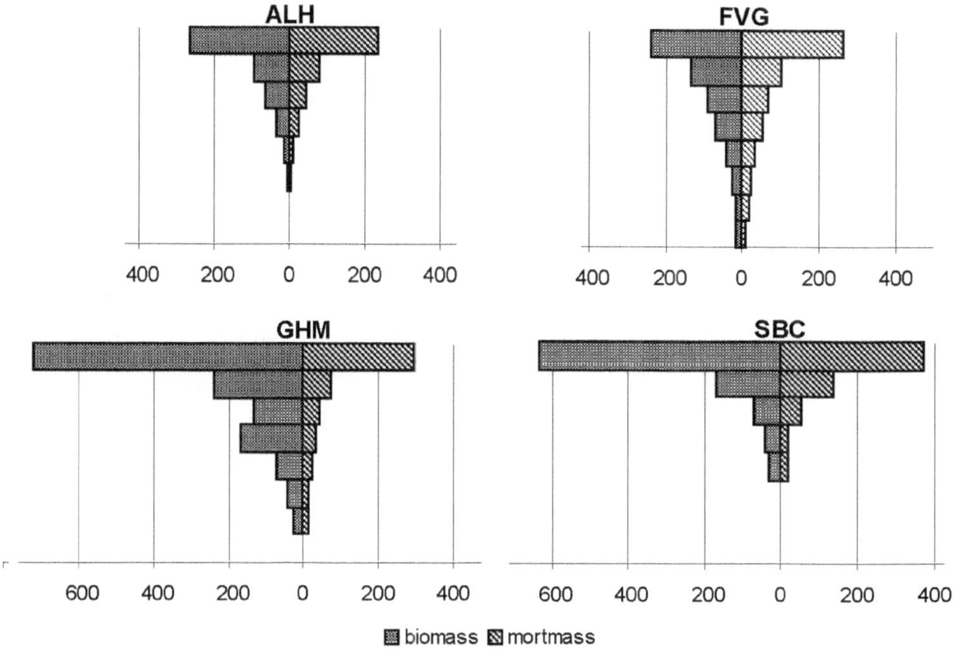

Figure 5.13 Vertical distribution of belowground biomass (g/m^2). Horizontal bars represent bio- or mortmass in soil layers of 4 cm each

It is interesting to compare the studied communities based on the ratios among different phytomass fractions (Table 5.5). Aboveground phytomass to belowground phytomass ratio is about 1 in ALH (due to considerable lichen biomass) and in FVG (due to great litter accumulation). The ratio is significantly lower in GHM an SBC (0.21 – 0.28). Aboveground biomass to belowground biomass ratio decreases from ALH (1.16) to SBC (0.14). Mortmass is the lowest in GHM in aboveground phytomass as well as in belowground phytomass (Table 5.4). By this is meant that the decomposition rate is relatively high in the community.

Potential belowground annual production is especially high in GHM. According to an estimation of the root decomposition rate (Onipchenko

1985) we assessed that the real root production was about 25% of the potential production. Most likely the real annual belowground production is about 10 – 30% of the belowground biomass. The values are similar to the data for Alaskan tundra (Shaver & Billings 1975), but significantly lower than in the lowland meadows and steppes. Low values of belowground production are in agreement with low "soil respiration" rates (Stepanov & Onipchenko 1989) and with very small annual rhizome increments of alpine plants.

Table 5.4. Belowground phytomass of the alpine communities (g/m^2, dry weigh, average and standart error, n = 10)

Fractions	ALH		FVG		GHM		SBC	
	av.	s.e.	av.	s.e.	av.	s.e.	av.	s.e.
Total phytomass	881	76	1201	62	1883	202	1536	118
Belowground mortmass	403	38	565	46	491	49	592	38
Belowground biomass	478	42	636	34	1392	177	944	128
Roots (< 0.5 mm in diameter)	275	13	417	22	518	27	616	66
Roots (0.5 – 2 mm in diameter)	58	10	206	15	180	35	138	15
Roots (> 2 mm in diameter)	50	20	1	1	379	114	1	1
Rhizomes, bulbs, corms	96	43	12	5	314	95	188	63
Potencial belowground annual production	128	33	143	28	354	71	120	14

Table 5.5. Ratios between different phytomass fractions (a/g – aboveground, b/g – belowground, v.p. – vascular plants)

Ratios	ALH	FVG	GHM	SBC
Total a/g phytomass : total b/g phytomass	0.89	1.01	0.28	0.21
A/g biomass : b/g biomass	1.16	0.55	0.23	0.14
A/g v.p. biomass : b/g v.p. biomass	0.23	0.48	0.23	0.14
A/g mortmass : a/g v.p. biomass	2.06	2.8	0.6	1.43
B/g mortmass : b/g v.p. biomass	0.84	0.89	0.35	0.63
A/g mortmass : b/g mortmass	0.57	1.52	0.39	0.31

5.2.5. Production estimation

We estimated the total annual net-production for the investigated communities to be about 150, 400, 550 and 200 g/m^2 for ALH, FVG, GHM, and SBC respectively. A moderate winter snowpack in the GHM seems favourable for plant production in the alpine zone. Plant productivity decreased in snow free communities experiencing deep soil freezing as well as in snow-bed communities with a short growing season (Onipchenko 1990).

Our data showed essential differences among the studied alpine communities in terms of their phytomass structure and production. It is interesting to note that the annual production varied more than the phytomass (3 – 4 and 1.5 times among the communities, respectively). Therefore, unproductive communities accumulate approximately similar amounts of phytomass as more productive meadows in the study area.

5.3. Spatial structure

5.3.1. Introduction

Spatial heterogeneity is a universal feature of natural plant communities. A number of causes may determine their horizontal spatial structure, namely environmental heterogeneity, some properties of plants (character of distribution, growth, life cycle), interactions between plants and random factors (Watt 1947, Rabotnov 1983). The principal role of the factors differs in different communities. In this chapter we consider spatial structure of alpine plant communities from two points of view. First, (Chapter 5.3.2.) we compare the heterogeneity of the 4 studied alpine communities (ALH, FVG, GHM, SBC) according to the so-called "mass-effect" (vicinity or spatial autocorrelation) approach. Second, we consider the spatial heterogeneity of soil properties within plant communities as a possible factor behind the community spatial structure. The following subchapters (5.3.3 – 5.3.6) deal with specific reasons of spatial structure in different communities. They include experimental investigations of the role of soil purity in ALH (chapter 5.3.3), the influence of *Festuca varia* on FVG spatial structure (chapter 5.3.4), gap formation in GHM (chapter 5.3.5), and microrelief and snow regime impacts in SBC (chapter 5.3.6).

5.3.2. Spatial heterogeneity of plant communities and soil properties according "mass-effect" approach

5.3.2.1. Aims and methods

Many researchers attribute the spatial pattern of species composition to environmental heterogeneity (Grubb 1977, Sohlberg & Bliss 1984, Chuyo 1985). If the environment is more or less homogeneous, the theory of mass effect may explain the apparent heterogeneity (Shmida & Ellner 1984, Wilson 1990, Mistral et al. 2000). It states that the pool of seeds and fruits brought in from neighbouring habitats can help support populations of rare species and increase species richness within a community. The direct measurement of the mass effect can be a complex matter. Shmida & Ellner (1984) suggested that the mass effect within a community exists if a significant difference in the floristic similarity between plots is found. In this case two neighbouring plots will be more similar than two remote ones.

Gradual soil property changes with space as well as slope processes may be the main reasons for mountain plant community heterogeneity. To investigate these suggestions we undertook the present study (Onipchenko & Pokarzhevskaya 1994, Onipchenko et al. 1998b).

The aims of this study were (1) to compare "floristic similarity – distance" curves among the alpine communities, (2) to obtain analogy curves for "soil property similarity – distance" relationships, (3) to estimate a role of orientation along versus across slope in determining floristic or soil property similarity between plots.

Five sample areas 5 × 5 m were laid out within each of the four alpine communities (Onipchenko & Pokarzhevskaya 1994). The sides of the areas were oriented along the slope. Each area was divided into one hundred square plots of 50 × 50 cm. The plots were arranged in horizontal and vertical rows. We checked for presence or absence of all vascular plant species in 52 plots in a checkerboard pattern within the sample areas.

One of the 5 × 5 m areas was chosen for the soil heterogeneity study in each community. The area was divided into 25 one m^2 plots. Species composition was noted for each plot and separately for the central ($500\,cm^2$) part of the plots. A soil sample at 10 cm depth, 11 cm in diameter was collected from the centre of each plot: 25 samples from each community were analysed (Onipchenko et al. 1998b). The following soil properties were studied (Onipchenko et al. 1998b): 1) pH in water and KCl solution, 2) K, Ca, Mg, mobile organic mater in water solution, 3) exchangeable K, Ca, Mg and mobile P, 4) total C (loss on ignition), N, and P, 5) nitrate and ammonium concentrations.

To calculate floristic similarity between the plots we used Sorensen distance as a measure. Czekanowski's Index was used for quantitative soil data after their transformation by dividing on the maximum value for each property.

5.3.2.2. Floristic similarities within the communities

The studied alpine communities showed different degrees of spatial "mass-effect" measured as spatial dependence of floristic similarity on the distance between plots (Fig. 5.14, Onipchenko & Pokarzhevskaya 1994). The inclination of the curve shows the degree of heterogeneity of the community structure. In view of the supposed connection between the heterogeneity of structure and the "mass effect", the steeper the curve, the more important the mass effect is for this type of community. Thus our curves allow us to estimate the degree of mass effect. A two-way ANOVA has shown the significance of the effect of community type and the distance between plots on the floristic similarity.

The ALH and SBC have the most gently sloping curves. These types of communities are thus the most homogeneous and the species of these communities are evenly distributed within them. The curve for SBC falls off more rapidly than that for the ALH.

The floristic richness of ALH plots is quite high, and there are many species with high frequency. These communities are the most homogeneous among four studied types, and the recent "mass effect" is negligible. We suggest the following explanation for this phenomenon: The species belonging to this community are equally distributed, because the low height and slight competitive ability of these plants set no obstacles for the settlement of species in this community. Still, this assertion is right perhaps only on the large scale, e.g., our sample area (see chapter 5.3.3). There were some observations that severe environmental conditions can influence the establishment of seedlings more than competition (Ryser 1990).

The FVG possesses the steepest curve, which shows the highest level of heterogeneity of horizontal structure. The curve for GHM lies slightly higher. The results allow us to conclude that the mass effect is best expressed in FVG and least in ALH.

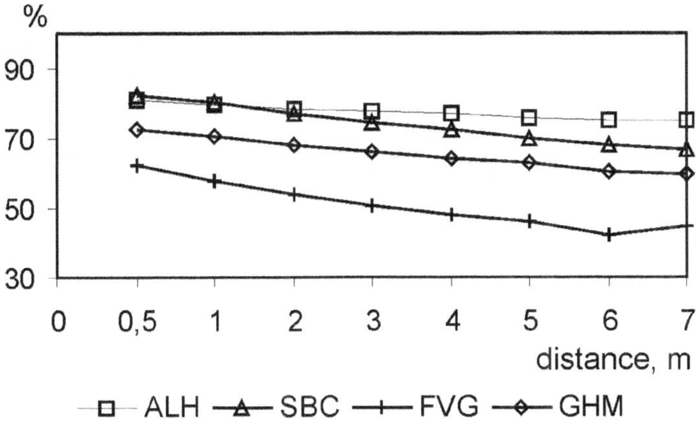

Figure 5.14 Floristic similarity (Sorensen Index, %) as a function of distance between plots. ALH – alpine lichen heath, FVG – *Festuca varia* grassland, GHM – *Geranium gymnocaulon – Hedysarum caucasicum* meadow, SBC – snowbed community. X – axis – distance (m) between plots (from Onipchenko & Pokarzhevskaya 1994 with permission of Geobotanical Institute SFIT).

FVG are productive communities, dominated by large tussocks grasses. They are the richest in species, but have only few species with high frequency. FVG are the least homogeneous among the studied communities, thus the mass effect is strongly expressed there. We suppose that the main dominants *Festuca varia* and *Nardus stricta*, producing a tough sod, prevent other species from settling within the community. It is possible that the input of diasporas from neighbouring areas, i.e., mass effect, can enable many species with low frequency to survive in FVG. These species, spreading vegetatively, form patches scattered in between *Festuca* and *Nardus* tussocks. Burrowing activity of rodents and other mammals (Fomin et al. 1989), although not very intensive, can help the survival of such species.

GHM is another example of an alpine community type with a high production and a considerable total number of species. Species with intermediate frequency prevail here. Burrowing activity of voles and other animals has a great effect on this community type (Fomin et al. 1989). The dominants *Geranium gymnocaulon* and *Hedysarum caucasicum*, producing a high phytomass, may prevent the settlement of other species. Although this community is rather heterogeneous, the degree of mass effect here is lower than that in FVG. We suspect, that intensive disturbance caused by digging animals lowers the impact of the dominant species, and facilitates the settlement of the territory by other species of plants.

SBC is the poorest in species, yet it is less homogeneous than ALH. We suggest that this fact can be explained as caused by a difference in the time of snowmelt during the vegetative season, which has an effect on the SBC structure (Chapter 5.3.6.).

As mentioned before, our large plots were oriented along the slope. We suspected that the distribution of diasporas (seeds, vegetative propagulae) depends on the slope processes. We expected therefore that most diasporas would move down the slope. Therefore, the degree of mass effect should differ in two directions: along the slope and across the slope. By comparing values of floristic similarity for vertical and horizontal rows of plots we could find out, if the mass effect in fact depends on the orientation. In this case two plots at different elevation would have different coefficients of similarity than two plots on the same level. However, we didn't find any significant differences (Onipchenko & Pokarzhevskaya 1994). The results are even more interesting, as the studied communities are situated on slopes of different steepness, ranging between 1 – 3 to 40 degrees.

5.3.2.3. Spatial structure of soil properties

Among other factors responsible for the heterogeneity of the community structure, we must mention the complexity of soils (Olsvig-Whittaker 1988). We calculated similarity between plots according to their soil properties

(Czekanowski's Index). This allowed estimation of the dependence of similarity on the distance between plots. In contrast to the pattern of floristic similarity (Fig. 5.14), no significant relationship was obtained (Fig. 5.15, Onipchenko et al. 1998b). With an increase in distance between the points of soil sampling from 1 to 7 m, the similarity in soil properties remained about the same. We observed the same pattern when calculating the similarity between plots along the slope and across the slope. Thus, slope processes have not lead to gradual soil property changes along slopes on the scale (1 – 5 m) studied within the communities.

Our data allowed us to compare soil property variability among the communities (Table 5.6). As we can see, higher mean values (soil nutrient concentrations) do not always coincide with higher variance values (Ca, SO_4, NO_3 for example). To estimate the relationship between mean and variance we use the ranks of the values (from 1 to 4).[2] analysis did not show a significant correspondence between the ranks of mean and variance values. Moreover, when we sum rank values for means we obtain the sequence of communities in the decreasing rank order:

FVG > ALH > GHM > SBC

The order of communities according to the sum rank values for variance is different:

GHM > SBC > ALH > FVG

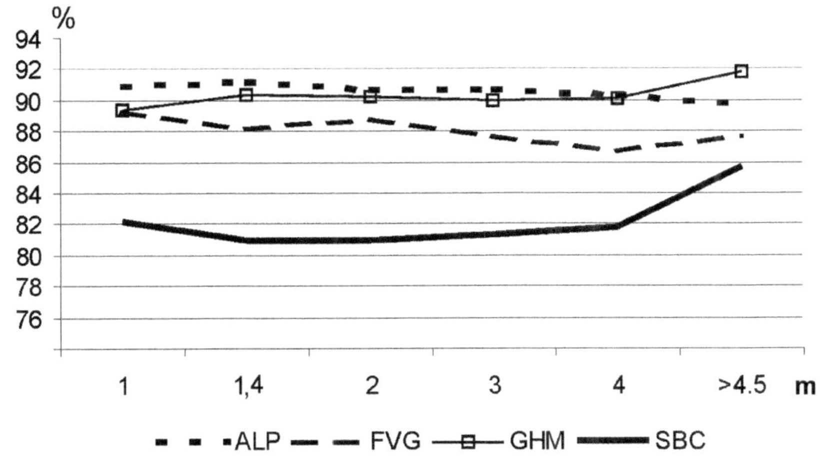

Figure 5.15 Soil property similarity (Czekanowski's Index, %) as a function of distance between plots. ALH – alpine lichen heath, FVG – *Festuca varia* grassland, GHM – *Geranium gymnocaulon – Hedysarum caucasicum* meadow, SBC – snowbed community. X – axis – distance (m) between plots (from Onipchenko et al. 1998b)

Table 5.6. Mean values and variances of soil properties. Soil samples were collected from each sq. m in plots of 5 × 5 m, n = 25. Units: mg/100 g for water extracted forms, exchangeable NH_4 and mobile P, mg-eq/100 g for exchangeable forms, % for total contents, loss of ignition, mg O_2/l for permanganate oxidation (from Onipchenko et al. 1998b)

Property	Mean				Variance			
	ALH	FVG	GHM	SBC	ALH	FVG	GHM	SBC
$pH_{(H2O)}$	5.58	5.05	4.86	4.76	0.0590	0.0074	0.0086	0.0540
$pH_{(KCl)}$	4.17	3.93	3.71	3.71	0.0893	0.0081	0.0046	0.0034
$K_{(H2O)}$	4.06	8.16	6.28	5.66	0.5064	3.7144	1.4816	1.3344
$Ca_{(H2O)}$	19.5	16.5	13.8	8.3	2.44	2.47	7.24	0.74
$Mg_{(H2O)}$	3.37	3.30	2.73	2.21	0.46	0.12	0.24	0.06
K_{exch}	0.69	1.29	0.87	0.83	0.0047	0.0572	0.0135	0.0146
Ca_{exch}	9.46	7.17	4.07	2.88	1.85	1.69	2.02	4.20
Mg_{exch}	1.65	1.82	1.00	0.76	0.0361	0.0997	0.0766	0.0236
P_{mob}	0.10	0.16	0.19	0.08	0.0001	0.0003	0.0004	0.0000
$SO_{4\text{-}H2O}$	2.06	3.43	2.55	2.11	0.64	0.36	1.14	0.61
C_{total}	9.52	9.59	9.10	9.34	0.521	0.868	1.786	3.631
N_{total}	1.03	0.90	0.85	0.95	0.0103	0.0063	0.0167	0.0164
P_2O_{5total}	0.28	0.31	0.29	0.33	0.0009	0.0009	0.0003	0.0012
loss on ignition	24.26	24.65	22.96	25.42	3.76	3.62	5.69	10.97
Oxidation	82.5	80.3	86.3	55.5	300	588	251	855
NH_{4exch}	2.43	1.90	3.16	2.84	0.1686	0.1114	0.2549	0.3120
NO_3	2.46	2.22	2.77	1.43	1.7930	0.5065	0.3109	0.2748

So, the communities of the upper parts of slopes (ALH, FVG) have higher mean values of soil nutrient concentrations but a lower spatial variability of the soil properties. In contrast, the communities on the lower parts of slopes and depressions (GHM, SBC) tended to have a higher spatial variability of the soil properties. Perhaps, a more prominent zoogenic disturbance regime in GHM is a reason of this phenomenon. Most interesting is the FVG position in the rows. This community is the first in the "mean" rank order and the last in the "variance" rank order. We can hypothesise that in spite of a higher nutrient concentration, dense and prostrate *Festuca varia* roots can equalise the soil properties by selective uptake from locally rich patches. Such selective uptake was shown for several species with Grime's competitive strategy (Jackson & Caldwell 1993, Grime 2001).

5.3.3. Alpine lichen heath spatial structure: hypothesis and experiments

5.3.3.1. Introduction

Alpine communities of humid mountains developing on windward slopes have fruticose lichens as the main components of aboveground biomass (Stanyukovich 1960, Gorchakovskii 1975, Grabherr et al. 1978, Wielgolaski 1980, Grabherr 1987b, Sedel'nikov 1988). Many communities have a fine spatial structure (Onipchenko 1984, 1985, 1994b, Grabherr 1987a, 1989), e.g., alpine lichen heaths (ALH).

The most common pattern of their spatial structure is the following: small lichen areas (up to 10 cm across) alternate with graminoid bunches (*Festuca ovina, Carex sempervirens, C. umbrosa*) or isolated shoots of vascular plants. Pure lichen or vascular plant patches of more than 20 cm in diameter do not occur in this community. There is a significant amount of thin roots in the soil under lichen patches (Fig. 5.16). We proposed the following hypothesis as an explanation of this structure (Onipchenko 1985, 1994b).

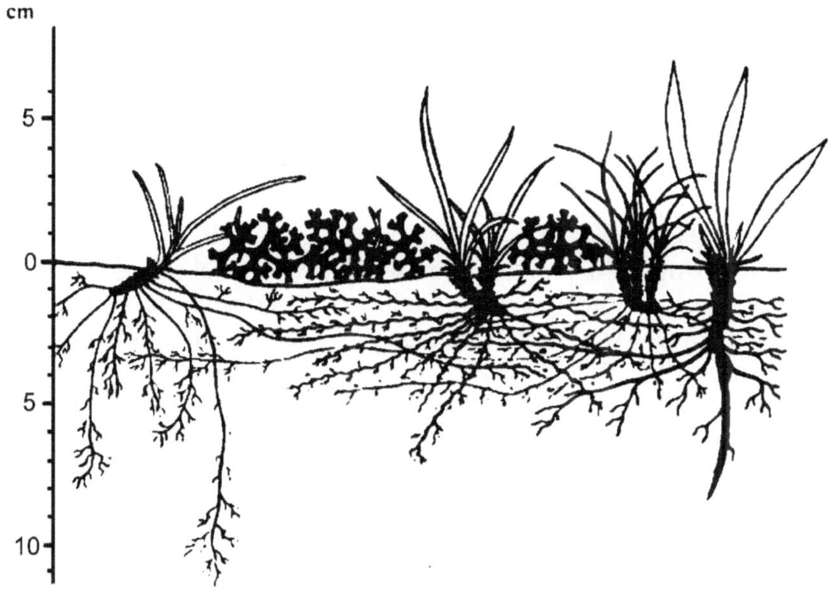

Figure 5.16 Structure of alpine lichen heath. Plants (from left to right): *Helictotrichon versicolor, Cetraria islandica, Carex umbrosa, Cetraria islandica, Festuca ovina, Plantago atrata.* (from Onipchenko 1985)

ALH soils are comparatively poor and shallow (Chapter 2). Because of this, vascular plants have to develop large root systems. These root systems occupy a larger area than the aboveground shoots, so open space remains in the aboveground layer of vascular plants. Fruticose lichens can occupy these free patches, because they receive their main nutrients with precipitation and atmospheric dust. The lichens constitute the main part of the aboveground phytomass, but they are not able to efficiently compete with vascular plants for nutrients. Vascular plants form the "framework" of the community. Because of frequent strong winds, fruticose lichens cannot develop without the "framework." Grabherr (1987a, 1989) suggested a similar hypothesis for the structure of the alpine sedge meadow (*Caricetum curvulae*).

Accepting the hypothesis, we might make the following assumptions (Onipchenko 1994b):

1. Removal of lichens probably should not change the other components of the plant community.
2. Enrichment of soil with nutrients (fertilization) should increase the aboveground vascular plant biomass and decrease the lichen biomass due to the competition for light.
3. Isolation of soil areas under lichen patches from roots (root cutting) should facilitate the settlement and growth of vascular plants on these areas in comparison with non-isolated (control) areas (Fig. 5.17).

To prove these assumptions a series of experiments was carried out (Onipchenko 1984, 1985, 1994b).

5.3.3.2. Methods

Four permanent plots of 1 m^2 were used for lichen removal. Using pincers, fruticose lichens were carefully removed from the plots by hand. The experiment was started in August of 1981. We attempted to remove all reappearing fruticose lichens from the experimental plots during the whole period of observation. We monitored shoot density on 16 small plots (25 × 25 cm, 4 plots per m^2) for 12 years. The density of shoots was compared with the initial data and with data for control plots.

The fertilisation experiments had nine variants: control, Ca, N, P, K, NP, NK, PK, NPK (Onipchenko 1994b). We used lime as a source of Ca (5 t per ha), ammonium nitrate as a source of N (90 kg N per ha), potassium chloride for K (60 kg K_20 per ha), and commercial double superphosphate for P (60 kg P_2O_5 per ha). We applied lime only once (May 1981), the other fertilisers once per year in the beginning of the growth season during three years (1981 – 1983). For each variant we used 5 plots with a size of 2 × 3 m. For biomass estimation we took 10 experimental cuts at ground level in 25 × 25 cm plots for each variant and each year. The number of shoots was counted for the plots and the weight of single shoots was calculated.

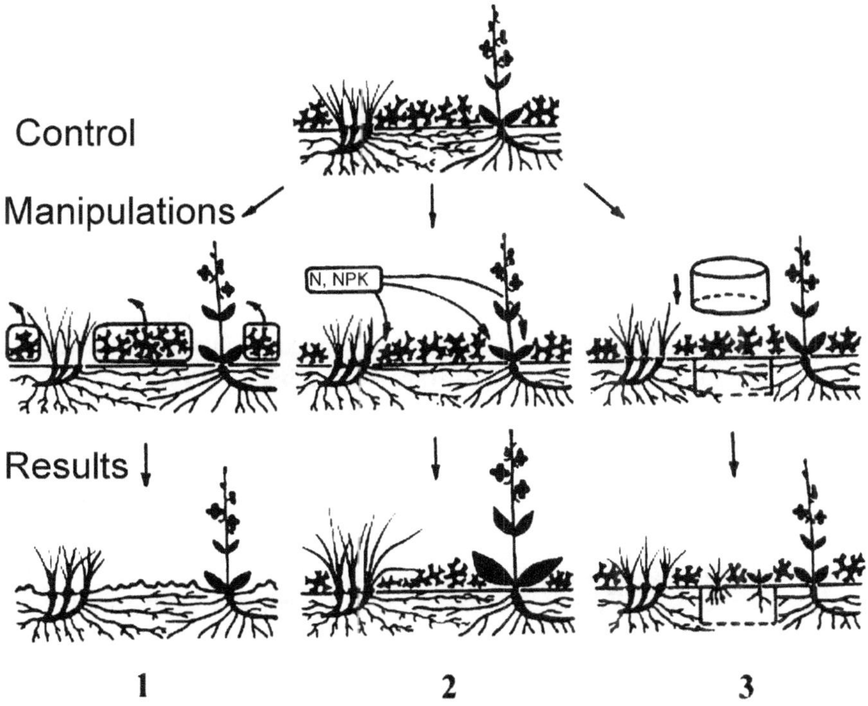

Figure 5.17 Schematic diagram of experiments: 1 - lichens removal experiment, 2-fertilisation, 3 – root cutting (from Onipchenko 1994b with permission of Geobotanical Institute SFIT)

For the root cutting experiment the sod under lichen patches was isolated from neighbouring vascular plant roots using metallic tubes (diameter 7 or 10 cm, depth 8 – 10 cm, 12 replications). The tubes were left in the soil to prevent lateral root growth (Fig. 5.17). Control plots had the same size (5 replications). We counted the number of shoots and seedlings at the end of the growth season for 20 years on experimental and control plots.

5.3.3.3. Results and discussion

Lichen removal experiments. There have not been any significant changes in vascular plant floristic composition on the experimental plots after lichen removal (Onipchenko 1994b). Vascular plant cover on experimental and control plots was about 40% and didn't change significantly after removal of lichens. The barren areas resulting from the removal of lichens haven't decreased in size and they haven't been overgrown by other plants. Sometimes frost "boiling" occurred on these sites. A considerable cover of *Cladonia* species (primary thalli) was found 11 years after the beginning of the experiment. It was not possible to remove the thalli without affecting the soil, so the experiment was finished after 12 years.

The shoot density of most species showed fluctuations from year to year. Some species decreased in shoot density during the experiment (*Gentiana pyrenaica, Helictotrichon versicolor*). It can be related to microclimatic changes or soil surface "boiling" since fine roots of these plants are located close to the soil surface (Onipchenko 1987b). *Eritrichium caucasicum* and *Luzula spicata* showed a relatively positive response to lichen removal (Onipchenko 1994b). Shoot density of *Vaccinium vitis-idaea* increased three- to five-fold from the first years to the last. This slow-growing evergreen dwarf shrub apparently can compete with fruticose lichens for space. The population density of the semiparasitic *Euphrasia ossica* rose sharply during the first years of the experiment. Species of *Euphrasia* show a positive reaction to disturbance, for example, *E. minima* is a very abundant species in the winter sports resorts of the Alps (Hofer 1981). Probably, the species have light-stimulated seeds (Grime 1981). *E. ossica* did not develop after artificial shading in ALH (Rabotnova et al. 1992). Probably, primary thalli of *Cladonia* negatively influenced the species during the last two years of the experiment.

In spite of some changes in the experimental plots after lichen removal, on the whole, the community structure and its floristic composition kept stable during the observation period. The results do not contradict the hypotheses being discussed.

Fertilisation experiments. Significant changes of plant biomass and composition were found only in the variants where N was brought in (Fig. 5.18, 5.19). P affected plant productivity only in NP and NPK variants, i.e., in the presence of N. The floristic composition didn't change during the 3 years of the experiment, but the aboveground biomass increased significantly especially for grasses and sedges. Lichen biomass slightly decreased in variants with NP and NPK, but the changes were not significant. Vascular plants formed dense covers in these variants. It seems that 3 years is too short a period for lichen cover degradation, because of the slow rate of lichen decomposition in the community (Voronina et al. 1986).

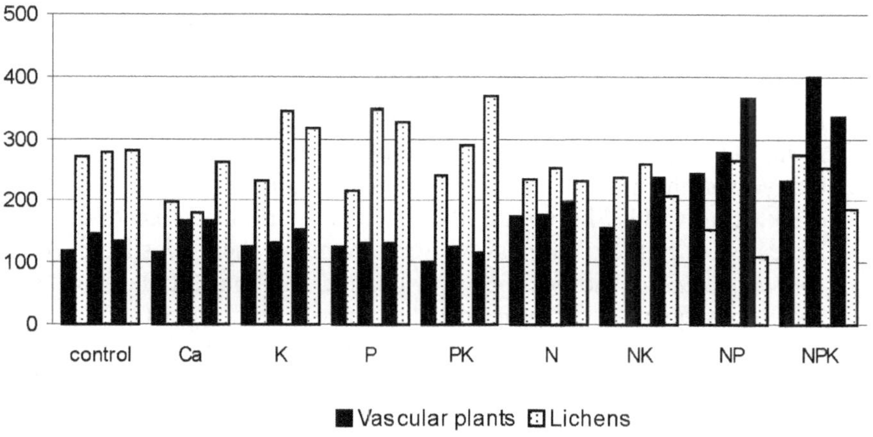

Figure 5.18 Vascular plant and lichen biomass dynamics during 3 years of the fertilisation experiment. The first pair of bars in each treatment stands for 1981 values, the second for 1982 and the last for 1983 (g per m^2, dry mass). (from Onipchenko 1994b with permission of Geobotanical Institute SFIT)

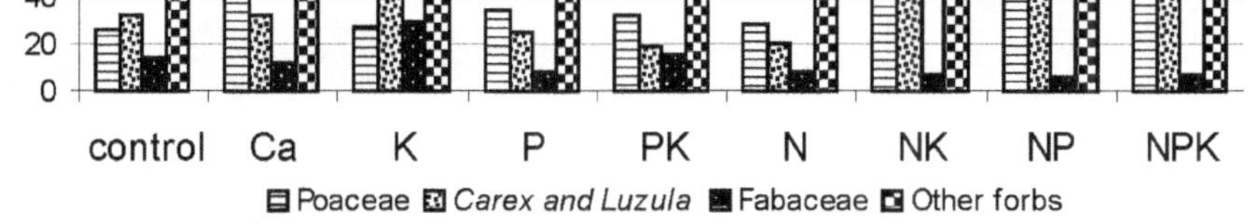

Figure 5.19 Average vascular plant biomass composition in different treatments of the fertilization experiment (g per m^2, dry mass). (from Onipchenko 1994b with permission of Geobotanical Institute SFIT)

Vascular plants had different responses to the addition of nutrients. Sedges and grasses (*Carex umbrosa, C. sempervirens, Festuca ovina, Helictotrichon versicolor*) increased their density of shoots as well as their size and weight on variants where N had been added (Table 5.7). A similar response of grasses and sedges to fertilization was observed in many experimental researches in arctic and alpine ecosystems (McKendrick et al. 1980, Shatvorjan 1981, Jonasson 1992). Most rosette forbs increased only their biomass without changing their density. Species as *Antennaria dioica* did not show any changes in shoot biomass or density. As a consequence, the proportion of grasses and sedges in the aboveground biomass was much greater for the variants with N than for the control (Fig. 5.19). It is interesting that the biomass of *Trifolium polyphyllum* (closely related to *Trifolium alpinum*) increased in the N variant. This species has not (or has a few) nodules on its roots and uses essentially mineral nitrogen forms from soils. The fact that the biomass of *Trifolium polyphyllum* did not increase in the PK variants is not typical for Fabaceae species (Rabotnov 1973).

Root cutting experiment. The density of shoots and seedlings in the experimental plots has slowly increased (Fig. 5.20), with some fluctuations, during the observation period. Significant statistical difference from the control plots was obtained only after 10 years. *Vaccinium vitis-idaea, Festuca ovina, Euphrasia ossica, Potentilla gelida* and *Primula algida* were the most active colonists of the experimental areas after root cutting (Table 5.8). Most of these plants germinated from seeds having originated both from the soil seed bank and from recent seed rain. *Vaccinium vitis-idaea* did

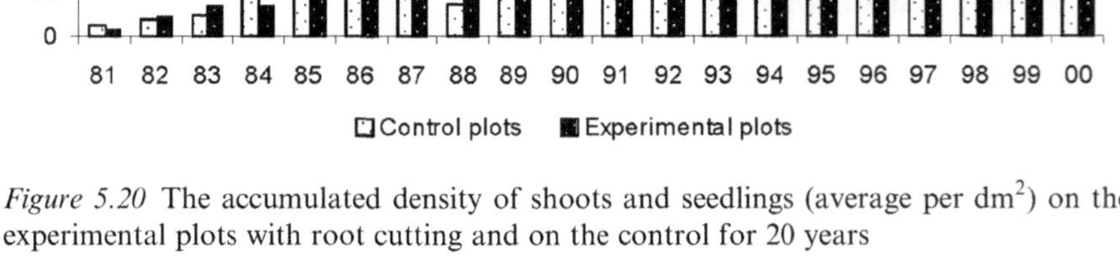

Figure 5.20 The accumulated density of shoots and seedlings (average per dm^2) on the experimental plots with root cutting and on the control for 20 years

not produce seeds in ALH, but it produced a well-developed underground shoot system for vegetative propagation. Apparently, it can develop new aboveground shoots from dormant buds after a "gap" has been detected.

Table 5.7. Shoot density in permanent plots and average shoot biomass of some species in the fertilisation experiment.* Number of shoots per 8 permanent plots of 25 × 25 cm. Abbreviations: n.d. - no data, m. - mean value, s.e. - standard error, Cont. – control (from Onipchenko 1994b with permission of Geobotanical Institute SFIT)

Species	variant	number of shoots*:				shoot	
		vegetative		generative		weight (mg per 1 shoot)	
		1980	1982	1980	1982	m.	s.e.
Helictotrichon versicolor	Cont.	78	60	0	1	54	5
	N	47	29	0	1	57	6
	NP	58	89	0	33	101	7
Carex umbrosa	Cont	243	230	12	5	60	7
+ *C. sempervirens*	N	166	276	9	14	77	7
	NP	325	588	41	79	88	8
Luzula spicata	Cont.	13	16	1	2	29	5
	N	9	16	1	2	29	6
	NP	25	14	3	9	98	15
Anemone speciosa	Cont.	10	14	6	3	266	22
	N	9	10	13	10	358	34
	NP	14	8	15	16	430	43
Campanula tridentata	Cont.	2	3	0	1	70	8
	N	8	2	6	7	79	11
	NP	8	13	1	2	111	13
Trifolium polyphyllum	Cont.	269	314	0	3	73	11
	N	187	216	2	12	116	8
	NP	0	0	0	0	112	17
Festuca ovina	Cont.	n.d.	180	2	12	19	2
	N	n.d.	186	0	16	24	3
	NP	n.d.	489	1	258	29	2
Antennaria dioica	Cont.	109	95	0	0	42	4
	N	145	89	0	0	33	3
	NP	96	83	0	0	43	3

5.3.4. *Festuca varia* grasslands

5.3.4.1. Introduction

The spatial structure of FVGs was studied from two points of view: 1) influence of dominant grass (*Festuca varia*) on the spatial structure of the community (Pokarzhevskaya 1998a), and 2) influence of the "island effect" on the species richness of FVGs (Sukhova et al. 1997).

Table 5.8. Dynamics of the shoot number of some alpine lichen heath species on experimental plots with root cutting and on control plots during 12 years (from Onipchenko 1994b with permission of Geobotanical Institute SFIT)

Year	81	82	83	84	85	86	87	88	89	90	91	92
control plots												
Vaccinium vitis-idaea	4	5	5	13	16	15	15	9	22	15	15	23
Potentilla gelida	0	0	0	0	0	0	0	0	0	0	0	0
Helictotrichon versicolor	0	0	0	0	0	0	0	0	0	0	0	0
Festuca ovina	0	0	0	0	0	0	0	0	0	0	0	0
Euphrasia ossica	0	0	2	4	4	11	0	1	1	0	1	1
Primula algida	0	0	0	0	0	0	0	0	0	0	0	0
experimental plots												
Vaccinium vitis-idaea	2	7	7	11	6	42	52	61	78	88	101	118
Potentilla gelida	2	3	4	2	3	6	5	5	6	6	6	7
Helictotrichon versicolor	0	1	1	2	6	5	3	1	1	3	4	4
Festuca ovina	0	0	0	1	3	5	4	10	11	17	32	44
Euphrasia ossica	0	0	9	7	1	17	14	11	13	38	3	7
Primula algida	0	1	3	1	5	6	4	4	4	5	7	5

First we discuss the influence of the dominant grass features on the community structure.

Festuca varia is the dominant in this community type. It is a dense tussock grass with obvious xeromorphic features. It produces numerous tillers, which form a large amount of litter after dying off. The root system of *Festuca varia* is well developed, especially in the upper soil layer (Onipchenko 1990). Thus the grass forms tussocks of different shapes and sizes, depending on slope steepness. Tussock forms vary from relatively large isolated "round" to smaller oblong, arranged in a sort of "rows" (Fig. 5.21, Pokarzhevskaya 1998a).

Tussock species have several features that allow them to be strong competitors, restricting clonal spread of other species. Such features include the litter they produce (Facelli & Pickett 1991), plant height (Elberse & Berendse 1993), growth form (Serebryakova 1971, Grime 1979), and high biomass (Epp & Aarsen 1989). In this way tussock species have an impact not only on the performance of single species but also on the structure of the whole community (Pokarzhevskaya 1998a).

In spite of the fact that *Festuca varia* is the dominant tussock grass, it doesn't eliminate other species from the community. These grasslands are relatively rich in species (Chapter 4). They also show small-scale heterogeneity.

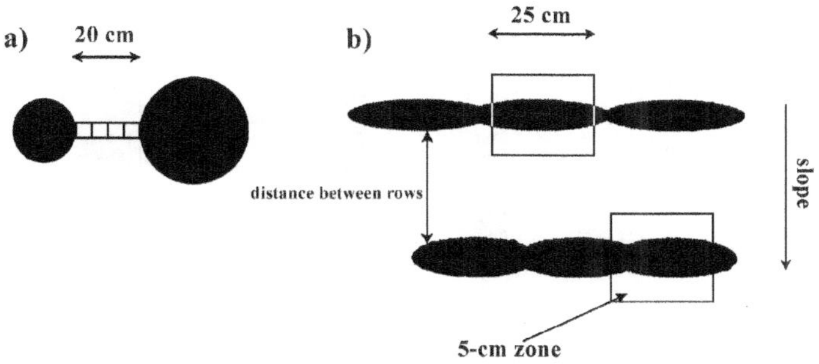

Figure 5.21 Scheme of the investigated plots on sites with (a) round tussocks, (b) rows. (from Pokarzhevskaya 1998a with permission of *Arctic, Antarctic and Alpine Research*)

5.3.4.2. Influence of Festuca varia on the spatial structure

A detailed study of the influence of different tussock forms of *Festuca varia* on the spatial structure of the community on a small scale was conducted by Pokarzhevskaya (1998a). Two questions were addressed:

How do *Festuca varia* tussocks of different shape vary in their influence on the community?

Which species do prefer to occur inside and between the tussocks?

For these aims tussocks of different shapes were studied. In the case of "round" ones belts of plots (transects) between tussocks were laid out. In addition, the presence and absence of species within tussocks were noted. For the investigation of *Festuca varia* "rows" presence and absence of species was recorded above and below the rows and within them (Fig. 5.21, Pokarzhevskaya 1998a).

The round *Festuca varia* tussocks of different sizes were inspected. They were situated at different distances from each other (Fig. 5.22). Shoots of different age categories were distributed equally over the tussock, and the litter amount was equally distributed around the plant. The number of phanerogam species outside the round tussocks was negatively correlated with litter cover and positively with distance to the neighbouring tussock (Pokarzhevskaya 1998a). *Nardus stricta, Anthoxanthum odoratum*, and *Leontodon hispidus* occurred less often in close vicinity of the tussock than at a distance. For other species the relation between occurrence and distance to tussock was not significantly different from random.

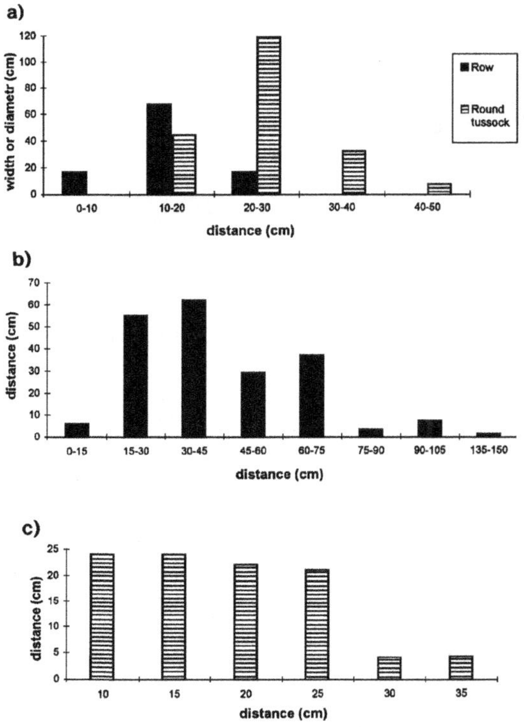

Figure 5.22 The distribution of tussock size and distances to the nearest neighbor: a) – row width and round tussock diameter; b) – distances to neighboring rows; c) – distances to neighboring round tussocks. (from Pokarzhevskaya 1998a with permission of *Arctic, Antarctic and Alpine Research*)

Only 10 species occurred inside the tussocks. They grew in 25% from the total number of the *Festuca varia* individuals. These species preferred to settle at the margins of tussocks. There was no correlation between tussock diameter and species occurrence. It is remarkable that the intruding species belonged to different morphological groups: from tap-rooted (*Taraxacum stevenii*) to long rhizomatous (*Campanula collina*) species.

In the case of tussock "rows", the row width and the distance between rows varied greatly (Fig. 5.22). There are different conditions above and below rows along the slope. A dense litter is formed below the row because *Festuca varia* shoots lie down under the snow cover. Above the row patches of bare ground occur. Most of the young vegetative shoots grow out in the upper and central part of the row, and reproductive shoots also prefer to grow insight the row. Thus the old shoots mostly formed the lower part of the row (Pokarzhevskaya 1998a).

A considerable number of phanerogam species (in total 56) was found between the *Festuca varia* rows. The most common species were *Campanula collina, Hedysarum caucasicum, Vaccinium vitis-idaea*, and *Viola altaica*. Species richness and composition were depended on the position along the slope. The number of species (55) above the rows was higher then below the rows (36). A stepwise regression showed that the number of species below the row was negatively dependent on row width. Species number above the row was positively dependent on the distance to the neighbouring row (Pokarzhevskaya 1998a). Species as *Anthoxanthum odoratum, Ranunculus oreophilus, Veronica gentianoides, Cetraria islandica*, and *Nardus stricta* preferred the above-row parts, species as *Daphne glomerata* and *Vaccinium vitis-idaea* were recorded more frequently below the rows, where shoot cover was densest and litter accumulated. Probably, the plants that can tolerate the shade provided by living and dead shoots can occur below the rows. The most abundant species within the rows were *Campanula collina* and *Hedysarum caucasicum*.

Pokarzhevskaya (1998a) concluded that *Festuca varia* restricts the species number and cover in the community on a small scale, but not enough to produce noticeable effects on the overall species composition in the studied grasslands. The results of this study show that the conditions outside a tussock are not always extreme for accompanying species and there is a possibility for establishment of seedlings in such conditions. But the picture inside the tussocks is different. Only few species can exist among the numerous, dense growing tillers of the grass.

Seed reproduction of *Festuca varia* was not recorded during the study. Even in years after abundant *Festuca varia* seed production, only a few seedlings were observed and none of them survived to the next season.

Very low or no seed reproduction has been described for the dominants of other alpine communities; these were *Carex curvula* (Erschbamer & Winkler 1995) and *Kobresia bellardii* (Bell & Bliss 1979). Despite the low seed reproduction the tussock growth form of *Festuca varia* allows it to dominate. The fact that a tussock can produce many tillers simultaneously leads to the expansion of the grass.

5.3.4.3. Influence of the island effect on the FVGs spatial structure

Community composition is determined by different factors. One factor is the general area of the communities. The basic laws of island biogeography were formulated in classical work of R.H. MacArthur and E.O. Wilson (1967) with an example of bird communities on oceanic islands.

Communities which occupy small areas in landscapes of another structure can be considered islands, for example, wood islands in steppe, forest bogs, mountain tops, etc. Due to the relatively sharp border with other communities, FVGs are suitable objects for an investigation of the island effect in alpine communities. A special investigation to determine the dependence of floristic saturation on different ecological parameters in FVGs, including the island effect, was conducted by Sukhova et al. (1997).

The more sharply restricted areas of FVGs were selected for the field study. All of selected patches were divided in two groups, namely "continents" (extending over more than 400 m^2) and "islands (less than 400 m^2). Plots of $2 \times 2\,m^2$ were laid out inside each studied patch. In these plots registration of all vascular plants species was conducted. Exposure, steepness of slope, cover of plants, mosses, stones were noted in each plot as well. Sizes of the islands and distance from each of them up to the proximate continent were recorded. Floristic richness (number of vascular plant species per plot) was compared for FVG "islands" and "continents". The dependence of floristic saturation and separate species frequency on ecological features was studied (Sukhova et al. 1997).

The mean floristic richness of FVGs showed no significant differences in "islands" and in "continents". They contained 27.7 and 28.0 species correspondingly (Table 5.9.). The same situation was observed for continents and islands separated by other closed plant communities. Continents and islands separated by screes showed insignificant differences between their mean quantities of floristic saturation (27.2 and 23.6 species per 4 m^2 respectively). Thus, there was only a weak tendency for "islands" towards a lower floristic richness for FVG areas separated by screes.

Table 5.9. Dependence of floristic richness on sizes, positions and surroundings of FVG patches. M – average floristic richness, vascular plant species/4 m^2; s.e. – standard error, n – number of observation; A.D. – pairs of classes with significant difference in floristic richness ($p<0.05$) (from Sukhova et al. 1997)

Property	Class number	Class characteristic		s.e.	n	A.D.
General sizes of massifs	1	"islands" (up to 400 m^2)	27.7	1.5	48	no
	2	"continents" (more then 400 m^2)	28.0	1.1	49	
Sizes of islands	1	less than 50 m^2	30.8	2.3	16	1—3
	2	from 50 to 120 m^2	28.2	1.7	16	2—3
	3	from 121 to 400 m^2	20.0	1.9	15	
Distances to "continents"	1	less than 50 m	25.9	2.2	16	2—3
	2	from 50 to 100 m	21.4	2.0	13	
	3	more then 100 m	31.2	2.0	17	
Surround vegetation	1	screes	23.6	2.3	14	no
	2	closed communities	29.1	1.8	34	

Floristic richness decreased significantly when the size of an island increased (Table 5.9). There were about 31 species per plot on island of less than 60 m^2 and 20 species on larger islands (more than 150 m^2).

There was no linear dependence of island floristic richness on the distance from continents (Table 5.9). This fact suggests that different factors affect the floristic saturation. Several differently directed factors may influence the floristic richness in FVG island, namely 1) island effect – decreasing amount of species in a smaller patches, 2) boundary effect – increasing species numbers in small plots, because there is migration of species from adjacent communities, 3) competitive effect of *Festuca varia* which expressed more strictly in large and dense patches, 4) age of the patches (large and old areas possess a high level of floristic richness, because they had more time for migration).

Probably, all these factors balance each other in the case of the FVGs surrounded by other dense communities. But in the case of FVGs surrounded by the mobile screes, the second factor renders less essential influence on floristic richness. This explains the tendency for the island effect found in this case (Sukhova et al. 1997). It was shown that the floristic richness was higher on slightly disturbed, more steep and less closed areas.

The studied alpine species showed different preferences of occurrence in "islands" and "continents". Six frequent species prefer islands (*Sedum tenellum, Gentiana biebersteinii, Hieracium lactucella, Senecio aurantiacus, Sibbaldia procumbens, Silene vulgaris*). The species possess effective colonisation abilities either by anemochorous seeds or by vegetative propagation. Species with deep or dense root systems (*Festuca ovina, Hedysarum caucasicum, Galium verum*) prefer large FVG areas ("continents") (Sukhova et al. 1997).

As a whole the "island effect" is not an important factor determining the floristic richness of FVG areas.

5.3.5. *Geranium gymnocaulon-Hedysarum caucasicum* meadow

The characteristic feature of GHMs is the intensive burrowing activity of different animals: small rodents, mainly voles *Pitymys majori* Thos. (Fomin et al. 1989), and large mammals like wild boars (*Sus scrofa* L.) and bears (*Ursus arctos* L.). The two latter species consume the roots of *Hedysarum caucasicum, Carum meifolium* and other herbs. Therefore, GHMs are permanently exposed to disturbances at different scales.

Several disturbances (or "gaps") can serve as regeneration niches in which seedling establishment and development of young plants take place (Grubb 1977). We studied the influence of the disturbance type on the composition of GHMs (Onipchenko & Rabotnova 1994).

Two types of disturbances were studied within GHM areas (Table 5.10):

1) mesogaps (with the typical size of 1000 – 10000 cm^2) formed as a result of the burrowing activity of large mammals. This type of disturbances is characterised by changes in the upper soil horizon, a sharp increase in solar radiation and a considerable shift in micro-climatic characteristics of the soil surface in comparison with undisturbed places. The age of the studied mesogaps was about 3 – 5 years.
2) microgaps (with the typical size of 100 – –1000 cm^2) formed as a result of vole's activity. In this case the availability of soil resources also increases, but insolation of the soil surface slightly changes, because of shading by neighbouring plants. The age of the studied microgaps was about 1 – 2 years (Onipchenko & Rabotnova 1994).

Large gaps can take over 10 years to grow over, while vole-burrowing places can easily recover in only a few (2 – 5) years. On the other hand, the latter often reappear after only a very short time.

Table 5.10. Some features of zoogenic gaps in alpine meadows (from Onipchenko & Rabotnova 1994 with permission of Geobotanical Institute SFIT)

Features	mesogaps	microgaps
Burrowing mammals	*Ursus arctos* L., *Sus scrofa* L.	*Pitymys majori* Thos.
Sizes of gaps, cm^2	1000 – 10000	100 – 1000
The periodicity of disturbances, years	>10	1 – 8
The depth of soil disturbances, cm	5 – 20	1 – 8
Size of sample plots, cm	25 25	10 10
Number of replications (n)	270	250
Floristic richness (number of species per plot, mean and standard error):		
gap plots	8.5 ± 1.8	5.0 ± 1.7
control plots	9.5 ± 1.3	7.3 ± 1.8

The experimental plots were laid out in meso- and microgaps. Control plots were placed near gaps on undisturbed places.

Floristic richness of control and disturbed plots did not significantly differ in either type of disturbances (Table 5.10). On the whole, 78 species of vascular plants were found in the plots under observation. The occurrence of the most abundant species is shown in the Table 5.11. According to the Ramenskii-Grime concept, species with a positive response to disturbances belong to explerents (or ruderals) (Ramenskii 1938, Grime 1979, 2001).

The studied alpine species fell into the following groups, according to their responses to different types of zoogenic disturbances (Onipchenko & Rabotnova 1994):

Table 5.11. Frequencies of alpine species under natural disturbances in GHM. Column "P" represents the significance level of differences according to the t-test: *** p >0.999, ** p >0.99, * p >0.95. (from Onipchenko & Rabotnova 1994 with permission of Geobotanical Institute SFIT)

Species	Frequency (%)						number
	control	mesogaps	P	contro 1	microgaps	P	of group
Agrostis vinealis	15.9	33.0	***	7.2	23.6	***	2
Anthemis cretica	17.0	37.4	***	10.0	11.2		1
Anthoxanthum odoratum	65.2	41.5	***	42.8	21.2	***	7
Campanula tridentata	13.3	14.8		6.8	8.4		4
Campanula collina	14.8	9.6		29.2	26.4		4
Cartex atrata	27.0	8.9	***	11.6	4.0	**	7
Carum meifolium	70.7	73.7		37.6	32.0		4
Catabrosella variegata	19.3	48.5	***	0.0	0.0		1
Corydalis conorhiza	47.8	34.4	**	0.0	0.0		5
Daphne glomerata	0.0	0.0		10.4	1.6	***	6
Deschampsia flexuosa	16.3	11.9		32.0	7.6	***	6
Draba hispida	0.0	0.0		5.2	14.4	***	3
Erigeron caucasicus	0.0	0.0		22.8	16.4		4
Euphrasia ossica	6.7	14.1	41*	10.0	7.6		1
Festuca brunnescens	40.7	42.2		6.4	27.6	***	3
Festuca ovina	5.2	12.2	**	0.8	0.8		1
Gentiana septemfida	0.0	0.0		8.8	1.2	***	6
Geranium gymnocaulon	89.6	26.7	***	91.2	34.0	***	7
Gnaphalium supinum	2.2	43.7	***	0.0	0.8		1
Hedysarum caucasicum	47.8	34.4	**	70.4	35.6	***	7
Leontodon hispidus	35.2	27.8		42.4	31.2	*	6
Luzula multiflora	25.2	11.1	***	5.6	4.0		5
Matricaria caucasica	28.9	51.1	***	11.2	48.4	***	2
Minuartia aizoides	21.5	35.6	***	9.6	9.6		1
Minuartia recurva	11.9	28.1	***	8.4	4.4		1
Nardus stricta	42.6	15.6	***	28.8	11.2	***	7
Pedicularis condensata	0.0	0.0		18.8	10.8	*	6
Phleum alpinum	53.0	14.4	***	27.6	14.4	***	7
Potentilla crantzii	29.6	13.0	***	5.2	4.4		5
Pulsatilla aurea	6.3	0.7	***	10.8	4.4	**	7
Ranunculus oreophilus	26.3	15.2	**	6.0	2.0	*	5
Rumex alpestris	43.0	10.4	***	21.2	13.6	*	7
Scorzonera cana	14.1	14.1		12.4	9.6		4
Sedum tenellum	3.3	38.1	***	1.6	5.6	*	1
Senecio taraxacifolius	10.0	1.1	**	0.0	0.0		5
Sibbaldia procumbens	24.8	48.5	***	5.6	11.6	*	2
Taraxacum stevenii	18.9	18.1		4.4	2.4		4
Veronica gentianoides	9.6	10.4		10.8	20.4	**	3

1. Typical R – species ("explerents"). Species increasing their frequency in the mesogaps and not responding to the disturbances at a smaller scale: *Anthemis cretica, Catabrosella variegata, Euphrasia ossica, Festuca ovina, Gnaphalium supinum, Minuartia aizoides, Minuartia recurva, Sedum tenellum.* It seems that the development of these species is mainly determined by the increase in insolation.
2. Explerents which positively respond to different scale disturbances. *Agrostis vinealis, Matricaria caucasica, Sibbaldia procumbens. Agrostis vinealis* possesses the highest degree of vegetative mobility, using every possible canopy gap to form new above-ground shoots. Other species from this group have a restricted vegetative mobility. They form a considerable soil seed bank and can rapidly resume from seeds in disturbed places (Semenova and Onipchenko 1991).
3. Species that increase in frequency only in microgaps and do not change (or only insignificantly) in mesogaps: *Draba hispida, Festuca brunnescens, Veronica gentianoides.* These species possess high tolerance to closed cover conditions and can indicate small scale disturbances.
4. Species that do not change in frequency at different scales of disturbance: *Campanula tridentata, Campanula collina, Carum meifolium, Erigeron caucasicum, Scorzonera cana, Taraxacum stevenii.* Most of the species have relatively large seeds and are tolerant to a wide range of ecological conditions: from open burrowing places to closed swards.
5. Species that decrease in frequency in the mesogaps and do not change (or only insignificantly) in microgaps: *Corydalis conorhiza, Luzula multiflora, Potentilla cratzii, Ranunculus oreophilus, Senecio taraxacifolius.* These species are adapted for growth in closed communities, but many of them can successfully resume in "small canopy breaks".
6. Species that decrease in frequency in "small canopy breaks" but do not change in mesogap plots: *Leontodon hispidus, Deschampsia flexuosa.* Besides these two species, *Daphne glomerata* and *Pedicularis comosa,* were not found in either mesogaps or control plots, and were included in this group.
7. Species that decrease in frequency both in meso and microgap plots. This is a large group of species, which dominates or has a high frequency level in undisturbed places in the community: *Anthoxanthum odoratum, Carex atrata, Geranium gymnocaulon, Hedysarum caucasicum, Nardus stricta, Phleum alpinum, Pulsatilla aurea, Rumex alpestris.* Yet, most of them have comparatively large seeds and can resume under disturbances, especially on microgaps. Therefore, for this group of species, natural zoogenic disturbances do not cause elimination but, on the contrary, can promote successful establishment of seedlings (Onipchenko & Rabotnova 1994).

5.3.6. Spatial structure of snowbed communities

The specific feature of SBCs is heavy snow accumulation (4 m and more). So, the vegetation period here is short but varies for different areas and in different years. Only a few species can survive under such conditions. Despite the extreme situation with snow depth, the soil parameters such as nutrients and moisture can be favourable for plants in SBC, compared to other alpine communities. Snow cover promotes storage of nutrients and melting water (Bowman 1992) and provides protection from desiccation by wind (Smith et al. 1995). Plants of snowbed communities are characterized by their short stature, but the vegetation cover reaches 65 – 90% (Pokarzhevskaya 1998b). Due to these plant features the communities look uniform and also the low species richness suggests that these communities are homogeneous.

An important explanation of vegetation cover diversity in alpine communities is heterogeneity in different edaphic factors (soil moisture, pH, substrate, nitrogen availability) and unevenness of snow distribution (Billings 1974, Acosta et al. 1991, Atkin & Collier 1992). Under the conditions of SBCs, it seemed that insignificant variation in topographic position can be very important for the length of the growing season or the amount of resources, and this will influence the species distribution and their functional characteristics (Billings & Mooney 1968). Thus there are a lot of different microhabitats within the small SBC areas (Pokarzhevskaya 1998b).

Pokarzhevskaya (1998b) used principal component analysis (PCA) to study the relationship between the microrelief and plant cover features. Results of this analysis show that three groups of pants exist in different habitats within SBCs:

1. *Carex pyrenaica, Corydalis conorhiza, Pedicularis nordmanniana, Sibbaldia procumbens* – species which appear first after snow melting. According to the PCA results they represent the relatively late-melting habitats.
2. *Campanula tridentata, Nardus stricta, Carum caucasicum, Anthoxanthum odoratum* – species of early to intermediate snow melting habitats. They also preferred slightly raised places.
3. *Potentilla crantzii, Catabrosella variegata* – species of early melting to intermediate sites. These species occur in the hollows with a large flow of melting water, which also brings considerable amounts of nutrients in the beginning of the growing season.

It is difficult to estimate the date of snow melt without long-term observations (Billings & Bliss 1959). But some special areas might be more convenient to study the relationships between topography and plant community composition. Such an area was found in the upper part of the valley Malyi Khutyi. The area (30 × 50 m) represents a very smooth gentle slope (about 5° steepness) exposed to the south-west. According to the preferent wind direction in winter (from the south) and the adjacent rock wall we presume that there is a steep snow depth gradient within the area (Fig. 5.23). Five belts of plots were laid out along the slope to estimate changes in species occurrences according to the microtopographic / snow depth gradient. Spearman rank correlation coefficients with Bonferroni correction (Legendre & Legendre 1998) were calculated applied to estimate the statistic significance of the dependence between species occurrence and plot position in the gradient (Table 5.12). Several groups of species were recognised among the abundant SBC species (Fig. 5.24 – 5.26).

Species of the first group occupy areas with the deepest snow. *Hyalopoa pontica, Sibbaldia procumbens, Taraxacum stevenii, Pedicularis nordmanniana* and *Carex pyrenaica* belong to this group (Fig. 5.24). The frequency of these species continually decreased along the plot series. *Carex pyrenaica* is known as a species of snowbed areas with the shortest growth season in different mountain regions in Northern Hemisphere (May et al. 1982).

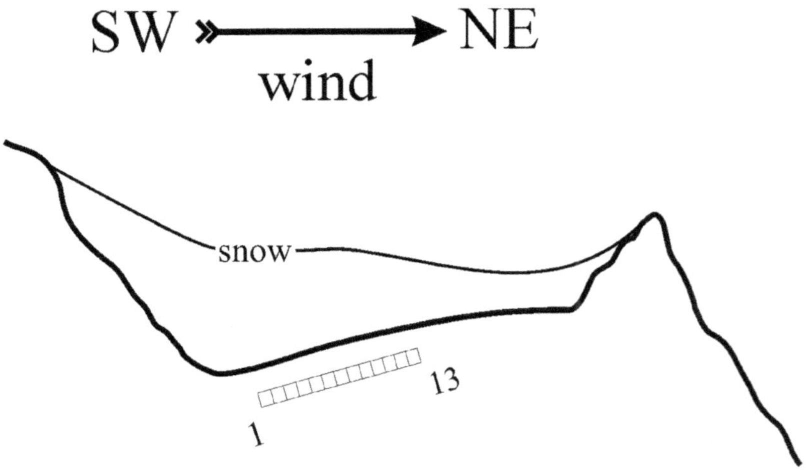

Figure 5.23 Cross section of the studied snow bed area in Malyi Khutyi.

Table 5.12. Spearman (rank) correlation coefficients between species occurrence and plot position in the snow depth gradient. R – correlation coefficient, t – values, p - significance levels (p – without and p' with Bonferroni correction). Number of plots (n) = 65. The smaller the plot number the thicker the snow cover.

Species	R	t(N-2)	p	p'
Sibbaldia procumbens	−0.507	−4.66	0.00002	0.0004
Taraxacum stevenii	−0.603	−6.00	0	0
Pedicularis nordmanniana	−0.595	−5.88	0	0
Carex pyrenaica	−0.754	−9.12	0	0
Agrostis vinealis	−0.454	−4.04	0.00015	0.0037
Catabrosella variegata	−0.052	−0.42	0.678	16.95
Gnaphalium supinum	0.334	2.81	0.0066	0.165
Corydalis conorhiza	−0.325	−2.73	0.0082	0.205
Phleum alpinum	−0.289	−2.40	0.0194	0.484
Luzula multiflora	0.375	3.21	0.0021	0.052
Carum meifolium	−0.035	−0.28	0.784	19.51
Briza marcowiczii	0.501	4.60	0.00002	0.0005
Hyalopoa pontica	−0.604	−6.02	0	0
Sedum tenellum	0.403	3.49	0.0009	0.022
Potentilla crantzii	0.032	0.25	0.802	20.05
Campanula tridentata	0.751	9.03	0	0
Matricaria caucasica	0.558	5.34	0.000001	0.00003
Carum caucasicum	0.571	5.52	0.000001	0.00003
Anthoxanthum odoratim	0.445	3.94	0.0002	0.0052
Minuartia aizoides	0.539	5.08	0.000004	0.0001
Carex atrata	0.129	1.03	0.306	7.65
Nardus stricta	0.169	1.36	0.178	4.46
Euphrasia ossica	0.371	3.17	0.002326	0.058
Gentiana pyrenaica	0.384	3.30	0.001609	0.040

The second group combines species preferring relatively deep snow, but decreasing slightly in the first plots with the deepest snow (*Corydalis conorrhiza, Phleum alpinum, Potentilla crantzii*). The two last species are common in alpine meadows as well.

Two species (*Catabrosella variegata* and *Carum meifolium*) represent the third group. They obviously have their maximum occurrence in the middle part of the snow depth gradient (Fig. 5.26).

Several species (*Luzula multiflora, Gnaphalium supinum, Briza marcowiczii, Sedum tenellum, Anthoxanthum odoratum*) are common in the plots with relatively shallow snow. They belong to the fourth group.

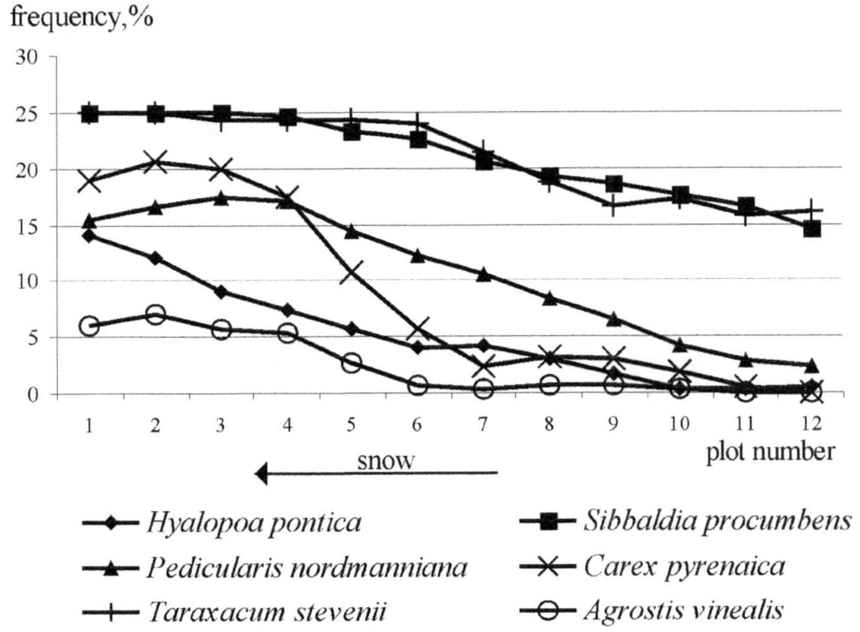

Figure 5.24 Smoothed curves of species occurrence for species preferring deeper snow cover.

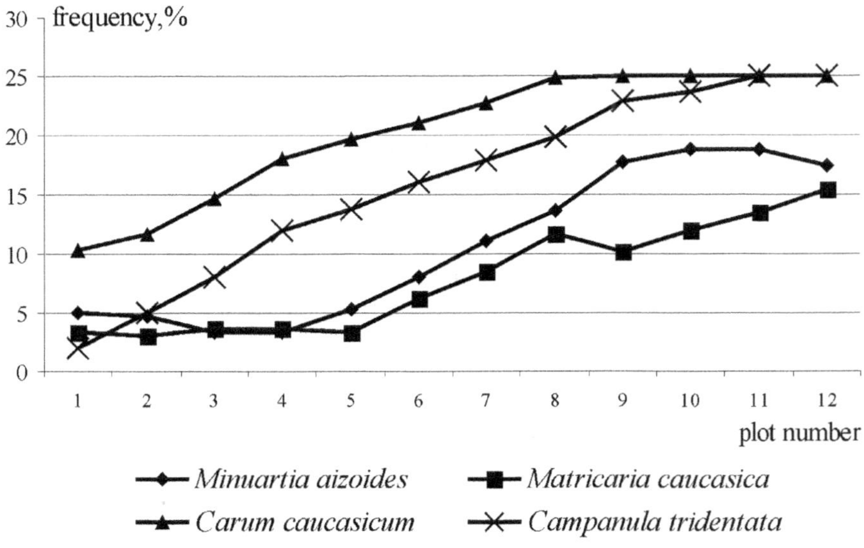

Figure 5.25 Smoothed curves of species occurrence for species preferring relatively shallow snow cover.

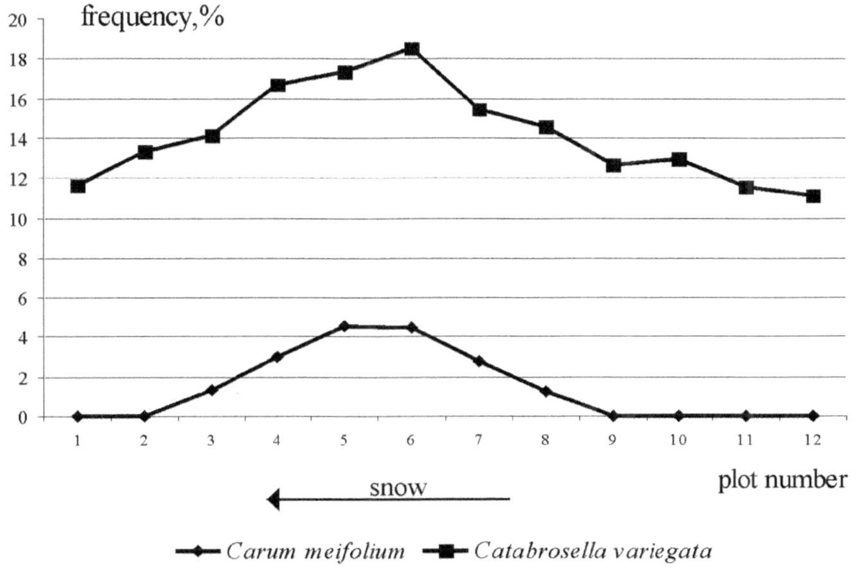

Figure 5.26 Smoothed curves of species occurrence for species preferring moderate snow cover or which are indifferent to the snow depth.

The abundance of species belonging to the fifth group increases continually when snow depth decreases (Fig. 5.25). This group combines species with wide distributions among different alpine communities (*Campanula tridentata, Matricaria caucasica, Carum caucasicum, Minuartia aizoides*). They can dominate in the areas with the least snow depth (but usually more than 3 m!) within snowbeds.

5.4. Soil-plant relationships

5.4.1. Introduction and methods

Plants can affect soil properties in different ways: by water and nutrient absorption (via roots or mycorrhizal fungi), by stimulating weathering processes, and by changing microbial activities through exudates, organic matter accumulation, etc. (Hobbie 1992, Strom 1995, Angers & Caron 1998; Berendse 1998; Kelly et al. 1998). There are only a few studies dealing with effects of individual plant species on soil nutrients (Wedin & Tilman 1993,

Reynolds et al., 1997). Virtually nothing is known about plants and communities from high mountains.

The four studied alpine communities differ significantly in soil properties as well (Chapter 2, 5.3.2.3, Grishina et al. 1993, Onipchenko et al. 1998b). There may be various reasons for the differences (parent rocks, weathering regime, activity of plants and others). It is conceivable that different species from these communities change soil nutrient concentrations and acidity. In our study of the plant communities and the spatial patterns in soil properties (Chapter 5.3.2.3.) some correlations between species abundance and soil properties were obtained (Table 5.13). To estimate the role of plants in the differentiation of the soil nutrient status, it is of interest to know if the direction of the changes agree with the patterns observed in nature.

Table 5.13. Correlation coefficients ($r \geq 0.6$, $p < 0.01$) between separate soil properties and the abundance of plants within alpine communities (from Onipchenko et al. 1998b)

Community	Index	Plant species	*r*
ALH	$Ca_{exchange}$	*Cladonia uncialis* (L.) Wigg.	0.69
	N_{total}	*Cetraria cucullata* (Bellardi) Ach.	−0.60
	C (loss on ignition)	*Primula algida* Adams	0.65
FVG	pH_{H2O}	*Festuca varia* Haenke	0.63
	pH_{H2O}	*Leontodon hispidus* L.	−0.70
	K_{H2O}	*Cruciata laevipes* Opiz	0.61
SBC	Ca_{H2O}	*Sibbaldia procumbens* L.	0.61
	NO_3	*Cladonia pyxidata* (L.) Hoffm.	0.61

We investigated the influence of 17 species from different alpine plant communities on soil nutrient concentrations and acidity (Onipchenko et al., 2000, 2001b). Four questions were addressed:

(1) Do alpine plant species have specific effects on soil properties under semi-natural conditions?
(2) Do species of the same family change soil properties in a similar way?
(3) Do the directions of the change caused by species from different communities coincide with the natural differentiation of soil properties among the alpine communities?
(4) Can we classify different species according to their ability to change soil nutrient concentrations?

We selected 17 species for our experiment, six monocots and 11 dicots, which are characteristic of one or more of the studied alpine communities (ALH, FVG, GHM and SBC) (Table 5.14).

A soil mixture from the upper layer (0 – 25 cm) of native alpine soils and pure sand was prepared (4:1) and homogenised by repeated sieving. Large

Table 5.14. List of studied species and communities in which they occur preferentially (ALH – alpine lichen heath, FVG – *Festuca varia* grassland, GHM – *Geranium gymnocaulon – Hedysarum caucasicum* meadow, SBC – snow bed community) (from Onipchenko et al. 2001b with permission of *Folia Geobotanica*)

Species	Abbreviations	Family	Communities
Anemone speciosa	*As*	Ranunculaceae	ALH
Antennaria dioica	*Ad*	Asteraceae	ALH
Campanula tridentata	*Ct*	Campanulaceae	ALH, SBC
Carex sempervirens	*Cs*	Cyperaceae	ALH
Carex umbrosa	*Cu*	Cyperaceae	ALH
Festuca ovina	*Fo*	Poaceae	ALH, FVG
Festuca varia	*Fv*	Poaceae	FVG
Geranium gymnocaulon	*Gg*	Geraniaceae	GHM
Hedysarum caucasicum	*Hc*	Fabaceae	GHM
Leontodon hispidus	*Lh*	Asteraceae	FVG, GHM
Matricaria caucasica	*Mc*	Asteraceae	GHM
Nardus stricta	*Ns*	Poaceae	FVG, GHM, SBC
Phleum alpinum	*Pa*	Poaceae	GHM
Scorzonera cana	*Sc*	Asteraceae	FVG
Sibbaldia procumbens	*Sp*	Rosaceae	SBC, GHM
Taraxacum stevenii	*Ts*	Asteraceae	SBC
Trifolium polyphyllum	*Tp*	Fabaceae	ALH

roots and stones (> 7 mm) were removed. The mixture was put in plastic boxes, which were buried in the ground so that the soil levels inside and outside were the same. Drainage holes in the bottom of the boxes allowed water to drain and prevented anaerobic conditions. Each plant species was grown from seeds in 10 replicates (boxes). Ten boxes with the same soil mixture but without plants were used as a control. Samples from the initial soil mixture were collected for analysis. Plants were grown for six years (1991 – 1996).

Chemical soil analyses included water soluble NO_3, NH_4, Ca, Mg, K, exchangeable Ca, Mg, K (1 M CH_3COONH_4), NH_4 (2% KCl), mobile (0.5 M CH_3COOH) and low mobile (0.1 M HCl) P, and pH (H_2O and 2% KCl).

5.4.2. Effects by species

Plants significantly affected all studied soil properties (Fig. 5.27). A complete list of species and soil nutrient concentrations in their monocultures is given in Onipchenko et al. (2001b).

The orders of species (from lowest to highest nutrient concentration) were different for different nutrients (Fig. 5.27). *Trifolium polyphyllum, Phleum alpinum, Anemone speciosa* and *Nardus stricta* did not change nitrate concentration in comparison with the initial soil. The lowest nitrate concentrations were found in soil with *Matricaria caucasica* and *Leontodon hispidus* (Onipchenko et al. 2001b).

Soil with *Hedysarum caucasicum* had the highest ammonium concentration, which did not differ significantly from the initial concentration. *Carex sempervirens, Taraxacum stevenii*, and *Sibbaldia procumbens* decreased exchangeable ammonium significantly in comparison with the control, but only the first species decreased water-soluble ammonium significantly.

Concentrations of mobile P were lower for soil under 4 grasses. Significantly (in comparison with the control) higher concentration of mobile P was found for *Hedysarum caucasicum* and *Leontodon hispidus* (Onipchenko et al. 2001b).

The pH of water solution was significantly lower for control samples than for 3 grasses (*Festuca ovina, F. varia, and Phleum alpinum*) as well as for *Carex umbrosa* and *Campanula tridentata* (Fig. 5.27). Most of the species (except *Antennaria dioica, Hedysarum caucasicum* and *Scorzonera cana*) decreased exchangeable acidity (pH – KCl) in comparison with the final control, but for only 3 species (*Phleum alpinum, Trifolium polyphyllum*, and *Festuca ovina*) was the decrease significant in comparison with initial conditions.

Concentrations of Ca and Mg were highest in initial soil samples (except exchangeable Mg) and much lower in final control samples. Most of the species were intermediate between those values (Fig. 5.27). It seems that different species prevent cation leaching to different extents. All the grasses had exchangeable Ca concentrations similar to the initial control. *Festuca varia* had the highest exchangeable Mg concentration, significantly higher even than the initial control.

Potassium shows a different pattern. Exchangeable K concentration was lowest in initial soil samples, but during the experiment the concentration increased under most of the species as well as in controls. Water-soluble K concentrations did not differ between initial and final control soil samples, but *Antennaria dioica* decreased water soluble K significantly in comparison with most of the other species (Onipchenko et al. 2001b).

Most of the species increased cation (Ca, Mg, and K) contents and pH values in comparison with the final control. This seems to be related to the prevention of cation leaching and to nitrification, leading to soil acidification after disturbance. Soils beneath monocultures and in the control boxes contained less Ca and Mg than the initial soil mixture. Nutrient losses by leaching may be the main cause of this decrease.

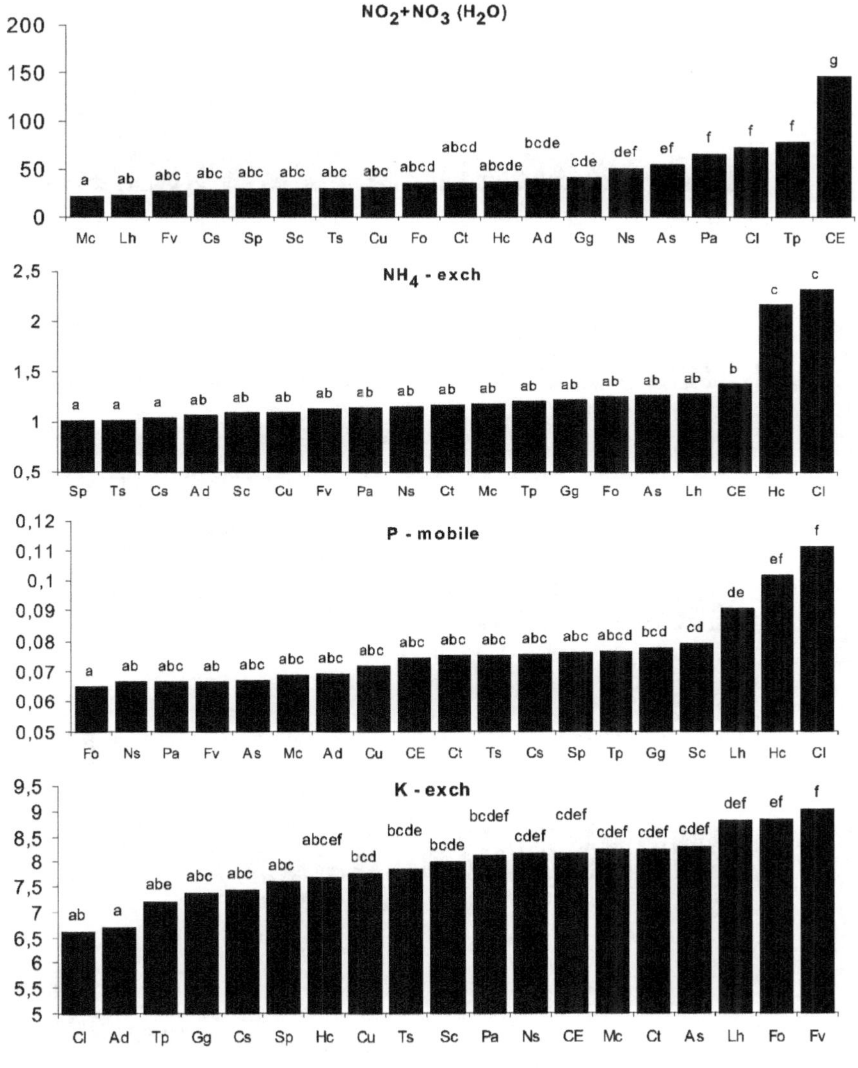

Figure 5.27a Soil nutrient concentrations for studied alpine species (see Table 5.14 for abbreviations) and initial (CI) and final (CE) control soil samples. Species are represented in order of increasing nutrient concentrations. Significant ($p<0.05$) differences between values are shown by different letters. All units except nitrates are in mg/100g, nitrates are in g/100g (from Onipchenko et al. 2001b with permission of *Folia Geobotanica*)

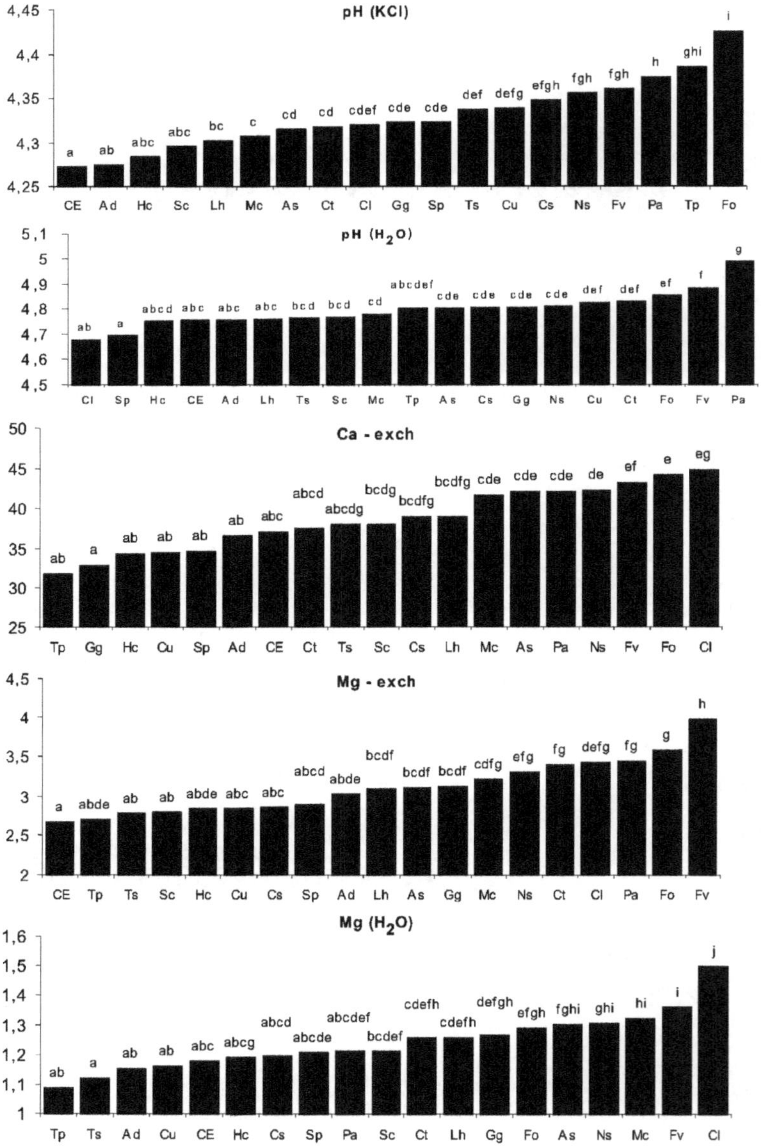

Figure 5.27b Soil nutrient concentrations and pH values for studied alpine species (see Table 5.14 for abbreviations) and initial (CI) and final (CE) control soil samples. Species are represented in order of increasing nutrient concentrations or pH values. Significant ($p < 0.05$) differences between values are shown by different letters. All units except pH are in mg/100g (from Onipchenko et al. 2001b with permission of *Folia Geobotanica*)

As a whole, control soil samples were more acidic and contained 2 – 7 times more nitrate than the soil under monocultures. The concentration of nitrates was much lower than ammonium. It seems that nitrification takes place very rapidly when plants do not use part of the ammonium in the soil, for example after any soil disturbance. Acidification and intensive nitrification are typical of disturbed soils without plants (Bormann & Likens 1979). On the other hand, plants that use nitrates tend to raise the pH in the rhizosphere relative to the surrounding soil (Nye 1981).

Our results provide further evidence for the considerable effects plant species have on soil nutrient concentrations, and for the species-specific nature of these effects. The results support the idea of biogeochemical specialization among plant species within communities. Tilman & Wedin (1991) and Wedin & Tilman (1993) have shown that grasses of a climax prairie community reduce soil nitrate concentrations much more strongly than early successional grasses. Our results show that the behaviour of nitrate is unique in this respect; of all the nutrients studied, only the final control concentration of nitrate was significantly reduced by all species. For every other nutrient, the final concentration was increased by some or all species.

According to the "colonisation – nutrient competition" hypothesis (Tilman 1990), there should be a trade-off between the colonisation ability of a plant and its ability to reduce soil nutrient concentrations. Even for nitrate, the only element dramatically reduced beneath all species, our results do not confirm this hypothesis. *Matricaria caucasica* lowered nitrate concentrations to a larger extent than other species, but this species mainly occupies open areas (gaps) in alpine communities and is a weak competitor in closed communities (Onipchenko & Rabotnova 1994). It produces many seeds, which can persist in the soil seed bank. Because of these traits, *M. caucasica* can be considered as a typical alpine ruderal species *sensu* Grime (1979) – (Onipchenko et al. 1998a).

5.4.3. Variation within and between taxa and samples

Nutrient concentration data were analysed using nested analysis of variance (Table 5.15.). In most cases, except pH (KCl), differences between samples accounted for most of the variance. Variance of pH and water soluble Ca was important at class level, with higher pH and Ca concentrations beneath monocots. Noticeable variance at family level was obtained for exchangeable forms of cations (Ca, Mg, K, NH_4) as well as for mobile P. Water soluble concentrations of K and NO_3 varied greatly at species level. In general, the influence of taxonomic position decreased in the order: sample > family > species > class. So family level is relatively important for determining soil properties (Onipchenko et al. 2001b).

Table 5.15 Percentage variance at each taxonomic level for soil nutrient concentrations (results of NESTED procedure) (from Onipchenko et al. 2001b with permission of *Folia Geobotanica*)

Soil property	Sample	Species	Family	Class
$NO_3 + NO_2$	52.9	29.3	17.8	0
NH_4 (ex.)	63.6	5.2	31.3	0
$NH_4(H_2O)$	46.8	10.9	42.3	0
P-low mob.	91.1	0.5	8.4	0
P-mob.	77.3	7.4	11.3	4.2
pH (KCl)	29.4	19.3	6.9	44.2
pH (H_2O)	49.9	12.4	6.3	31.4
Ca (ex.)	64	0	36	0
Ca (H_2O)	77	8.7	0	14.3
Mg (ex.)	44.1	7.4	48.5	0
Mg (H_2O)	54.9	20.6	24.5	0
K (ex.)	75.6	11.7	12.7	0
K (H_2O)	73.2	26.8	0	0
Mean	61.5	12.3	18.9	7.2

Significant differences ($p<0.05$) between families were found for pH(H_2O), exchangeable Ca and Mg, marginal differences ($p<0.1$) for pH(KCl), mobile P, and both ammonium forms (Table 5.16). Grasses had significantly higher cation contents in soils (Ca, Mg). Fabaceae tended to increase ammonium concentration. There was a non-significant tendency for both Asteraceae and Cyperaceae to reduce the nitrate concentration.

The results of this experiment suggest the occurrence of a degree of geochemical "specialization" among taxonomic groups. Species belonging to Fabaceae tended to decrease cation contents (Ca, Mg and K) in comparison with other species and increase nitrogen concentrations (NH_4, NO_3). It is well known that Fabaceae increase soil nitrogen concentration through nitrogen fixation (Carr et al. 1980, Bowman et al. 1996). However, among our Fabaceae species only *Hedysarum caucasicum* had *Rhizobium* nodules while these were practically absent in *Trifolium polyphyllum* roots. Consistent with this observation, ammonium content was much higher under *Hedysarum*, but *Trifolium polyphyllum* did not differ from other species in soil ammonium concentration. These results contrast with observations on other alpine species of *Trifolium* (*T. alpinum* in the Swiss Alps, Gigon 1999; *T. dasyphyllum* in the Colorado Rocky Mountains, Thomas & Bowman 1998) which have high *Rhizobium* nodulation and nitrogen fixation activity.

Table 5.16. Mean soil nutrient concentrations under plants from different families and results of one-way ANOVA (species as replications, df = 3, 9). Families with one species were excluded from the analyses. Significantly different values are indicated by different letters (from Onipchenko et al. 2001b with permission of *Folia Geobotanica*)

	Fabaceae	Asteraceae	Poaceae	Cyperaceae	p-value
$NO_2 + NO_3$	56	28	44	29	0.1568
NH_4 (ex.)	1.68 a	1.12 b	1.16 b	1.06 b	*0.0804*
$NH_4(H_2O)$	1.48 a	0.86 b	0.82 b	0.74 b	*0.0828*
P – low mob.	4.54	4.90	4.45	4.84	0.3477
P – mob.	0.089 a	0.077 ab	0.070 b	0.074 ab	*0.0675*
pH (KCl)	4.33 ab	4.30 a	4.38 b	4.34 ab	*0.0555*
pH (H_2O)	4.78 a	4.77 a	4.88 b	4.82 ab	*0.0213*
Ca (ex.)	33.0 c	38.6 b	42.9 a	36.7 bc	*0.0008*
Ca (H_2O)	8.6	8.7	9.1	9.0	0.3369
Mg (ex.)	2.77 a	2.98 a	3.58 b	2.85 a	*0.0034*
Mg (H_2O)	1.14	1.21	1.29	1.18	0.1166
K (ex.)	7.4	7.9	8.5	7.6	0.1877
K (H_2O)	3.76	3.72	3.79	3.49	0.8636

In spite of low or nonsignificant differences between families we have to note that Cyperaceae (*Carex* spp.) tended to reduce ammonium contents, while Asteraceae and Cyperaceae tended to reduce nitrate concentrations. These results support the suggestion that species of Asteraceae prefer nitrate to ammonium as a source of N (de Graaf et al. 1998). Grasses tended to reduce the phosphorus. Leishman (1999) found that *Festuca ovina* reduced the phosphorus concentration in soil more than *Leontodon hispidus*, which is consistent with our results. It is interesting to note that *Anemone speciosa* (Ranunculaceae) was close to grasses in terms of its geochemical effects (see below). The reason for this may be related to fact that this species has the highest rate of mycorrhizal infection (Baikalova & Onipchenko 1988, Onipchenko & Zobel 2000), which probably enables it to absorb phosphate very effectively.

5.4.4. Comparison with natural community soil properties

It was not possible to estimate unambiguously the difference in soil properties between species from different communities, because several species were common in two or more communities (Table 5.13). Our artificial soil mixture originated from soils of GHM and FVG, the two most productive communities, but its fertility was reduced by adding soil from lower horizons and sand. To estimate the role of plants in soil differentiation between communities we calculated Spearman rank correlations between average nutrient concentrations in native soils (see Chapter 2) and average concentrations for species which are dominants or common in the communities (Onipchenko et al. 2001b). The rank order coincided completely for K contents (both exchangeable and water-soluble form). It is necessary to note that with four points, only perfect correspondence is statistically significant. Close (but not significant) rank order was found for exchangeable Mg and pH(H_2O). There was therefore only weak correspondence between properties of native soils and nutrient concentrations for the species abundant in those communities. As a whole we can conclude that experimental soils and native soils do not correspond.

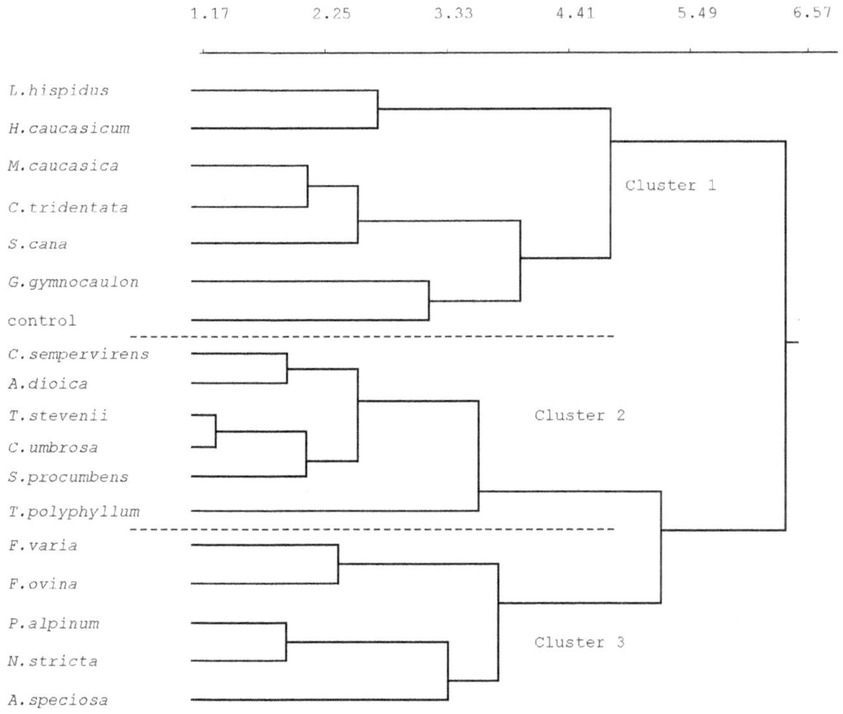

Figure 5.28 Dendrogram of species cluster analysis (Complete linkage procedure). Dotted lines show borders between 3 main clusters (see Table 5.17) (from Onipchenko et al. 2001b with permission of *Folia Geobotanica*)

But some of our observations accord with the natural pattern of species correlation with soil properties. *Sibbaldia procumbens* prefers areas which are richer in $Ca(H_2O)$, *Festuca varia* occurs more often on less acid soil, but *Leontodon hispidus* prefers more acid areas within communities (Table 5.13). Soils under monocultures of the species changed in the same direction (Fig. 5.27). So the plant species can really influence some soil properties within communities.

5.4.5. Cluster analysis of the species

Three groups of species were distinguished in the cluster dendrogram on the basis of the differences in soil properties (Fig. 5.28). Generalized changes of soil properties by species of these clusters are given in Table 5.17. Influence of the "cluster" factor on soil nutrient concentration was significant for most nutrients except for ammonium and nitrates.

Table 5.17. Relative soil nutrient concentration under plants of different clusters (L – low, M – medium, H – high) (see Fig. 5.28) (from Onipchenko et al. 2001b with permission of *Folia Geobotanica*)

Clusters and species	NH_4	NO_3	P	Ca	Mg	K
Cluster 1						
Campanula tridentata						
Geranium gymnocaulon						
Hedysarum caucasicum	*H*	*L*	*H*	*L/M*	*M*	*M/H*
Leontodon hispidus						
Matricaria caucasica						
Scorzonera cana						
Cluster 2						
Antennaria dioica						
Carex sempervirens						
Carex umbrosa	*L*	*M*	*M*	*L/M*	*L*	*L*
Sibbaldia procumbens						
Taraxacum stevenii						
Trifolium polyphyllum						
Cluster 3						
Anemone speciosa						
Festuca ovina						
Festuca varia	*M*	*H*	*L*	*H*	*H*	*M/H*
Nardus stricta						
Phleum alpinum						

Group 1 includes forbs from productive alpine grasslands (FVG and GHM) which had the least effect on soil properties (note that the final control soil also belongs to this group). Higher soil NH_4- and P-concentrations are typical for the first group. The second group is formed by forbs and sedges from the extreme habitats, i.e. the snow-free alpine lichen heath and snowbed communities. Species of this group had lower NH_4-, Mg-

and K-concentrations than others. The four grass species, together with *Anemone speciosa*, formed the third group for which low P- and high Ca- and Mg-concentrations are typical.

As a whole, our results showed that individual plant species can change soil properties in different manners. As a rule, differences among taxonomic groups were bigger than those between individual species from different communities. This may be an indication of some complementarity between species with respect to their effects on soil properties. But this is a question for future research.

5.5. Seasonal dynamics

5.5.1. Introduction

Seasonal dynamics of alpine vegetation was a very attractive topic for many researchers. Main features of high-mountain phenology are: 1) reduction of growth periods due to lower temperature in high altitude areas; 2) high variability of phenological stages according to local weather and snow accumulation; 3) flower bud building a year (or even two years!) prior to flowering (Narinyan 1959b, Mark 1970, Gamtsemlidze 1979, Kudo 1991, etc.). But not much is known about comparative phenological patterns within the alpine belt in the Caucasus. Plant phenological types are poorly studied aspects of alpine seasonal dynamics. Phenological divergence among alpine plants may serve as an important mechanism of species coexistence. An experimental approach for this is given in Chapter 7.3, but the main aim of the current chapter is to describe the phenological patterns of four studied alpine communities (Alpine lichen heath - ALH, *Festuca varia* grassland - FVG, *Geranium – Hedysarum* meadow - GHM, snow bed community – SBC, see Chapter 1.4 for detailed description) and to characterise plant phenological types in these communities.

5.5.2. Methods

Phenological observation were made in 1981-1983 (ALH), and 1984-1989 (FVG, GHM, SBC) on Mt. M. Khatipara on permanent plots (Onipchenko 1983, Onipchenko & Ustinova 1993). The interval between observations was 3-7 days. Studied species number was different in the four communities, but the species formed more than 95% of aboveground biomass of these communities (Table 5.18).

Table 5.18. Number of species studied for their phenology and its proportion in aboveground vascular plant biomass (%)

Community	Number of species	Proportion of biomass
ALH	25	> 98%
FVG	32	97,4%
GHM	24	97,9%
SBC	17	96,5%

The following phenological phases were observed: 1) green shoots before flowering, 2) bud appearance, 3) flowering, 4) maturation of seeds (green fruits), 5) seed ripening (dispersion). Averaged phenological curves (number of species in different phenological phases in the course of the vegetation season) were obtained after 3-years observation. The curves were smoothed using the sliding mean approach (Golubev 1969). Phenological periods were separated according to prevailing phases (Gorchakovskii & Bulatova 1979, Onipchenko 1983).

5.5.3. Phenological curves and periods

Most of the phenological curves are unimodal (Fig. 5.29.). We distinguished five periods of seasonal dynamics.

Spring period lasts from snowmelt to prevailing species in flower. Due to practical snow absence in ALH, the beginning of this period was determined here from the date of upper soil horizon thawing.

Early summer is a period with prevailing species in flower.

Late summer is a period with prevailing species in their seed maturation phase.

During the **autumn** most of the species have ripe fruits and disperse their seeds. Leaves of most of the plants lose their green colour during this period. A few species (*Hieracium* spp., *Gentiana biebersteinii*) are still in flower during the autumn.

Winter period begins with the establishment of a permanent snow cover. It usually happens at the end of October or the beginning of November. Snowfalls can take place from the second half of August to the first half of June, but the autumn and spring snow cover usually disappears within several hours or days.

The phenological curves for species in flower are similar for the four studied communities (Fig. 5.30). The maximum was about 50% of all species under observation. The position of the maximum was later for communities with deeper snow and later dates of snow releasing. But the difference between the dates of the maximum in flowering species (5-15 days) differ less between communities than the difference between the dates

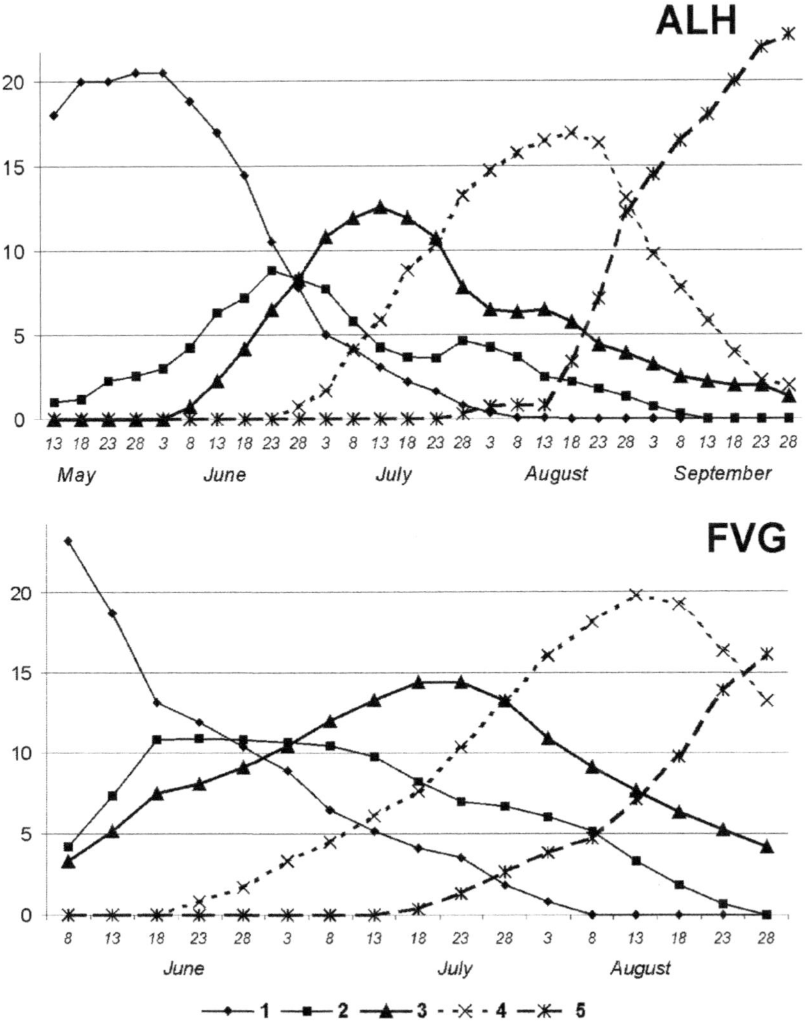

Figure 5.29a Phenological curves for alpine communities (ALH – alpine lichen heath, FVG – *Festuca varia* grassland). Ordinate: the number of species in a particular phenological phase. Abscissa: date. Phenological phases: 1) green shoots before flowering, 2) bud appearance, 3) flowering, 4) maturation of seeds (green fruits), 5) seed ripening (dispersion). (from Onipchenko 1994c)

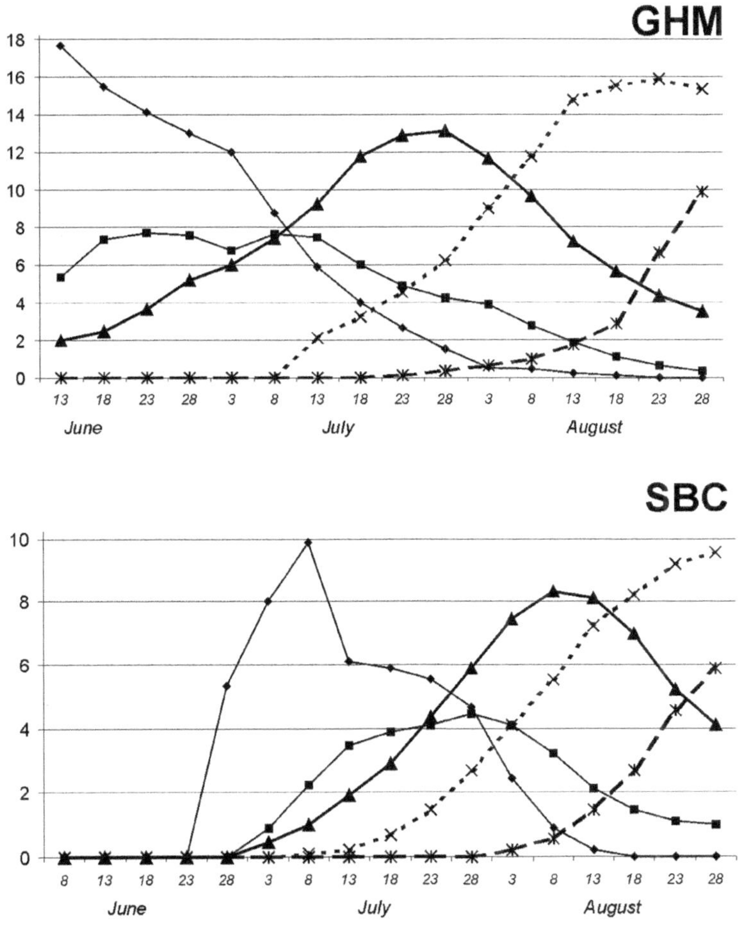

Figure 5.29b Phenological curves for alpine communities (GHM – *Geranium-Hedysarum* meadow, SBC – snow bed community). Ordinate: the number of species in a particular phenological phase. Abscissa: date. Phenological phases: 1) green shoots before flowering, 2) bud appearance, 3) flowering, 4) maturation of seeds (green fruits), 5) seed ripening (dispersion). (from Onipchenko 1994c)

Table 5.19. Phenological periods in the alpine communities

Period		ALH	FVG	GHM	SBC
Spring	Mean dates	End of April – June, 25	End of May – July, 3	June, 5 – July, 10	July, 1 – July, 25
	Species in flower	*Fritillaria collina Anemone speciosa*	*Viola altaica, Fritillaria collina*	*Campanula tridentata, Carex atrata, Sibbaldia procumbens*	*Gagea fistulosa, Sibbaldia procumbens, Potentilla crantzii*
Early summer	Mean dates	June, 25 – July, 25	July, 3 – August, 5	July, 10 – August, 5	July, 25 – August, 12
	Species in flower	*Anemone speciosa, Campanula tridentata*	*Veronica gentianoides, Carum caucasicum, Ranunculus oreophilus, Myosotis alpestris*	*Carum meifolium, Geranium gymnocaulon, Hedysarum caucasicum, Erigeron caucasicus*	*Gnaphalium supinum, Minuartia aizoides, Taraxacum stevenii, Pedicularis nordmanniana*
Late summer	Mean dates	July 25 – August, 25	August, 5 – August 25	August, 5 – September, 5	August, 12 – September 10
	Species in flower	*Oxytropis kubanensis, Minuartia circassica, Aster alpinus*	*Festuca varia, Leontodon hispidus, Campanula collina*	*Leontodon hispidus, Deschampsia flexuosa, Festuca varia, Festuca brunnescens*	*Hyalopoa pontica*
Autumn	Mean dates	August, 25 – October	August, 25 – October	September, 5 – October	September, 10 – October
	Species in flower	*Gentiana septemfida, Gentiana biebersteinii*	*Gentiana septemfida, Gentiana biebersteinii*	*Gentiana septemfida*	–

of snow melting (20-25 days). This observation supports the idea that the later the snow release takes place the quicker the vegetation development occurs (Baloyan 1987) because later development takes place under more favourable temperature conditions. The FVG curve is less steep than the curves for the other communities, so the species in this community flower less synchronously than in other communities.

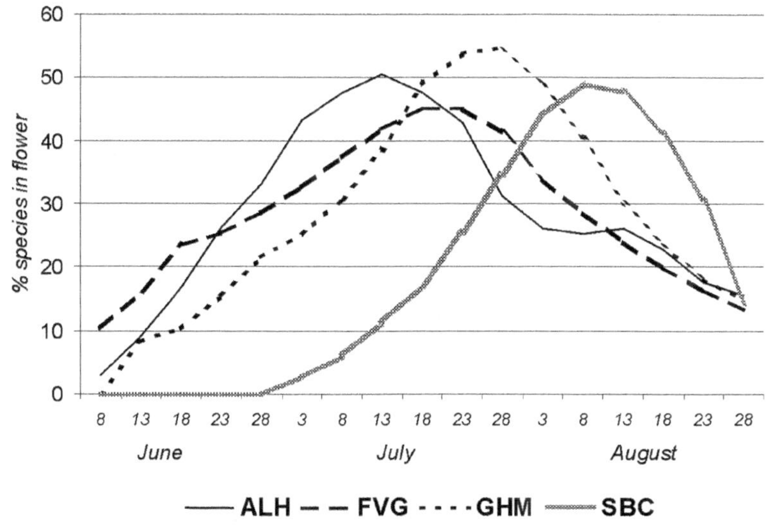

Figure 5.30 Phenological curves for species in flower

For each studied community we can distinguish the full set of the seasons according to the mentioned criteria (Table 5.19).

Similar phenological curves were obtained for alpine communities in the mountains in Central Asia (Lynov 1986, Petrova 1990), the Polar Ural (Igosheva 1984) and Armenia (Ziroyan 1983).

5.5.4. Plant phenological types

According to Borisiva's (1972) classification we distinguish 5 phelological types among the alpine species studied:

1) **Evergreen plants**. The leaves live more than one year and most of the leaves are green during winter (*Vaccinium vitis-idaea, Antennaria dioica, Gentiana pyrenaica* etc.).
2) **Summer-winter green plants**. Either basal parts of the leaves (*Festuca varia, Nardus stricta*) or whole leaves of the late generation (*Sibbaldia procumbens, Viola altaica*) are green during winter.
3) **Spring-summer-autumn green plants**. Their leaves are green during at least three phenological periods but they die in autumn or winter. Most of the alpine plants belong to this group (*Anemone speciosa, Trifolium polyphyllum* etc.).

4) **Spring-summer green plants**. The plants have relatively short vegetative periods (spring and early summer). Leaves die before (*Gagea fistulosa, Corydalis conorhiza, Fritillaria collina*) or during (*Cerastium purpurascens, Polygonum bistorta*) the late summer period.
   Spring ephemeral plants (*Gagea fistulosa, Corydalis conorhiza*) start to grow under the snow and develop aboveground roots during snowmelt (Narinyan 1959a). The roots grow into the snow and adsorb nutrients from melting water. After snowmelt, dry crusts of such dead roots surround the young shoots (Fig. 5.31).
5) **Summer-autumn green plants**. This group is represented by annual species which develop mainly during late summer and autumn (*Euphrasia ossica, Gentiana biebersteinii*).

Figure 5.31 Dry crust of dead aboveground roots surround young shoots of *Corydalis conorhiza* in an alpine snowbed community

We assessed the proportion of species of different phenological types separately according to species number and aboveground biomass (Fig. 5.32.). Most of the alpine species in all four studied communities belong to the spring-summer-autumn green plant group. The biomass proportion of these plants is the highest in GHM. On the other hand, the role of summer-winter green plants is very important in FVG, where the main dominants

(*Festuca varia, Nardus stricta*) belong to this phenological type. It is necessary to note that alpine plants of this group (in contrast to evergreen alpine shrubs) can support photosynthetic activity during winter time when snow cover is not deep (Nakhutsrishvili 1975).

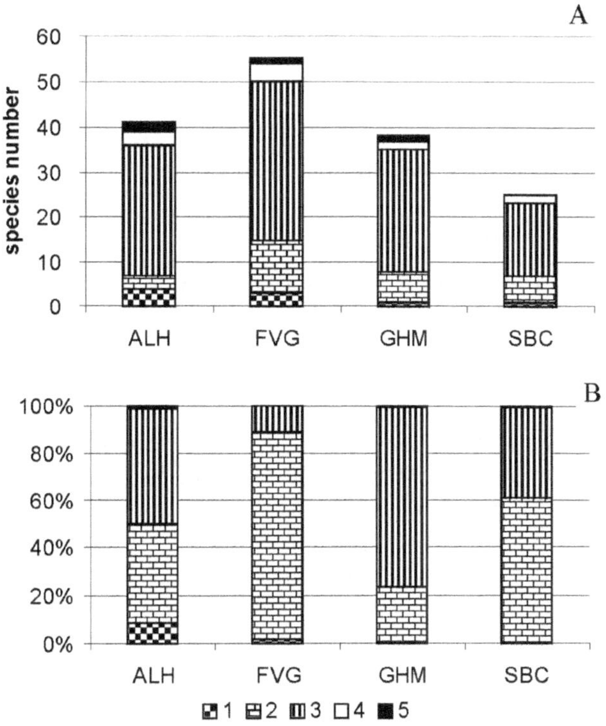

Figure 5.32 Proportion of different phenological types in studied communities. A – number of species, B – % of total aboveground biomass; 1 – evergreen plants, 2 – summer-winter green plants, 3 – spring-summer-autumn green plants, 4 – spring-summer green plants, 5 – summer-autumn green plants

The role of evergreen plants is noticeable in ALH. This community develops on poor soils and has the lowest production among studied alpine communities (Chapter 5.2). So, our observations support the idea of increasing leaf longevity as an adaptation to resource scarcity (Monk 1966, Chapin 1980, Moore 1980, 1984, Diemer et al., 1992). The significant role of spring-summer-autumn green plants in the most productive GHM community is consistent with this idea as well. Plants with (partly) wintergreen leaves are common in temperate alpine communities in different mountain regions: the Colorado Rocky Mountains (Bell 1974), Tien Shan (Lynov 1982), North Ossetia (Popov 1983). The role of plants, which have a short period of vegetation (spring-summer green and summer-autumn green plants), is negligible in all four studied communities. Annual and ephemeral plants are relatively rare in humid arctic and alpine areas (Mark & Bliss 1970, Bliss 1971, 1985, Marchand & Roach 1980, Reynolds 1984).

5.6. Fluctuations and trends in the dynamics of alpine lichen heath and snowbed communities

5.6.1. Introduction

Alpine plant communities are more sensitive to climatic changes than lowland ones, and vertical migrations of phytogeographical belts for 200 – 400 m took place in the Caucasus during the Holocene (Tushinskii 1957, 1962, Shcherbakova 1973). Therefore alpine ecosystems have attracted significant scientific interest as models for studying plant response to global changes such as CO_2 and increasing temperature (Chapin & Körner 1994, Körner 1995, Diemer & Körner 1998, Grabherr et al. 1995, Schappi & Körner 1996, 1997, Körner et al. 1997). However, there are few direct observations of vegetation dynamics during recent decades (Khramtsov 1982, Lesica & Steele 1996).

Dynamics and spatial structure of plant communities may be interrelated. For model communities with a competitive hierarchy, Lehman & Tilman (1998) showed that less-abundant species had more temporal variation and more aggregated spatial distributions. But there is no empirical evidence for these ideas.

Alpine communities are greatly influenced by different types of natural and antropogenic disturbances. Revegetation successions take a long period of time. That is why the study of recovery of disturbed alpine plant communities is of both theoretical and practical interest (Willard & Marr 1971, Brown et al. 1978, Urbanska 1986, 1995). But there are only few works based on the direct long-term field observations (Gadzhiev 1979 Frank & del Moral 1986, Roxburgh et al. 1988, Tsuyuzaki 1991). Wild boar digging occurs rather often in high altitude areas and therefore it may serve as a suitable model of animal and human impact on alpine communities.

The aims of the present chapter were to compare dynamics of two contrasted communities (snowfree alpine lichen heath and snowbed community) during 14 years of permanent plot observations in the Teberda Reserve. We considered the following aspects of inter-year dynamics: 1) comparison of temporal variability of shoot density between the communities; 2) correlation between spatial and temporal variability; 3) separation of groups of species according to their dynamic behaviour; 4) correlation between climatic parameters and species abundance; 5) linear trends; 6) short-term cycles in species dynamics and autocorrelation functions. As a special aspect we studied revegetation dynamics of alpine lichen heath after wild boar digging.

5.6.2. Methods

Two transects 0.5 × 5 m were established in each community. They were divided into 40 plots of 25 × 25 cm. The total number of plots was 80 for each community. We counted shoots (vegetative and generative) as well as juveniles on the plots from 1986 to 1999. Counting took place at approximately the same time each year, at the end of July or early August for ALH and in the middle of August for SBC.

Climatic parameters were obtained from the nearest meteorological station (Teberda, 1328 m a.s.l.). We used monthly mean temperature, mean minimal (mean daily minima) and mean maximal (mean daily maxima) temperatures and sum of precipitation. Data for the current year (from January to June) and the previous year (from January to December) were used.

We found some patches with a total area of about 5 square meters dug up by wild boars (*Sus scrofa* L.) in the autumn of 1980. Wild boars dug up the turf horizon of soil extracting apparently large roots of *Carum caucasicum, Campanula tridentata* and *Trifolium polyphyllum*. Most of the soil surface was deprived of plants and that led to the surface frost activity damaging seedlings and young plants. Five permanent plots of 25 × 25 cm were set up on these patches. Twenty undisturbed plots of the same size were taken as a control. Data have been collected annually during 20 years since 1981 in July and August (Onipchenko & Golikov 1996). The percent cover of vascular plants, lichens and bryophytes was recorded, as well as the percentage of bare soil surface. The number of juveniles and the density of vegetative and generative shoots were calculated for each vascular plant species.

Due to different distribution features of the studied species, we use nonparametric (Spearman) correlation coefficients to estimate the relationships between different species dynamics and the possible influence of weather on shoot number dynamics. We analysed four correlation matrices. They represented correlation coefficients 1) among ALH species, 2) among SBC species, 3) between common ALH and SBC species 4) among ALH, SBC and weather conditions. Correlation diagrams were drawn to show significant positive correlations between species. To assess the direction and degree of change in annual shoot counts we tested for presence of a linear trend using a standard technique (Zar 1999). To estimate possible temporal autocorrelation we calculate autocorrelation functions (relationships between correlation coefficients and lags) for each studied species (Onipchenko et al. 2001).

5.6.3. Variation in shoot density

Shoot density varied for all studied species in both communities (Table 5.20, 5.21, Fig. 5.33). Mean temporal variation coefficients were significantly bigger for generative shoot numbers than for total shoot numbers in both communities (ALH: 74.0% for generative shoots and 27.4% for total shoot numbers, $p=0.026$; SBC 82.9% and 42.1%, $p=0.001$). In ALH species shoot density varied significantly less than that for SBC ($p=0.035$), but there were no differences in generative shoot variations between the communities. All studied ALH species showed variation in total shoot number of less than 60%, but 5 out of 20 species in SBC exceeded this value (Onipchenko et al. 2001).

A significant positive correlation coefficient was obtained between spatial and temporal variation for the combined data ($n=60$, $r=0.34$, $p < 0.001$) and for total shoot numbers in SBC ($n=20$, $r=0.52$, $p < 0.05$). Two cushion plants (*Arenaria lychnidea, Minuartia circassica*) were among the obvious exceptions to the overall trend: the species had very aggregated spatial distributions but they were very stable temporally (Table 5.20).

As a whole, dominants (ALH: *Anemone speciosa, Campanula tridentata, Carex umbrosa, Festuca ovina, Antennaria dioica, Trifolium polyphyllum*, SBC: *Gnaphalium supinum, Minuartia aizoides, Sibbaldia procumbens, Taraxacum stevenii*) were more stable than less abundant species in both communities. Mean variation coefficient for ALH dominants was 18% versus 32% for nondominants, $p=0.0052$, corresponding values for SBC species were 22% and 47%, $p=0.0017$.

So, we obtained 3 main results: 1) the temporal variation for total shoot number was lower than for generative shoots, 2) species in SBC fluctuated more than in ALH, and 3) spatial and temporal variation were positively correlated for most of the species, except in cushion plants. Great interannual variation in generative shoot numbers has also been shown in long-term observation in ALH (Golikov & Onipchenko 1999), as well as for *Bromus variegatus* in subalpine meadows (Khramtsov 1982). Between – year variation in seed production is higher than variation in plant density in alpine areas (Lesica & Steel 1996, Logvinenko & Onipchenko 1999).

Table 5.20. Parameters of spatial and temporal variability of total shoot numbers. Spatial variability parameters were counted for 80 plots (25 × 25 cm) in 1998 year. Std – standard deviation, CV – variation coefficient. (from Onipchenko et al. 2001 with permission of *Oecologia Montana*)

Species	temporal variability (n = 14)			Spatial variability (n = 80)		
	mean	std	CV,%	mean	std	CV,%
	Alpine lichen heath (ALH)					
Agrostis vinealis	46	23.6	52	0.43	1.6	381
Anemone speciosa	212	34.3	16	2.93	2.2	76
Antennaria dioica	289	56.6	20	4.73	10.6	224
Arenaria lychnidea	116	24.8	21	1.41	8.2	579
Campanula tridentata	1522	240.3	16	18.54	10.3	55
Carex umbrosa	1541	374.9	24	16.23	9.3	57
Carum caucasicum	493	126.3	26	6.34	3.7	59
Erigeron caucasicum	19	6.1	32	0.31	1.0	308
Eritrichium caucasicum	358	128.4	36	2.98	4.2	140
Festuca ovina	7384	1838.1	25	109.3	53.0	48
Gentiana pyrenaica	997	319.5	32	10.99	13.1	119
Helictotrichon versicolor	798	100.6	13	10.46	10.3	98
Luzula spicata	105	21.2	20	1.23	3.4	276
Minuartia circassica	40	13.6	34	0.71	5.2	728
Pedicularis caucasica	18	9.5	54	0.20	0.9	467
Pedicularis comosa	50	17.0	34	0.24	0.5	204
Trifolium polyphyllum	353	22.2	6	2.30	5.8	253
Veronica gentianoides	21	6.9	33	0.19	0.7	371
	Snow bed community (SBC)					
Agrostis vinealis	777	179.7	23	10.05	15.0	150
Anthoxanthum odoratum	602	309.9	51	13.26	25.6	193
Carex atrata	119	18.5	15	1.50	3.9	257
Carex oreophila	1700	335.8	20	16.31	16.3	100
Carum meifolium	507	408.9	81	9.39	16.0	170
Catabrosella variegata	1852	424.0	23	15.09	12.1	80
Corydalis conorhiza	453	469.3	104	4.71	7.5	160
Festuca brunnescens	180	141.6	79	4.86	17.5	360
Festuca ovina	243	97.5	40	3.98	8.3	210
Gentiana pyrenaica	19	15.0	81	0.50	2.6	521
Gnaphalium supinum	2772	553.9	20	32.01	63.0	197
Hyalopoa pontica	23	14.0	62	0.24	0.8	354
Luzula multiflora	244	89.0	36	3.25	10.1	310
Minuartia aizoides	6827	1881.9	28	66.89	43.3	65
Nardus stricta	7885	2863.2	36	100.41	86.9	87
Pedicularis nordmanniana	687	279.9	41	9.20	11.5	125
Phleum alpinum	699	276.4	40	11.11	14.7	132
Potentilla crantzii	1122	275.5	25	14.80	25.0	169
Sibbaldia procumbens	6915	1376.3	20	86.09	43.0	50
Taraxacum stevenii	2771	520.5	19	27.84	16.2	58

Table 5.21. Parameters of spatial and temporal variability of generative shoots. (see Table 5.20 for abbreviations). (from Onipchenko et al. 2001 with permission of *Oecologia Montana*)

Species	temporal variability (n=14)			Spatial variability (n=80)		
	mean	std	CV,%	mean	std	CV,%
	Alpine lichen heath (ALH)					
Anemone speciosa	65	12.8	20	0.93	1.2	130
Campanula tridentata	346	104.0	30	6.24	3.9	63
Carex umbrosa	38	46.5	121	0.19	0.5	283
Carum caucasicum	118	18.4	16	1.65	1.5	88
Eritrichium caucasicum	22	13.0	60	0.04	0.2	510
Festuca ovina	130	109.0	84	1.68	1.9	112
Gentiana pyrenaica	90	37.0	41	0.81	1.4	174
Helictotrichon versicolor	22	30.5	136	0.28	0.7	259
	Snow bed community (SBC)					
Anthoxanthum odoratum	41	39.6	97	1.78	3.8	215
Carex oreophila	203	67.8	33	2.08	3.3	159
Carum meifolium	18	13.4	76	0.34	0.8	231
Catabrosella variegata	64	90.9	141	0.03	0.2	628
Corydalis conorhiza	39	37.2	94	0.28	0.6	231
Gnaphalium supinum	94	107.1	114	0.79	1.7	218
Luzula multiflora	35	24.6	70	0.30	1.2	402
Minuartia aizoides	116	138.1	119	0.33	0.8	234
Nardus stricta	85	112.3	132	4.70	9.4	199
Pedicularis nordmanniana	141	85.9	61	2.26	2.9	127
Phleum alpinum	51	31.8	62	0.89	2.0	221
Potentilla crantzii	35	27.1	78	0.10	0.4	439
Sibbaldia procumbens	180	74.6	41	4.18	4.3	102
Taraxacum stevenii	210	86.5	41	1.51	2.1	142

Abundant nonclonal (*Anemone speciosa, Campanula tridentata*) as well as clonal (*Festuca ovina, Carex umbrosa, Carex oreophila*) and cushion (*Arenaria lychnidea, Minuartia circassica*) species showed relatively low variation in temporal shoot density. A low rate of temporal variation was observed for *Carex curvula* in the Alps (Erschbamer et al. 1998).

Our results support the idea that the dynamics and spatial structure of plant communities may be interrelated. The behaviour of SBC species was more consistent with the competitive hierarchy model of Lehman & Tilman (1998) than that of the ALH species. We may speculate that a competitive hierarchy plays a more important role in SBC than in ALH, where facilitation between dominants and subordinate species has been shown (Aksenova et al. 1998).

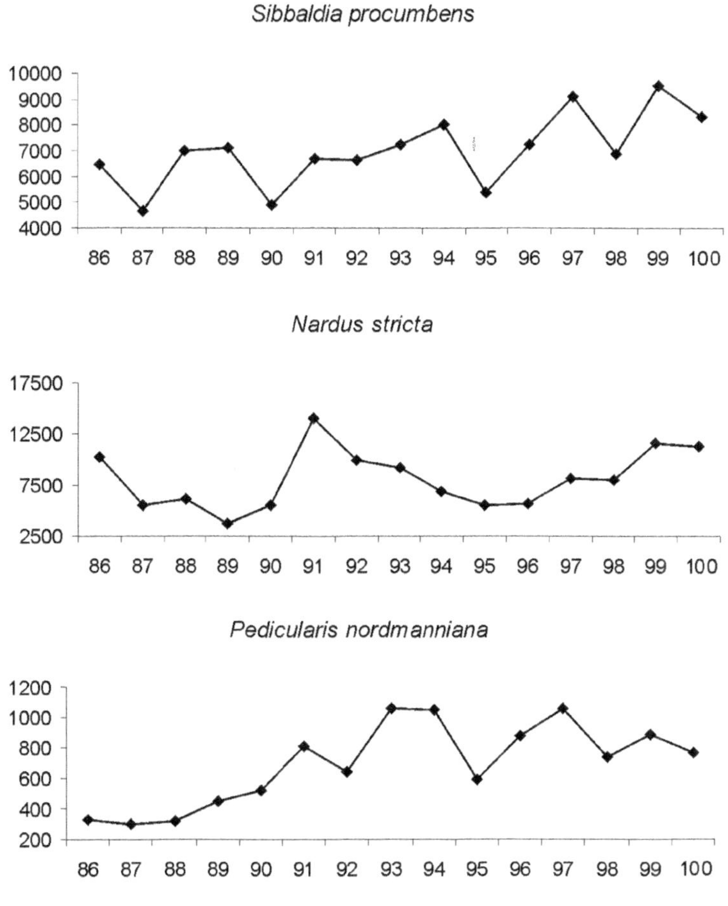

Figure 5.33. Examples of dynamics of 3 alpine species from SBC. Abscissa – years of observation (1900+), ordinate – total number of shoots on permanent plots ($5\,m^2$) (from Zakharov et al. 2001).

5.6.4. Species groups according to their dynamic behaviour and weather conditions

Analysis of the correlation matrix allowed as to distinguish groups of species with similar dynamic behaviour (Onipchenko et al. 2001). Several small correlated groups were obtained in ALH (Fig. 5.34). *Eritrichium caucasicum, Gentiana pyrenaica, Agrostis vinealis*, and *Luzula spicata*

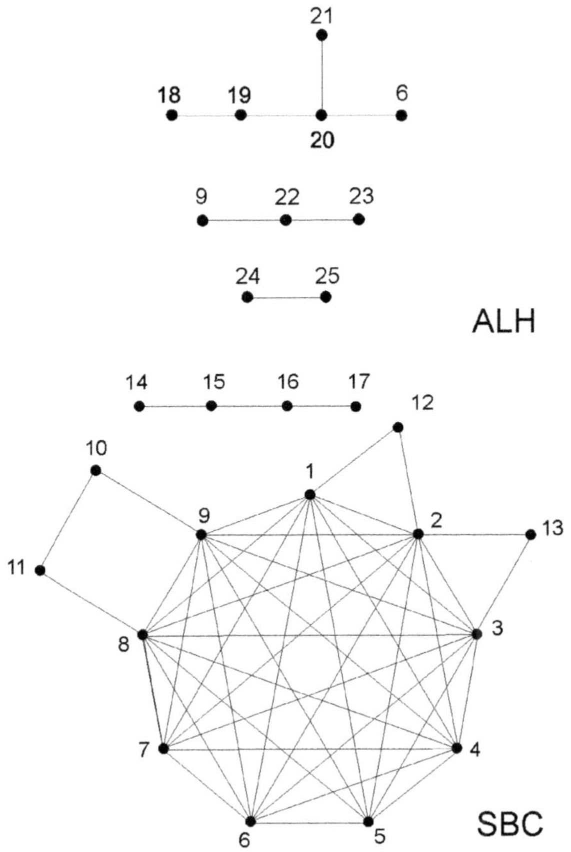

Figure 5.34. Correlation diagrams for groups of species with similar dynamic behaviour. Values of positive Spearmen correlation coefficient (significance level $p<0.01$) are indicated by solid lines within communities only. Communities: ALH – alpine lichen heath, SBC – snow bed community. Species: *1 – Carum meifolium, 2 – Pedicularis nordmanniana, 3 – Luzula multiflora, 4 – Anthoxanthum odoratum, 5 – Corydalis conorhiza, 6 – Gentiana pyrenaica, 7 – Festuca brunnescens, 8 – Phleum alpinum, 9 – Festuca ovina, 10 – Potentilla crantzii, 11 – Nardus stricta, 12 – Hyalopoa pontica, 13 – Sibbaldia procumbens, 14 – Carex oreophila, 15 – Taraxacum stevenii, 16 – Minuartia aizoides, 17 – Gnaphalium supinum, 18 – Veronica gentianoides, 19 – Agrostis vinealis, 20 – Eritrichium caucasicum, 21 – Luzula spicata, 22 – Minuartia circassica, 23 – Erigeron alpinus, 24 – Arenaria lychnidea, 25 – Carum caucasicum.* (from Onipchenko et al. 2001 with permission of *Oecologia Montana*)

formed the largest group at a significance of $p=0.01$. *Festuca ovina, Minuartia circassica* and *Erigeron alpinus* formed another group, while *Carum caucasicum* and *Arenaria lychnidea* were positively correlated with each other. But between groups there were many positive correlations for the species at a lower level of significance ($0.05>p>0.01$), thus the species of those three groups did not show major differences in dynamics. Completely different shoot dynamics was shown by *Trifolium polyphyllum*. This species had negative correlations ($p<0.05$) with 3 species of the second group (*Minuartia circassica, Festuca ovina, Erigeron alpinus*).

A completely different picture was obtained for SBC where most of the species showed positive correlations with each other (Fig. 5.34). Nine species had significant positive correlations with the other eight (*Festuca ovina, Carum meifolium, Pedicularis nordmanniana, Luzula multiflora, Anthoxanthum odoratum, Corydalis conorhiza, Gentiana pyrenaica, Festuca brunnescens, Phleum alpinum*). Four other species comprised a second group in SBC (*Carex oreophila, Taraxacum stevenii, Minuartia aizoides, Gnaphalium supinum*). It is interesting to note that there were no significant negative correlations among SBC species at $p<0.01$. SBC species had more interrelated dynamics than ALH species.

There were three species (*Agrostis vinealis, Festuca ovina* and *Gentiana pyrenaica*) common to ALH and SBC. We calculated the Spearmen correlation coefficients for shoot numbers of the species between the two communities. A significant positive correlation coefficient was obtained for *Festuca ovina* ($r = 0.705$, $p<0.01$) only.

A few significant correlation coefficients were obtained between shoot numbers and current or previous weather conditions (Table 5.22.). Several SBC species (*Anthoxanthum odoratum, Luzula multiflora, Phleum alpinum*) showed a positive correlation with minimal temperature in the previous October. A warm autumn can facilitate shoot development of these species in the following year. Precipitation in May of the previous year correlated positively with the number of generative shoots of *Anemone speciosa*. Generative shoots of *Eritrichium caucasicum* correlated negatively with precipitation in April of the current year.

5.6.5. Trends in species dynamics

Inter-year dynamics of the communities can be divided into two components: fluctuations (non directional changes, see Miles 1979) and successional (directional) trends.

Rabotnov (1972) classified fluctuations in grasslands into 3 main types: 1) "hidden" – composition and structure of communities are stable, only quantitative changes in shoot dynamics take place; 2) "oscillations" – several

Table 5.22. Significant Spearman correlation coefficients (CC, $p < 0.001$) between shoot numbers and weather parameters. Abbreviations: Sh. – shoots: G - generative, S - sum of vegetative, generative and juvenile shoots; Com: SBC - snow bed community, ALH - alpine lichen heath, temp. – temperature (from Onipchenko et al. 2001 with permission of *Oecologia Montana*)

Species	Com	Sh	Weather parameter	CC
Anthoxanthum odoratum	SBC	G	Mean temp. in previous year October	0.81
Anthoxanthum odoratum	SBC	G	Mean minimal temp. in previous year October	0.83
Corydalis conorhiza	SBC	G	Mean temp. in current year January	−0.76
Luzula multiflora	SBC	S	Mean minimal temp. in previous year October	0.77
Phleum alpinum	SBC	S	Mean minimal temp. in previous year October	0.76
Sibbaldia procumbens	SBC	G	Precipitation in current year April	0.79
Anemone speciosa	ALH	G	Precipitation in previous year May	0.79
Eritrichium caucasicum	ALH	G	Precipitation in current year April	0.78
Eritrichium caucasicum	ALH	G	Mean temp. in previous year December	−0.76
Helictotrichon versicolor	ALH	S	Mean temp. in current year February	−0.83

dominants alternate according to climatic variations between years; 3) "demutations" – cycles include local disturbance and recovery. As both studied communities kept their structure and composition through the whole period of observation, we may consider the fluctuation as "hidden" according to this classification.

Several species of ALH and SBC showed a significant linear trend for the period of observation (Onipchenko et al. 2001). A positive trend (significant increase of shoot number) was found for 8 species in SBC (*Anthoxanthum odoratum, Carum meifolium, Corydalis conorhiza, Festuca brunnescens, Festuca ovina, Gentiana pyrenaica, Hyalopoa pontica, Luzula multiflora*), most of them (except *Corydalis conorhiza* and *Hyalopoa pontica*) more typical for more productive alpine meadows. One common SBC species (*Catabrosella variegata*) demonstrated a significant decrease in population density (Onipchenko et al. 2001). Far less (5 from 19 studied species, *Arenaria lychnidea, Campanula tridentata, Erigeron alpinus, Festuca ovina, Minuartia circassica*) ALH species showed an increase in shoot numbers. Only one species (*Trifolium polyphyllum*) showed a decrease in population density in ALH.

As a whole, about half of the studied species showed a progressive increase in shoot density during the study period. This trend was more common among SBC species, and the species concerned are more widespread in alpine meadows than in SBC (Onipchenko 2002). Moen and Oksanen (1998) considered snow bed communities in Sweden as secondary communities, developed under strong grazing. Our SBC were grazed by domestic ungulates until 1943 and they still show some signs of former overgrazing (high abundance of low palatable species - *Sibbaldia procumbens, Nardus stricta*). The increase of abundance in meadow species may therefore be the result of secondary succession after the ending of grazing.

Other possible explanations of the successional trends may be connected with current climate change. We have found a positive trend for minimal October temperature for the Teberda meteorological station during the last 19 years (Blinkova & Onipchenko 2001). But only few species demonstrated a positive correlation with the October temperature (Table 5.22). The lists of the species with such correlations and linear trends overlap only slightly.

Most of the SBC species that had a concordant dynamic behaviour (Fig 5.34.) had a significant linear trend and increased their shoot numbers during the observation period, so the successional trend is more obvious in this community. These results are consistent with the hypothesis that the snowfree alpine communities in the Caucasus are more stable than snowbed communities for short time periods. The same is true for longer periods (centuries and millennia) (Pavlova & Onipchenko 1992, Blinnikov 1994).

5.6.6. Autocorrelation analysis

According to their autocorrelograms all species can be divided into two groups. The first group (most of the species in both communities) showed no significant correlation coefficient for any temporal lag (from 1 to 10 years). The other group included species with more conservative dynamics, namely they had significant positive correlation coefficients for a lag of 1 year. That is, next year's number of shoots depended on the current year's shoot density. Nine species (3 from ALH and 6 from SBC) belonged to this group (Table 5.23), suggesting that relatively more species from the snow bed community demonstrated such conservative dynamics.

Table 5.23. Significant autocorrelation coefficients for a lag of 1 year. R – correlation coefficient, *t*-criteria value (n = 13)

Species	Community	R	*t*
Anthoxanthum odoratum	SBC	+0.61	2.27
Festuca brunnescens	SBC	+0.70	2.62
Festuca ovina	SBC	+0.59	2.20
Gentiana pyrenaica	SBC	+0.54	2.04
Hyalopoa pontica	SBC	+0.65	2.42
Pedicularis nordmanniana	SBC	+0.53	2.00
Carex umbrosa	ALH	+0.58	2.17
Erigeron alpinus	ALH	+0.66	2.45
Gentiana pyrenaica	ALH	+0.52	1.94

So, the short period of observation (14 years) did not allow us to evaluate cyclic dynamics of alpine species. Cycles, if they exist, appear to be longer than 10 years. Further observations are necessary to study this aspect of alpine plant population dynamics.

5.6.7. ALH revegetation after wild boar digging

5.6.7.1. Species richness

We recorded a total of 20 vascular plant species on the disturbed plots during the whole period of observation. The number of species recorded annually increased from 10 in 1981 to 16 in 1994, as well as the average number per plot (from 5.2 to 10.2, respectively). The t-test shows that the difference between the disturbed and control plots is statistically significant at the 1% level in the two first years and at the 5% level in the following four. By year 10 the average number of vascular plant species on the disturbed plots was approximately the same as on the control (about 9.5 species per plot) one (Fig. 5.35). Floristic richness per plot increased due to the gradually increasing frequency of species registered from the very beginning of our observation (*Luzula spicata, Carum caucasicum*) as well as the arrival of new species (*Eritrichium caucasicum, Veronica gentianoides, Euphrasia ossica, Pedicularis comosa, Alchemilla caucasica*). All of the recorded species are typical of alpine lichen heaths. There appeared no species alien to this type of community. This proves the statement that the species composition of plant communities in the high altitude areas does not change during revegetation successions (McMahon 1982). Severe environmental conditions prevent diaspores of other species from invading disturbed plots.

5.6.7.2. Plant cover

Overall the revegetation of alpine lichen heath has been proceeding slowly: the traces of digging are clearly distinguishable even 15 years after the disturbance had happened. The bare soil area has reduced considerably (from 71% to 5% on average) during the time of our observation, whereas the cover of vascular plants has increased from 2% to 32%, getting close to the control one. At the same time the cover of fruticose lichens has reached only 27%. It makes up approximately half of the control value. The cover of bryophytes on the disturbed plots has exceeded a few times the control one that was never more than 1% (Fig. 5.36). In all probability it is connected with the sudden denudation of a considerable piece of soil surface after wild boar disturbance. It is supposed that bryophyte species (particularly

Polytrichum juniperinum and *Pohlia nutans*) demonstrate the ability for quick colonisation of bare soil areas.

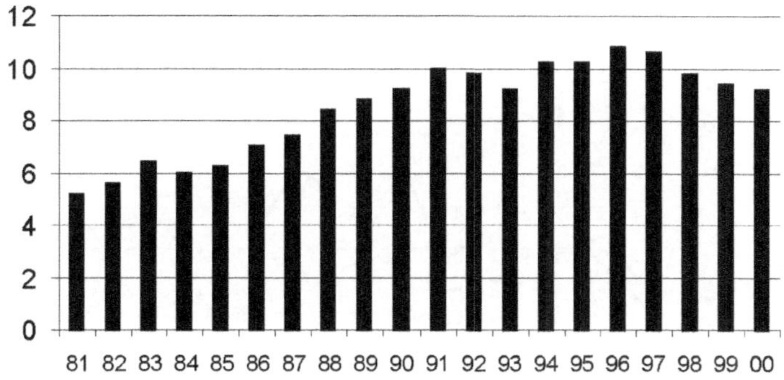

Figure 5.35 Dynamics of floristic richness on wild boar diggings in alpine lichen heath. Abscissa – years of observation (1990+), ordinate – mean number of vascular plant species in a plot of 25 × 25 cm

Thus, there is a marked difference between the components of alpine lichen heath in the rate of recovery. The slowest to recover are fruticose lichens. Apparently this is due to the low rate of their growth, especially of *Cetraria islandica, Cladina mitis* and *Thamnolia vermicularis*. On the other hand, because of frequent strong winds, fruticose lichens cannot develop in the alpine lichen heaths without the "framework" of the community, formed by vascular plants whose floristic composition does not change significantly after removal of lichens (Chapter 5.3, Onipchenko 1985, 1994b). It is interesting that the investigation of a New Zealand high-alpine cushion field has yielded similar results: 11 years after mechanical disturbance the lichen cover dominated by *Cetraria islandica, Thamnolia vermicularis* and *Alectoria nigricans* had not recovered (Roxburgh et al. 1988).

5.6.7.3. Shoot density

In the first year of our observation the shoot density of most species on the disturbed plots was significantly lower than that of the control (Onipchenko & Golikov 1996, Fig. 5.37). In later years a drastic increase was observed for some graminoids – *Festuca ovina, Luzula spicata, Carex* spp. (*C. umbrosa, C. sempervirens*) – due to the intensive vegetative growth of the plants remaining after the disturbance. It is suggested that these species demonstrate the "explerent" strategy properties sensu Ramenskii (1938), namely a quick clonal growth and colonisation ability (Chapter 6.3)

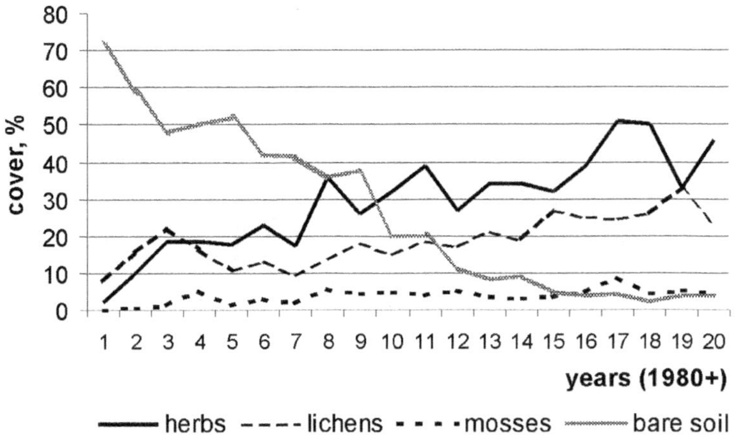

Figure 5.36 Dynamics of plant and bare soil cover after boar digging

There are some observations (Shiffers 1953, Tsareva 1978, Brown & Johnston 1979) that *Luzula spicata* is one of the first plants to occupy the disturbed sites in high altitude areas. It is therefore recommended for the restoration of alpine communities.

Seed reproduction is the main factor promoting the restoration of taprooted polycarpics, such as *Carum caucasicum* and *Campanula tridentata*. Their seedlings have appeared on the dug plots in the following fashion: 1) from the soil seed bank (Semenova and Onipchenko 1994); 2) from the adult individuals remaining on site after disturbance; 3) from the nearest surroundings. That is why in the population structure of these species on the disturbed plots juveniles are significantly more than on the control plots. On the other hand, the proportion of generative plants is considerably less than in the control.

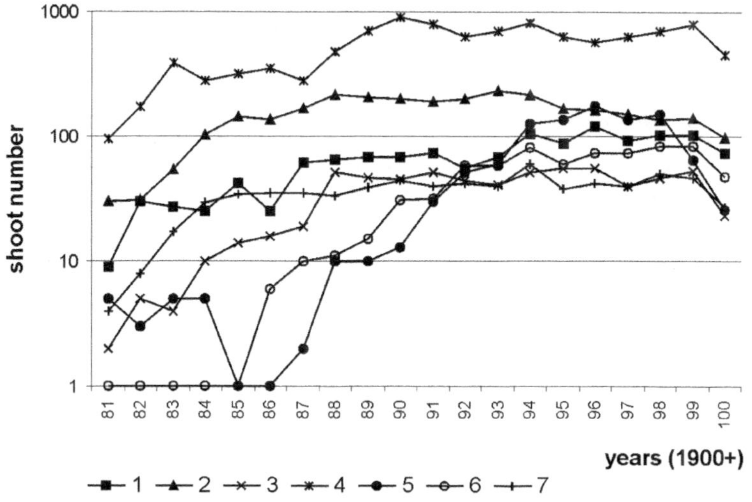

Figure 5.37 Shoot number dynamics for 7 abundant species after boar digging. Species: 1 - *Campanula tridentata*, 2 - *Carex umbrosa*, 3 - *Carum caucasicum*, 4 - *Festuca ovina*, 5 - *Gentiana pyrenaica*, 6 - *Helictotrichon versicolor*, 7 - *Luzula spicata*

The shoot density of *Anemone speciosa* has been recovering very slowly. Apparently, small seed yield capacity and the lack of vegetative reproduction of this species can explain this fact. The same holds true for *Eritrichium caucasicum*.

We can conclude that the species composition does not change considerably during local revegetation in the alpine zone. The most abundant species of alpine lichen heath remained dominants on the disturbed plots. Disturbance changes only the relative role of some groups of species. We proposed a number of characteristic features to distinguish stages of revegetation of alpine lichen heath from the undisturbed communities (at least during several ten-year periods): 1) presence of some patches of bare soil surface; 2) low coverage of fruticose lichens; 3) greater abundance of graminoids; 4) low abundance of *Anemone speciosa*.

5.7. Successions and historical dynamics during the Holocene

5.7.1. Introduction

Palaeoecological investigations in the alpine ecosystems are complicated by the lack of adequate proxies. Modern alpine communities are by definition located above the treelines and are thus unsuitable for dendrochronological analysis. Likewise, in the alpine areas, few lakes exist suitable for pollen analysis. We used pollen analysis. We used pollen and phytolith analysis from soils to infer the alpine communities' history in Teberda.

From the historical-floristic point of view, the western slope of the Caucasus is part of the larger Mediterranean region while the eastern part is included in the Turan region that extends eastward towards the Caspian and into Central Asia (Grichuk 1982). Within this region, the history of vegetation of the higher mountains received very little attention. Most palaeoecological reconstructions of the high-mountain Caucasus was done out of necessity at a broad regional level, primarily through pollen analysis of medium-sized lakes or treeline reconstructions (Makhova & Patyk-Kara 1961, Khapaev 1976, Serebryanny et al., 1980, Sokolova 1982, Margalitadze & Kimeridze 1985, Knyazev et al. 1992). Local reconstructions of individual communities remain elusive.

In our work at Teberda Station it was important to attempt local reconstructions of community histories to be able to compare their historical development with the observed modern-day composition and spatial pattern. While modern analysis of plant community structure, population biology of plants, and experimental methods (Chapters 5, 6, and 7) can teach us a lot about underlying mechanisms of species' coexistence, it is imperative to check these observation against a solid historical background.

5.7.2. Methods

We use two different proxies: analysis of pollen and spores in relatively undisturbed aggrading alpine soils and phytolith analysis of the same soils to investigate historical changes in plant composition at our research site during the mid to late Holocene (beginning ca. 7,000 cal. BP).

Pollen analysis of lake cores is a much more common, standard technique, while analysis of moss cushions or soil samples can be more problematic because of poor preservation of pollen, differential preservation and high likelihood of mixing of the soil record (Minckley & Whitlock 2000). Nevertheless, pollen analysis of mineral soils was demonstrated to provide good results in some cases (Berezina & Tyuremnov 1969, Savina 1982). Alpine soils of our area are promising in that respect: they have moderately low pH, high humus content, and low microbiological activity, all favouring pollen preservation. More importantly, bioturbation is low in all our communities, except the *Geranium* meadows (GHM), and soils apparently grow upwards by accumulating organic matter and dust that comes with wind and precipitation (Pavlova & Onipchenko 1992).

Soils samples were obtained from 8 soil profiles for pollen and 4 of the same profiles for phytoliths from the four main alpine communities found in the study area and described in greater detail in Chapter 4 of this volume (ALH, FVG, GHM and SBC).

Both pollen and phytolith records from soils were compared against modern collections of plant and soil surface samples. Modern pollen spectra were gathered from 8 pinch samples of soil per community from 25 – 100 m^2. Separation of pollen and spores was done by the standard Grichuk method that involves heavy liquid flotation and is very similar to the standard Faegri et al. (1989) method. Pollen grains were tallied on temporary slides in glycerine under a light microscope by counting minimum of 300 grains. Further details are provided in Pavlova & Onipchenko (1992).

Phytolith analysis is a relatively new method of palaeoecological studies, although Russian soil scientists have used it since the mid 1930s (Tyurin 1937). Phytoliths are silicified replicas of plant cells, which are morphologically distinct, abundant, and durable in soils, loess, cave deposits and other environments (Piperno 1988). The phytoliths' main strength is their durability under a wide range of depositional conditions and their possibility of identifying plant communities, and sometimes individual taxa, based on matching palaeoassemblages with modern analogues. In addition to individual shape counts, phytolith concentrations can be used to infer presence/absence of forested vegetation, as well as to indicate presence of buried A soil horizons (Wilding & Drees 1971, Blinnikov et al. 2002) and as such can be a useful supplement to pollen.

We had to build our own reference collection of phytoliths from plants and soils of Teberda (Blinnikov 1994, Volkova et al. 1995). The geographically closest and only other example of a phytolith collection is described in Kiseleva (1992) from North Ossetia subalpine grasslands and it supports our main observations.

Extraction of phytoliths from plants was done by dry oxidation for a few hours in a muffle furnace at 550°C. A modified wet oxidation and gravity sedimentation method of Piperno (1988) was used for extraction of phytoliths from soils. Phytoliths were extracted from the silt fraction of soil (5 – 100 microns, 20 – 50 g of soil per sample). After soil fractionation and removal of clay and sand, the organics were removed by a concentrated

(70%) nitric acid with a pinch of potassium perchlorate added per 100 ml beaker and then heated for 1 hour. Carbonates were removed by washing the residue in 10% hydrochloric acid. Opal fraction (phytoliths) were separated from quartz and other heavier minerals using flotation in a zinc bromide solution (specific gravity of 2.3 g/cm^3). Concentration of phytoliths was calculated based on the ratio of the total dry weight of the opal residue to the total dry weight of the initial sample. Information about vegetation composition was obtained based on identification of individual phytolith shapes from fossil samples under a light microscope and matching the paleoassemblages against a reference collection from modern plants and soils by using a key developed earlier (Volkova et al. 1995).

5.7.3. Pollen analysis results and discussion

5.7.3.1. Modern pollen spectra

Most of this section is described based on earlier work published in Russian (Pavlova & Onipchenko 1992). All analysed soil profiles and surface soil samples contained abundant well-preserved pollen enabling identification. Not surprisingly, in modern spectra a lot of pollen was coming from lower elevations being brought in by wind. Other researchers experienced the same problem at high elevation sites (Janssen 1981, Randall et al. 1986, Minckley & Whitlock 2000). In our area *Pinus, Artemisia* and Chenopodiaceae pollen and fern spores are the most common types from the vegetation below the alpine zone. In all communities the amount of arboreal pollen was greater than 50%, with the highest recorded for the heath (ALH) near the ridgeline and the lowest in the snowbed communities (SBC) (Fig. 5.38). SBC and much of GHM is still covered with snow in May-early June when *Pinus* pollination is at its peak. Spore concentrations in all samples was below 5%.

In the modern spectra the highest proportion of non-arboreal pollen belongs to Apiaceae, Poaceae (Graminae), Caryophyllaceae, Asteraceae and Geraniaceae (largely from *Geranium gymnocaulon* in GHM). The first four families are expected to dominate the pollen spectra in the arcto-alpine environments (Fig. 5.39). FVG pollen is dominated by Poaceae, SBC – by Asteraceae, and GHM by *Geranium*. ALH pollen contains a relatively high proportion of Apiaceae and Caryophyllaceae.

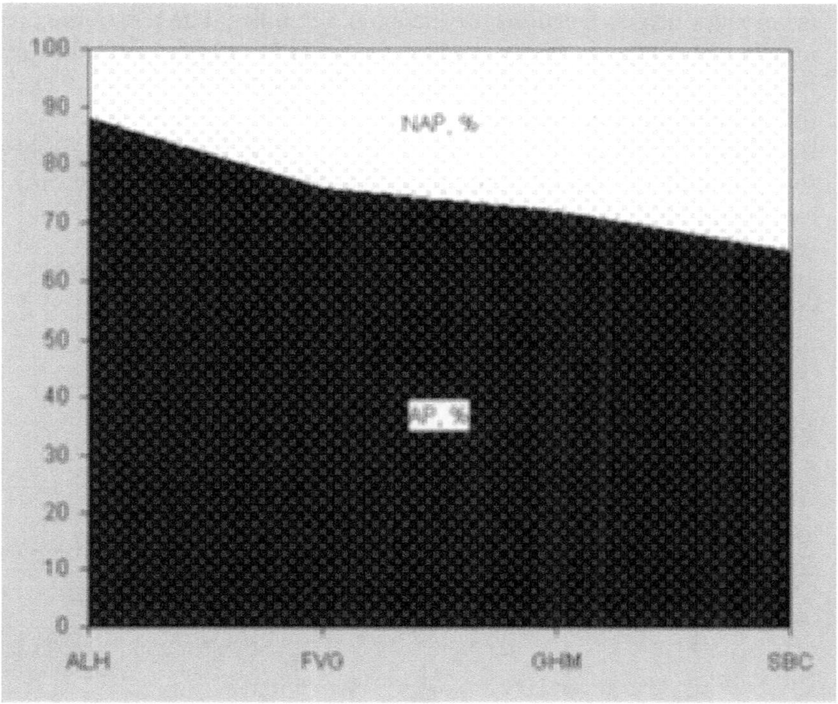

Figure 5.38 Change in proportion of arboreal (AP) vs. non-arboreal pollen (NAP) in modern samples from soils under 4 alpine communities. ALH – alpine lichen heath, FVG – *Festuca varia* grassland, GHM – *Geranium gymnocaulon* – *Hedysarum caucasium* meadow, SBC – snowbed community.

The results of the pollen analysis from 8 profiles are represented in pollen diagrams created in Tiliagraph and arranged from the driest (ALH-Upper) to the wettest communities (SBC-Lower) (Fig. 5.40 – 5.47). A small percentage of unidintified pollen in each sample was included in the nonarboreal pollen category. In every community, "upper" sites tend to have a lower amount of snow and are consequently drier than the lower sites.

5.7.3.2. Alpine lichen heath pollen

Alpine lichen heaths appear to have changed little over the period of soil development (Fig. 5.40, 5.41). A date of 2160 ± 80^{14}C BP was available from the 10 – 25 cm interval in this profile, and the increasing presence of *Tilia, Quercus* and *Ulmus* in the middle of the profile at 6 – 10 cm depth suggests that the soil genesis included at least the mid-Holocene "optimum" ca. 5000 – 6000 yrs BP. The top 0 – 2 cm from the upper ALH profile with a high level of grass pollen resemble modern FVG spectra as established by modern analogue matching by calculating the Euclidean distance between non-arboreal pollen proportions for the main 5 families. The rest of the profile matches well the modern spectra from ALH: similar proportions of Poaceae, Apiaceae, Caryophyllaceae and Asteraceae pollen. Based on the pollen diagram, we conclude that the ALH site was never covered by meadow or forest vegetation and that this community likely formed at the beginning of the Holocene from species capable of surviving snowless conditions.

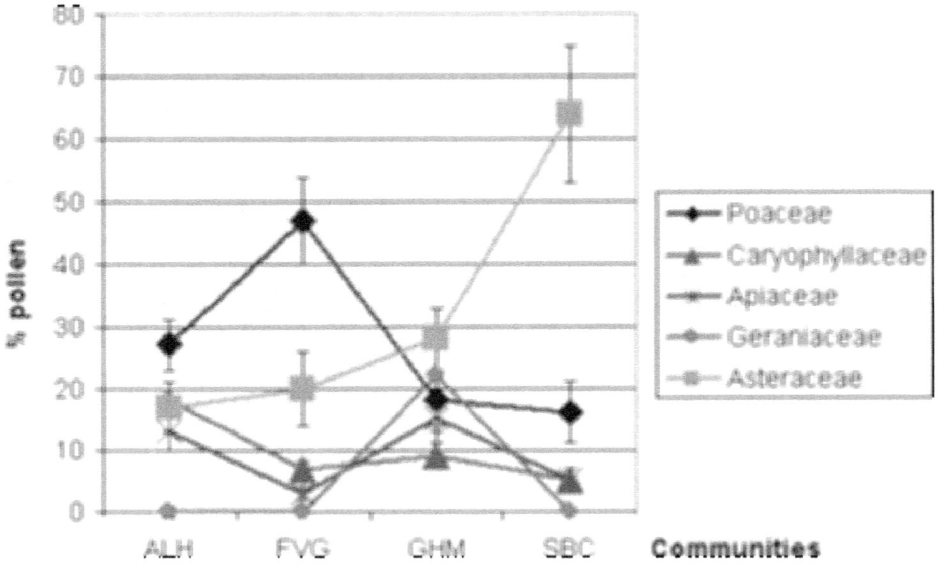

Figure 5.39 Proportion of the main five families in the total non-arboreal pollen in modern samples from soils under 4 alpine communities. Vertical bars are ±S.E. (n=8).

5.7.3.3. *Festuca grassland pollen*

The upper and lower FVG profiles exhibit similar trends (Fig. 5.42, 5.43) in terms of tree pollen, but are different in terms of local NAP pollen. The best match for the top of the upper FVG profile (0.2 cm) comes from the modern ALH, while the rest of the profile suggests the presence of a FVG community down to about 30 cm depth implying a multi-century period of grassland development. This is largely based on high levels of Poaceae pollen accounting for over 50% of the NAP in these samples. Below 30 cm depth the modern analogue match becomes SBC due to a very high proportion of Asteraceae pollen implying wetter conditions than today. The Upper-FVG was radiocarbondated at $3610\pm80\,^{14}C$ yrs BP near the bottom at the depth of 25 – 35 cm (Grishina et al. 1987). We believe that the real age of

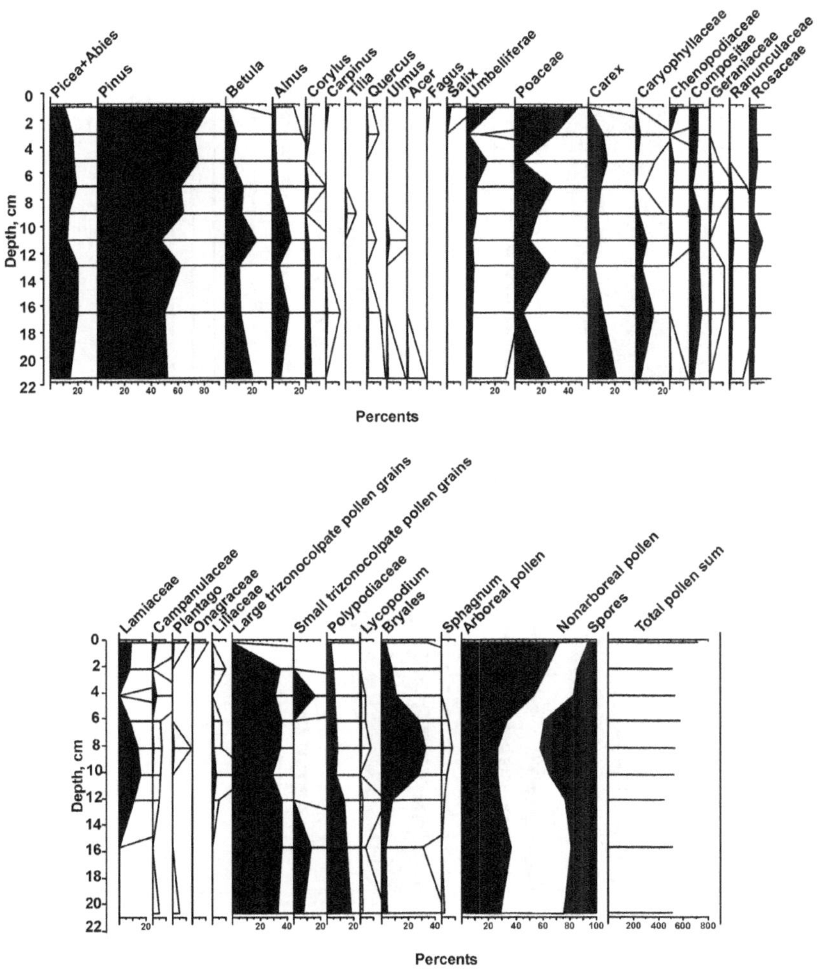

Figure 5.40 Pollen percentages and total pollen counts from ALH-Upper soil profile, Mt. Khatipara alpine station, Russia. Stippled area represents ×10 exaggeration for rare taxa. Total pollen sum represents the total number of pollen grains counted. Redrawn with modifications from Pavlova and Onipchenko (1992).

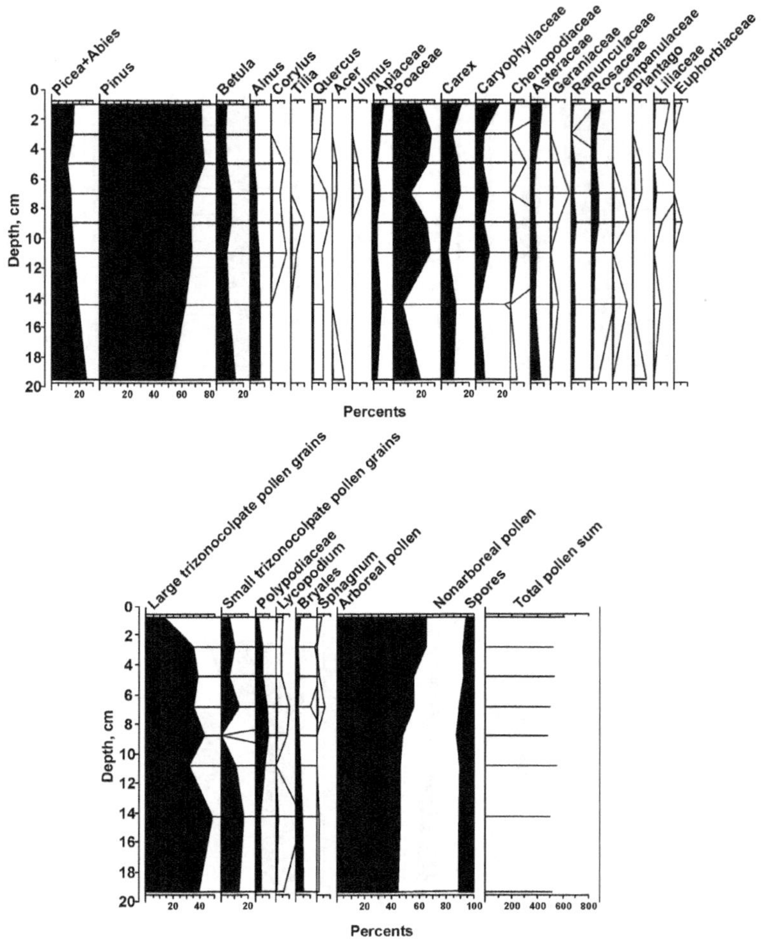

Figure 5.41 Pollen percentages and total pollen counts from ALH-Lower soil profile, Mt. Khatipara alpine station, Russia. Stippled area represents ×10 exaggeration for rare taxa. Total pollen sum represents the total number of pollen grains counted. Redrawn with modifications from Pavlova & Onipchenko (1992)

the profile from this modern soil with active biological processes is closer to 5000 – 6000 cal. years BP.

The lower FVG profile shows greater variability. It is located near a temporary brook on a mountain slope near a patch of *Geranium* meadow community and a wet meadow, and thus may have a somewhat mixed source for its pollen. The closest match for the top two samples from this profile was a modern SBC community. Most of the profile matches equally well modern GHM and FVG pollen implying moderately wet conditions, certainly wetter than the FVG-Upper site, while the bottom portion of it below 20 cm is more similar to an ALH/FVG transition implying drier conditions at that time. Both profiles agree on the distribution of tree pollen: *Pinus* accounts for almost 80% of all AP near the top, while *Tilia, Corylus* and *Quercus* become more prominent at the depths of 25 – 30 cm (likely corresponding to the Holocene Atlantic optimum).

5.7.3.4. Geranium meadow pollen

Unlike the previous two communities, GHM spectra suggest that the current *Geranium-Hedysarum* vegetation is relatively recent at these sites. *Geranium* pollen is very scarce in both GHM profiles, while a high level of Poaceae pollen suggests the historical presence of FVG-like grassland at the present-day GHM sites (Fig. 5.44, 5.45). Both profiles are similar to each other, with the wetter GHM-Lower profile having a greater percentage of *Geranium* pollen in the profile. This profile was radiocarbonated at 2950 ± 80^{14}C yrs BP at the depth of 19 – 30 cm (Grishina et al. 1987). The sample from 22 cm depth has no analogue in that it contains a very high proportion of Caryophyllaceae pollen (50% of NAP) and Polypodiaceae and *Sphagnum* spores implying wet conditions at the time. We interpret the history of this community as largely being a grassland, with *Geranium* and other moisture-preferring species coming *en masse* not earlier than ca. 500 years ago.

5.7.3.5. Snowbed community pollen

Both snowbed spectra (Fig. 5.46, 5.47) were similar to each other. The upper half of both profiles (16 – 25 cm depth with the date of 3630 ± 80^{14}C yrs BP, Grishina et al. 1987) has spectra typical for modern SBC. The lower portion of both profiles resemble a mix between modern-day GHM and SBC with a lower proportion of Asteraceae pollen and some *Geranium* present. Mikeladze (1960) suggested that SBC in the Caucasus originate as open assemblages on wet sites and gradually develop a carpet-like appearance. Our results suggest that SBC have gone thorough a meadow stage.

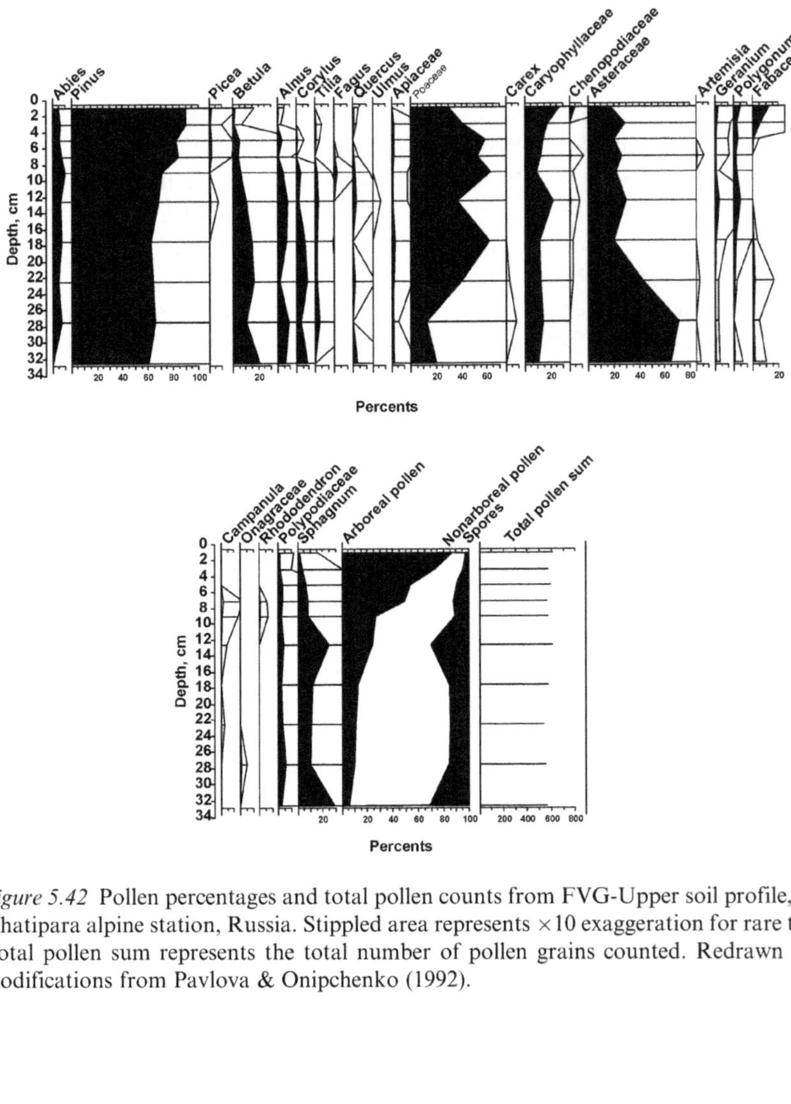

Figure 5.42 Pollen percentages and total pollen counts from FVG-Upper soil profile, Mt. Khatipara alpine station, Russia. Stippled area represents ×10 exaggeration for rare taxa. Total pollen sum represents the total number of pollen grains counted. Redrawn with modifications from Pavlova & Onipchenko (1992).

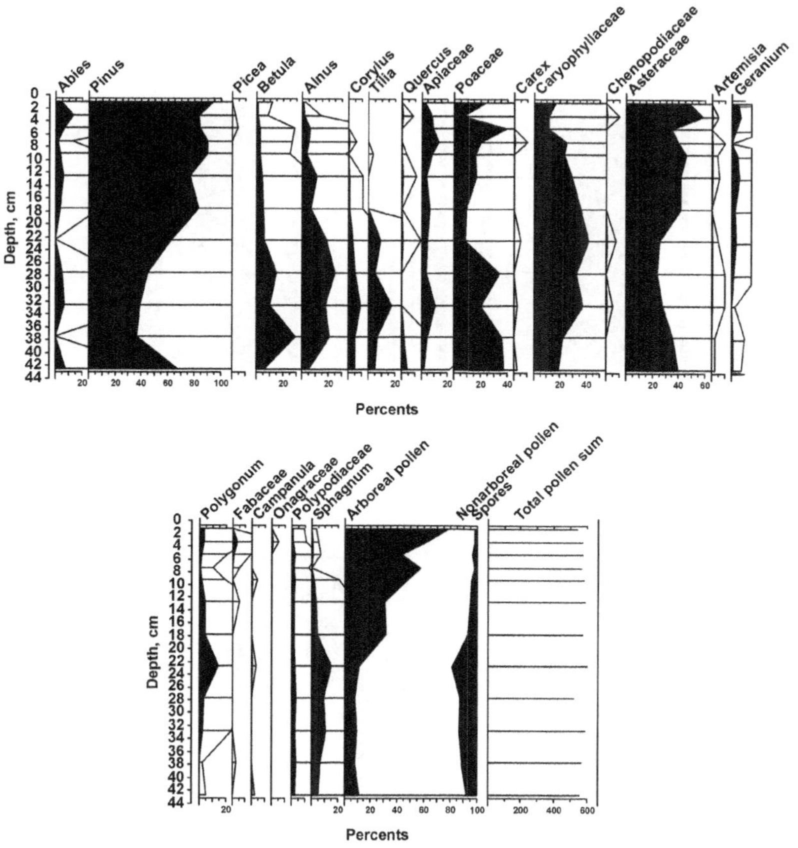

Figure 5.43 Pollen percentages and total pollen counts from FVG-Lower soil profile, Mt. Khatipara alpine station, Russia. Stippled area represents ×10 exaggeration for rare taxa. Total pollen sum represents the total number of pollen grains counted. Redrawn with modifications from Pavlova & Onipchenko (1992)

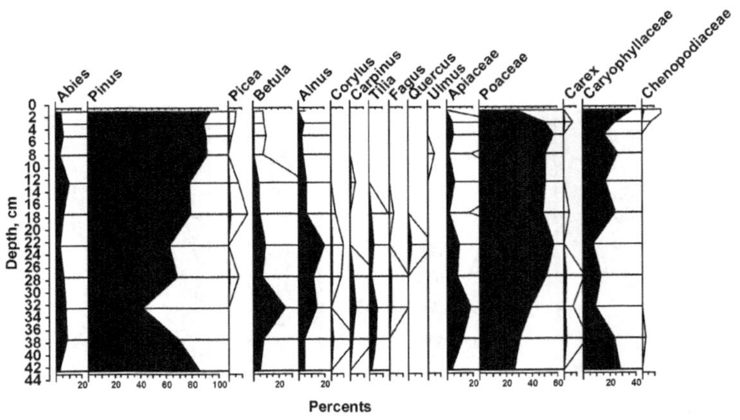

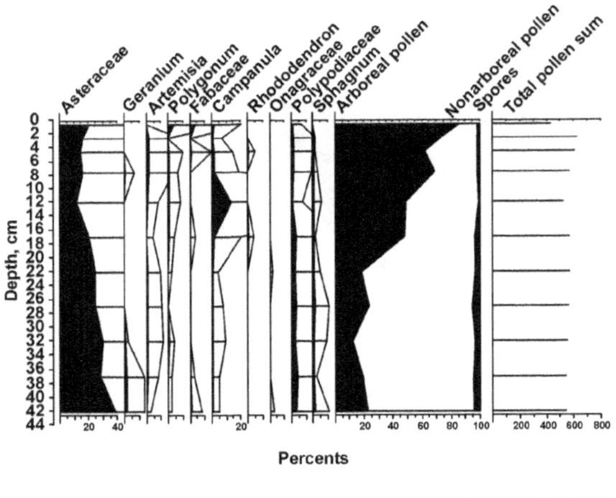

Figure 5.44 Pollen percentages and total pollen counts from GHM-Upper soil profile, Mt. Khatipara alpine station, Russia. Stippled area represents ×10 exaggeration for rare taxa. Total pollen sum represents the total number of pollen grains counted. Redrawn with modifications from Pavlova & Onipchenko (1992)

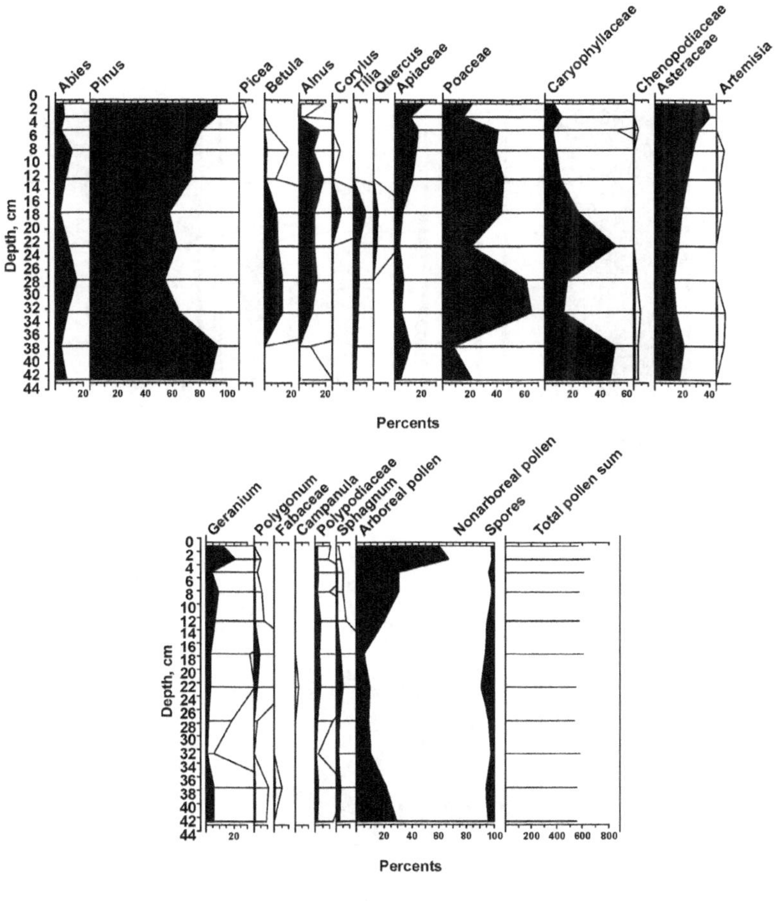

Figure 5.45 Pollen percentages and total pollen counts from GHM-Lower soil profile, Mt. Khatipara alpine station, Russia. Stippled area represents ×10 exaggeration for rare taxa. Total pollen sum represents the total number of pollen grains counted. Redrawn with modifications from Pavlova & Onipchenko (1992)

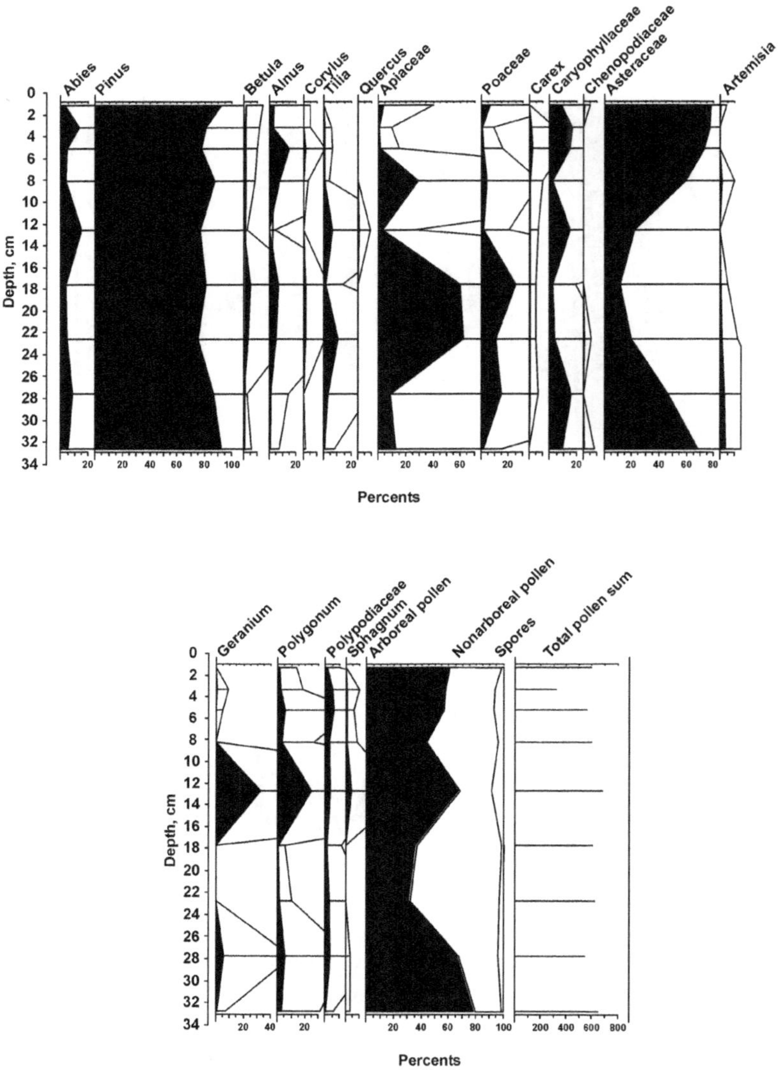

Figure 5.46 Pollen percentages and total pollen counts from SBC-Upper soil profile, Mt. Khatipara alpine station, Russia. Stippled area represents $\times 10$ exaggeration for rare taxa. Total pollen sum represents the total number of pollen grains counted. Redrawn with modifications from Pavlova & Onipchenko (1992)

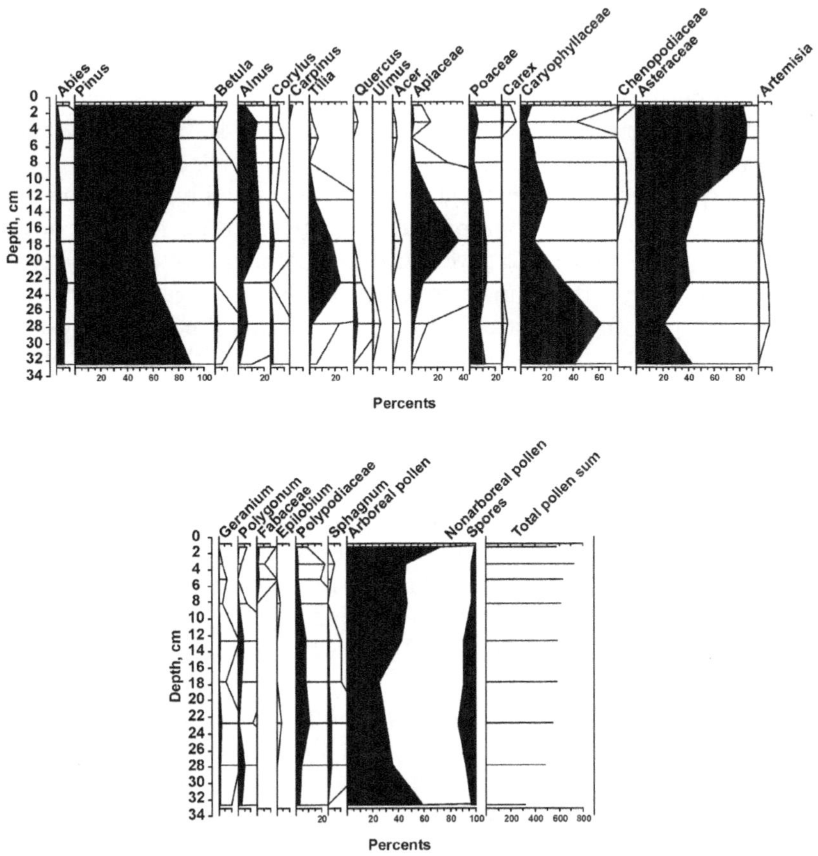

Figure 5.47 Pollen percentages and total pollen counts from SBC-Lower soil profile, Mt. Khatipara alpine station, Russia. Stippled area represents ×10 exaggeration for rare taxa. Total pollen sum represents the total number of pollen grains counted. Redrawn with modifications from Pavlova & Onipchenko (1992)

5.7.4. Phytolith analysis: results and discussion

5.7.4.1. Modern phytolith data

Our phytolith palaeodata from four of the same profiles as used for the pollen study were reported earlier (Blinnikov 1994), but a more comprehensive modern analogue study (Volkova et al. 1995) made us to adjust our interpretation of palaeovegetaion reconstructed from phytolith spectra. Presumably, phytoliths provide us with a more local record than pollen (Piperno 1988). E.g., we did not encounter any tree phytoliths in any of the phytolith samples collected in the same general area as the pollen samples with the treeline being a few hundred meters down the slope. Since our original work was done, a wealth of new phytolith research has taken place (Runge 1998, Meunier & Colin 2001) making it necessary to revise some of our old terminology of morphotypes.

Despite many attempts at developing phytolith nomenclature (Twiss et al. 1969, Brown 1984, Piperno 1988, Mulholland 1989), there is as yet no uniform nomenclature code, although some attempts are being undertaken by the Society for Phytolith Research. We designed our own classification system, modified from Twiss et al. (1969) and Fredlund & Tieszen (1994) in describing Poaceae (grass) morphotypes, and followed Piperno (1988) and Bozarth (1992) in describing non-Poaceae morphotypes.

Twenty-seven morphotypes were distinguished in Blinnikov (1994), but in the present chapter we only use the aggregate data recognizing only 8 morphotype groups which can all be easily distinguished in the alpine soils: (i) rondels (formerly known as "hats", types 1 – 4 and 12 in Blinnikov 1994); (ii) straight-edge plates (types 5 – 8, 10 ibid); (iii) wavy-edged plates (types 9, 11, 13 – 16 ibid); (iv) long cells (17 – 18); (v) *Nardus* phytolith (branched rod type, #19 ibid); (vi) *Carex* phytoliths (cone-shaped #27 ibid); (vii) trichomes (#20 – 23 ibid) and (viii) other forms (Fig. 5.48). In this section we provide the phytolith data recalculated based on this classification system, therefore, they differ slightly from the previously published diagrams (Blinnikov 1994). The community identification was based on the published phytolith key (Volkova et al. 1995).

It is important to recognize the tremendous potential of silica phytoliths that can provide valuable information about past communities even outside of the grasslands having "classic" Festucoid, Panicoid and Chloridoid phytoliths (Twiss et al. 1969). All our communities contain 100% Festucoid grasses, all of which are C_3 species.

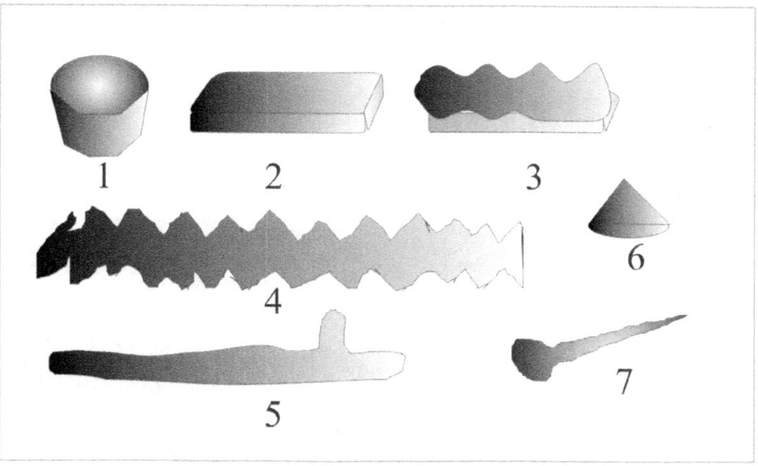

Figure 5.48. Phytolith morphotype groups recognised in this study: 1 rondels, 2 - straight-edged plates, 3 - wavy-edged plates, 4 - long cells, 5 - *Nardus* phytolith (branched rod type), 6 - *Carex* phytoliths (cone-shaped), 7 - trichomes

5.7.4.2. ALH community phytoliths

Only the ALH-Upper profile was analysed for phytoliths, as a more typical location within the ALH community (Fig. 5.49). The phytoliths were much more frequent in the silt (10 – 50 m) than in the fine sand fraction (50 – 250 m), and only the former is considered here.

In the context of the aggrading stable soils at our research sites (see 5.7.2 above) we assume that the changing frequencies of phytoliths in the soil horizons Z4. reflect changes in vegetation composition. This has been demonstrated to be the case in the aggrading loess palaeosols of the Columbia river basin (Blinnikov et al. 2002). In Teberda, however, we are dealing with unburied modern soils, and in case of GHM – a community with a high rate of burrowing activity. Nevertheless, there are obvious differences in terms of phytolith frequencies among the samples from different depths. Based on the phytolith key of Volkova et al. (1995), the top sample (0 – 2 cm) and the two near the bottom may be tentatively assigned to ALH. These samples have a moderate proportion of rondels (16 – 35%), straight plates (18 – 25) and a particularly high proportion of wavy plates (26 – 30%), while they lack diagnostic branched-rod phytoliths of *Nardus*, common in *Geranium* meadows or snowbeds. The rondels in these samples have been most likely contributed by *Festuca ovina*, while wavy plates may have come from *Helictotrichon versicolor*. The presence of small amounts of sedge phytoliths and trichomes is noteworthy, because *Carex* and trichome-producing species (e.g., *Antennaria dioica*) are common in the heaths.

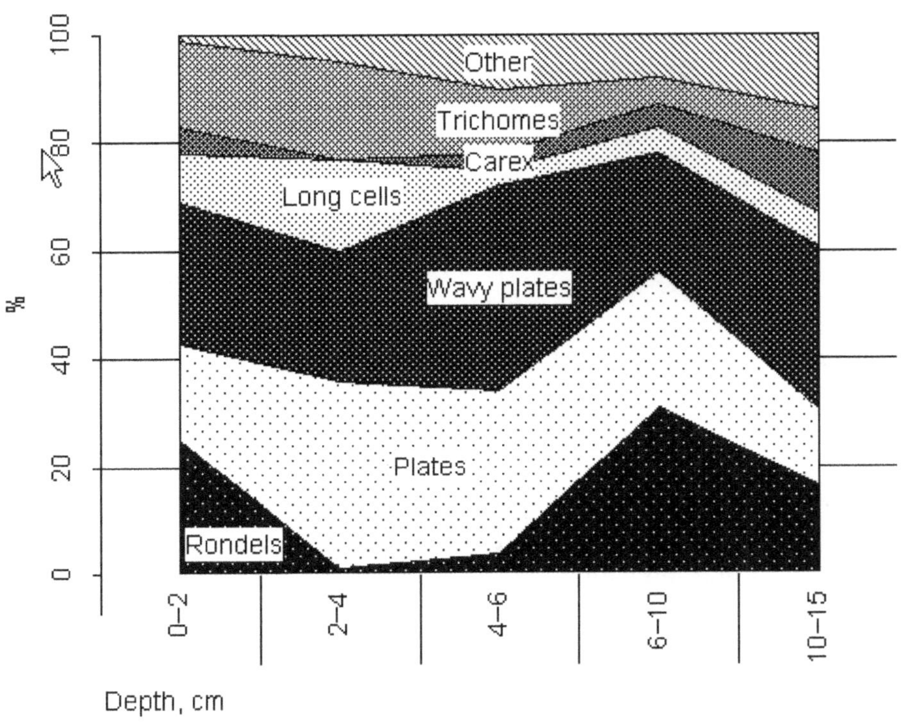

Figure 5.49. Phytolith diagram of the ALH-Upper profile.

The very low rondel proportion in the other two samples is hard to explain and these samples do not have a clear modern analogue in our database, which also includes alpine bog and subalpine meadow and pine forest phytolith samples. Typical surface assemblages from our area today all have at least 20% or more of rondels, which are particularly frequent in *Festuca* genus (80 – 90.5% of all phytoliths produced by fescues in our area are rondels) but can also be found in many other Festucoid grasses. Lack of rondels, which are very durable phytoliths, cannot be explained by variation in preservation rate either. Based on our observations elsewhere, rondels are the least likely to be dissolved in the soil profile (confirmed also by Fredlund & Tieszen 1994).

The original interpretation (Blinnikov 1994) of the decrease in rondels at moderate depths in ALH was that it was a period when grasses providing rondels played a smaller role than composites providing trichomes, implying cooler wetter conditions than today. Tushinskii (1962) describes a "vernau" glaciation stage (ca. 500 – 800 cal BP) from the study area that can be correlated with the Little Ice age when the conditions perhaps were Z4. cooler and wetter. The only available date from the ALH-U profile comes from below this layer at 10 cm depth with the age estimated at 2610±80 ^{14}C yrs BP (IGAN), but the calendar age is likely at least 2000 years older.

Our conclusion is that the ALH had a rather stable history based on phytolith analysis with the conditions being somewhat more conducive to the growth of Asteraceae and less conducive to the growth of grasses, in particular, *Festuca*. This is not directly supported by the pollen data from the same two depths, however (Fig. 5.40). The pollen analysis conclusion that ALH has remained a rather steady community is supported by phytolith analysis (i.e., no phytolith samples could be interpreted as indicative of any other modern alpine community from the research area, except ALH).

5.7.4.3. FVG community phytoliths

The FVG community produces very abundant modern phytoliths with rondels always representing >50% based on the modern study from our area (Volkova et al. 1995). The FVG-Upper soil profile (Fig. 5.50, see Fig. 5.41 for the pollen diagram from the same profile) contains over 50% of rondels in every sample. Very significantly, however, the 3 samples between 2 and 8 cm of depth contain branched rods, diagnostic of *Nardus stricta*. This species increases in response to grazing in the modern alpine communities of the Caucasus and may indicate a period of higher sheep grazing in the past. The research area has been free from livestock grazing since at least the 1940s. Savinetskii (1992) provides independent evidence for an abrupt increase in domesticated animals grazing in the Caucasus at ca. 2000 BP.

Decrease in rondels, increase in wavy forms and appearance of branched rods between 2 and 8 cm depth in the FVG-Upper profile has been interpreted as evidence of a community type more similar to modern-day GHM (mesophytic meadows with *Nardus stricta*). Pollen analysis (Fig. 5.41) suggests that *Geranium* Z4. was present at the site in the same period.

Thus, it is very probable that the FVG community has always had a large share of *Festuca varia* present, but more recently it went through a stage resembling modern-day *Geranium* meadows with a greater share of forbs.

5.7.4.4. GHM community phytoliths

The top of the GHM-Lower profile looks similar to the present-day GHM spectra, but the layer immediately below suggests a community intermediate between a FVG grassland and a GHM meadow (Fig. 5.51). This would fit the interpretation of the pollen diagram for the site (Fig. 5.45). Further down the profile the community looks more like a GHM meadow again, although the diagnostic branched rods of *Nardus* disappear entirely below 6 cm of depth. *Nardus* again likely indicates a recent period (since a few centuries) of severe grazing.

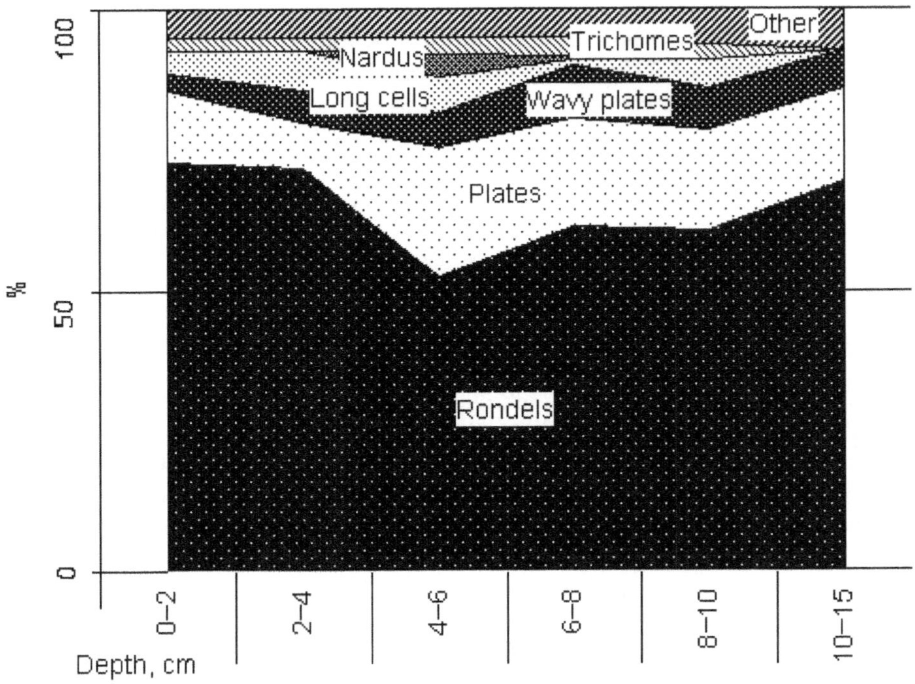

Figure 5.50. Phytolith diagram of the FVG-Upper profile

The bottom portion would match modern-day ALH heaths better than a modern GHM, but these two communities can largely be told apart based on the presence/absence of branched rods of *Nardus* according to the key (Volkova et al. 1995). Our interpretation (Blinnikov 1994) suggests that these were hypothetical "parent" communities intermediate in character between ALH and GHM today, which definitely would imply colder and drier conditions. It is important to remember also that due to the high burrowing activity of voles in GHM, there spectra are the least reliable.

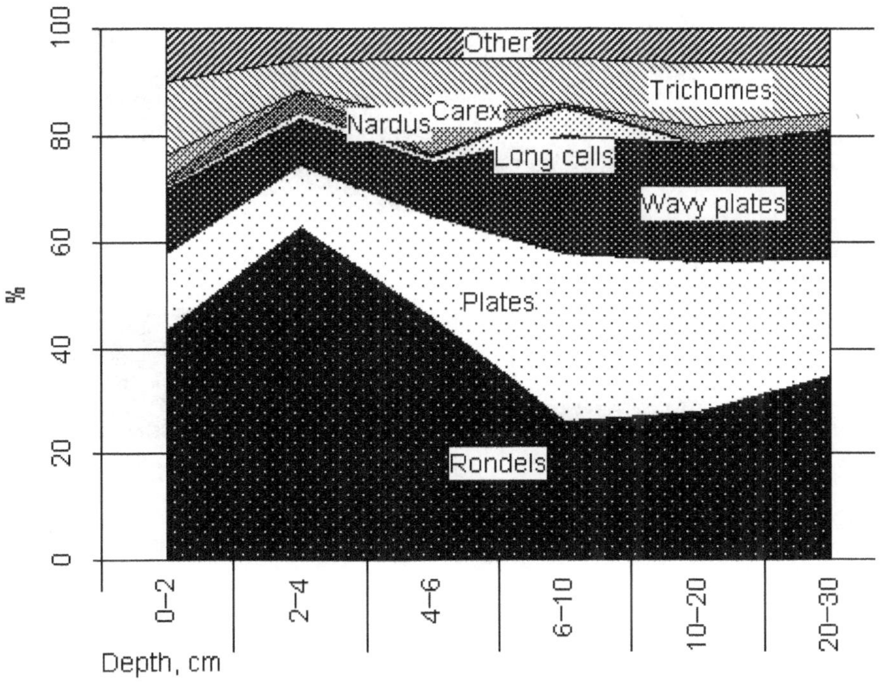

Figure 5.51. Phytolith diagram of the GHM-Lower profile

5.7.4.5. SBC community phytoliths

The deepest two samples from the SBC-Lower profile (Fig. 5.52) differ strongly from the upper layers. All the upper layers were interpreted as "classic" snowbeds due to the low prevalence of rondels and strong presence of *Nardus* rods. The lower samples were interpreted as a non-analogue community similar in character to the "parent" forb-grassland as the one described for the bottom portion of GHM (see above). This community likely existed in the area during the drier stage (2nd xerothermic) of Tushinskii (1962) ca. 2,000 – 2,500 cal BP. Preservation of phytoliths was adequate throughout the profiles, and we could easily tell most of the forms apart.

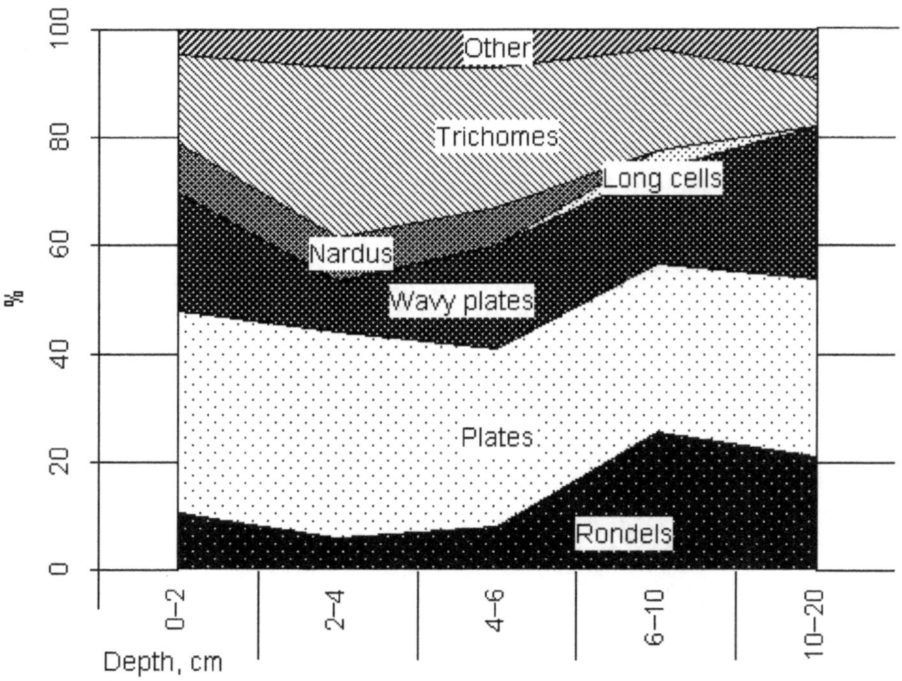

Figure 5.52. Phytolith diagram of the SBC-Lower profile

5.7.5. Conclusions

Application of phytolith and pollen analysis together can provide additional insight into the reconstruction of vegetation as was demonstrated here and also in the Columbia Basin of Washington State (Blinnikov et al. 2001). Pollen and phytolith results broadly agree on the interpretation of the communities' histories during the mid- to late-Holocene. The ALH communities have not changed much over this period. The ALH sites certainly never supported meadows or forested vegetation. Phytolith data suggests that part of the community history was a wetter period a few hundred years ago perhaps coinciding with the Little Ice Age. The FVG community likewise has been rather stable, in particular, the more typical and drier upper site. *Festuca* has always played an important role in the community. The ptrsence of *Nardus stricta* in the recent past suggests a period of more intensive grazing starting ca. 2000 BP, which agrees with data from other parts of the Caucasus. That was also a wetter period, again, perhaps, reflecting the Little Ice Age, or a preceding similar episode. Interestingly, the GHM community has had a xeric episode in its history and generally appears to have emerged relatively recently (the past few hundred years) from a hypothetical non-analog "parent" community that shared some features with present-day lichen heaths. The snowbed sites (SBC) have supported a snowbed community for the past few thousand years, but previously supported a non-analog, more xeric community similar to the "parent" GHM described above.

Chapter 6

Population biology of alpine plants

6.1. Age structure and population adaptations in three non-clonal alpine perennial plants

6.1.1. Introduction

Extreme climate conditions (i.e., a cold winter and a short summer with significant fluctuations in temperature between day and night) is the main stress factor in alpine habitats. Consequently, alpine plants must adapt their age population structure and growth strategy to the fluctuating severe environment. In general, the aim of alpine plants' adaptations is to minimise the risk of a sudden local extinction under more or less continuously severe external conditions.

In this chapter we explore possible mechanisms of adaptations at the population level of three non-clonal alpine species *Anemone speciosa* Adams ex G. Pritz. (Ranunculaceae), *Campanula tridentata* Schreb. (Campanulaceae) and *Carum caucasicum* (Bieb.) Boiss. (Apiaceae) (Fig. 6.1).

We used the concept of discreet phases in plant populations' development (Rabotnov 1950, 1964, Uranov 1975, Uranov et al. 1977, Gatzuk et al. 1980, White 1985, Zaugolnova et al. 1988) to estimate life span. The life cycle (ontogeny) of a plant may be described as a sequence of discreet developmental stages. Each stage may be characterized by physical time (years, months, etc.) and by "biological time" inherent to the organism's life cycle (Robbins 1957). Division of the continuous

V.G. Onipchenko (ed.), Alpine Ecosystems in the Northwest Caucasus, 181–222.

ontogenetic process into developmental stages allows to obtain a coherent scale of biological time for different species with different life spans.

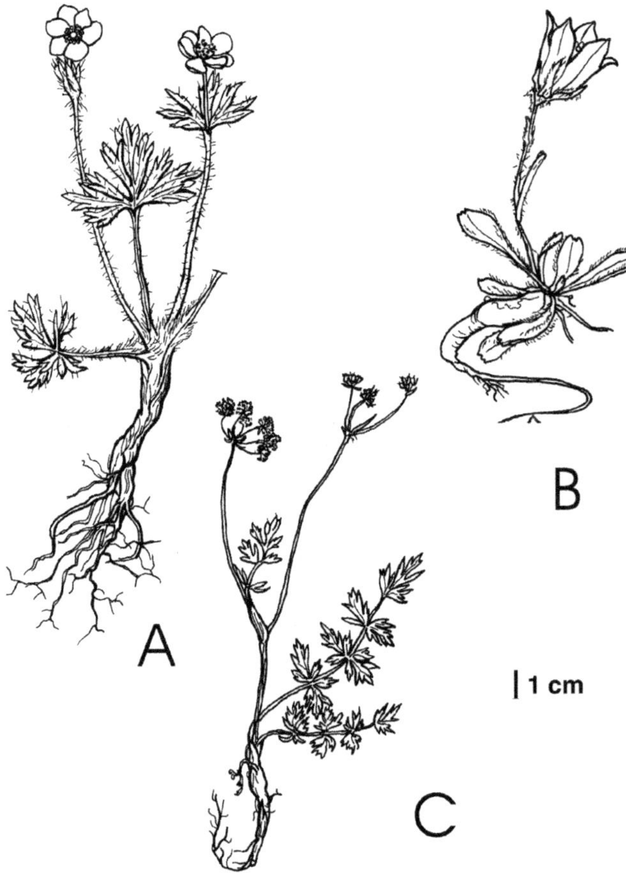

Figure 6.1. Studied plant species (generative stage) A - *Anemone speciosa,* B – *Campanula tridentata* C. - *Carum caucasicum* (Figure by E.Semenova).

Continuous processes of individual plant development may be subdivided into several stages based of structural indicators that reflect functional importance. There are the presence vs. absence of embryonic, juvenile or mature morphological features; the ability of an individual to reproduce sexually or to propagate vegetatively; the ratio between live and dead matter as well as growing and non-growing plant parts.

The ontogeny of a plant organism is frequently subdivided into four periods and nine stages (Rabotnov 1950, Uranov 1975): 1. latent period (seed); 2. pre-reproductive period (seedling, juvenile, immature, virginal); 3.

reproductive period (young, mature, and old reproductive); 4. post-reproductive (senile) period.

The actual number and sequence of stages may differ from the list above. The main result of such a description is that we can mark the development of the plant as a sequence of age stages and describe a population as a set of individuals belonging to certain age stages with their corresponding distribution, so called, "age spectrum" (Sizov et al. 1999a,b).

The splitting of the whole ontogenesis into a few age stages gives us the possibility of using the proportion of individuals recorded at each stage at each time period as a vector which describes the internal age-stage structure of the population for analysing its inner dynamics and conditions of population stability. We use a simple matrix model for the analysis of the population dynamics and, most importantly, for the analysis of the sensitivity of different parameters which describe the dynamics of changing age-stages and the role of different developmental stages of the plant population in the maintenance of stability conditions.

6.1.2. Materials and methods

Data were collected in the alpine zone of Mount Malaya Khatipara, Teberda Nature Reserve. Field observations were carried out in the alpine lichen heath (ALH), which was described in detail in the preceding chapters.

Three non clonal typical alpine species were chosen as the objects of our study, namely *A. speciosa, C. tridentata*, and *C. caucasicum*. Individual marked specimens of the species were observed on permanent plots (200 x 25 cm) from 1980 to 1999. Observations were made annually at the end of August. Presence of flowers and ontogenetic stage (juvenile, vegetative, generative) of the specimens were recorded. All individuals from permanent plots were monitored individually. Field investigations were made to minimize external disturbances, e.g., we did not attempt digging to check the root system development. For the purpose of this study, plants that had flowers before the start of the observations but did not flower during the observation period are classified as vegetative.

Four age stages were monitored: seedlings (with existing cotyledons), juvenile (with juvenile leaves), vegetative (with adult leaves but without generative shoots), and generative (with generative shoots).

The total numbers of marked generative plants were 168, 335 and 1029 for *A.speciosa, C.tridentata* and *C.caucasicum* respectively. Since no excavations were made, juveniles could also include some small old plants. The number of ripened seeds was counted for every generative *A. speciosa* and *C. caucasicum*. If an individual plant was missing in a certain year, it was recorded as "conditionally dead". If this plant was found during the next year, it was then marked as a plant which had a "secondary dormant" age-state.

A single rosette was considered an individual plant. Two - three percent of the adult *C. caucasicum* population may have 2-3 rosettes per single root system, but most have only one. Mature individuals of *C. tridentata* may have 2-3 rosettes per single root system in 30% of the cases. This should be considered in the discussion of the results.

6.1.3. Age-state composition

We can see the dynamics of the age-state composition during the period of investigation in figures 6.2-6.4.

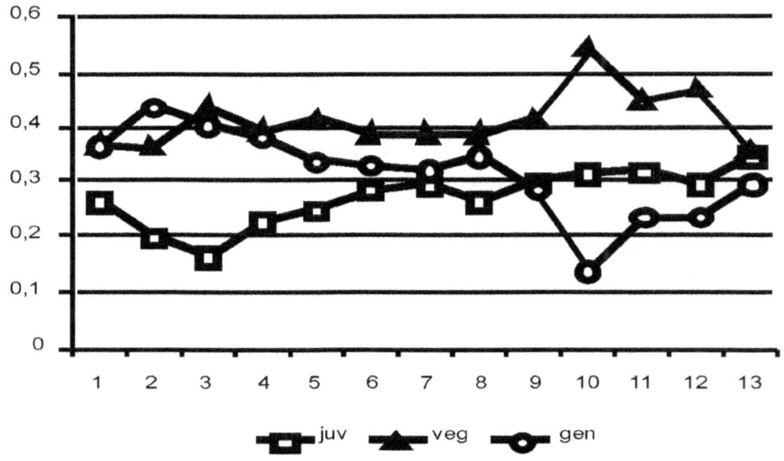

Figure 6.2. Age-state composition of *Anemone speciosa* populations during the first 13 years of investigation (juv – juvenile, veg – vegetative, gen – generative). Horizontal axis corresponds to the years of investigation, vertical axis – proportion of the total number of plants.

The amount of vegetative and generative plants remains approximately constant. It can be easily seen that the numbers of vegetative and generative individuals seem to be fluctuating in synchrony with each other, which is probably due to the climatic differences among the years. On average, the highest proportion of juveniles for the whole observational period was found for *C. caucasicum* (0.33), the smallest for *C. tridentata* (0.24), with *A. speciosa* having the intermediate value (0.28).

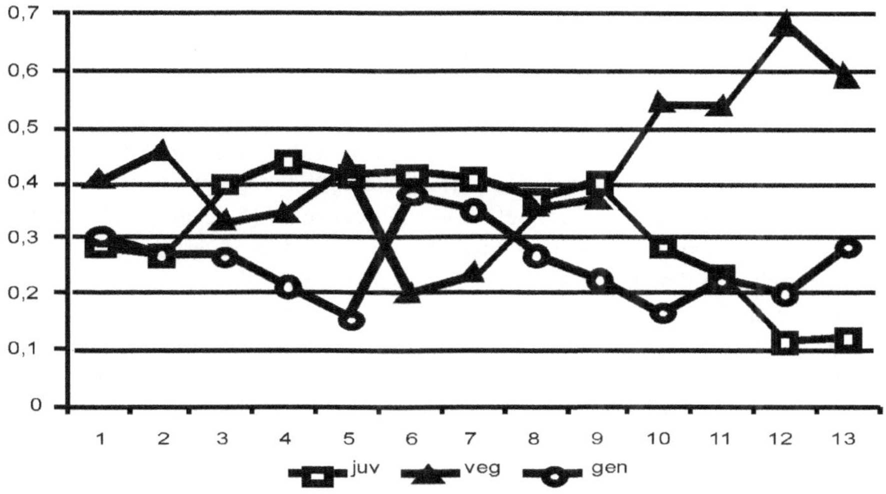

Figure 6.3. Age-state composition of *Campanula tridentata* populations during the first 13 years of investigation. See Fig. 6.2 for abbreviations.

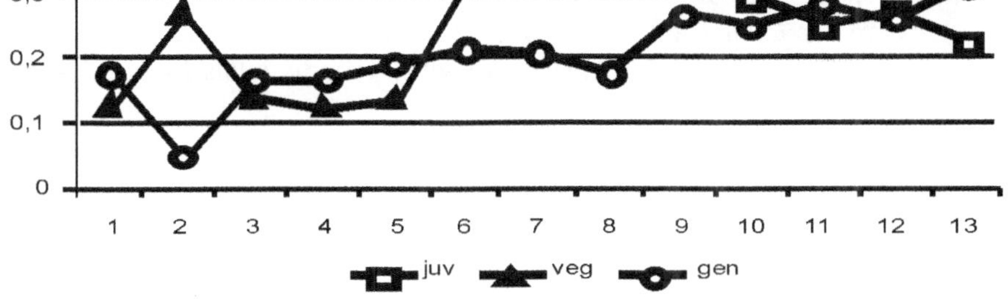

Figure 6.4. Age-state composition of *Carum caucasicum* populations during the first 13 years of investigation. See Fig. 6.2 for abbreviations.

6.1.4. Life spans of age stages

We calculated life spans of age stages based on established life tables (Sizov et al. 1999b). Since field investigations were made without observing the root system, we knew nothing about the history of the observed individuals prior to the beginning of the investigation. At the same time, many plants have continued their life cycle upon the conclusion of the study. Therefore, our data are both left- and right-censored (Nelson 1982). To avoid left censoring we assumed that the start of the investigation was the beginning of that age stage in which the plant was registered at the beginning of the study. For processing of the right censored data we use the approach proposed by Chiang (1984). It is based on the assumption that the probability of death or transition into the next stage in a cohort is constant after the observations are completed. We believe that Chiang's method is useful in situations where no data on the distribution of death or transition probability after the period of observations are available.

6.1.4.1. Juvenile stage

The transition of juvenile individuals into the vegetative state is an important stage in their life because it means a considerable increase in the likelihood of survival. However, the life span of the juvenile stage of individuals varies. We suppose that the variability in microhabitats and annual weather fluctuations determine the variability. Such non-synchronous transitions from one age-state to another and its likely role in the population dynamics were previously discussed by Zhukova (1986) and Zhukova & Komarov (1990).

Figure 6.5 shows that approximately 50% of all juvenile plants of the studied species attain the vegetative state within the first five years. However, about 10% of individuals turn into the vegetative state only after 10 years, and some individuals of *A. speciosa* stay in the juvenile state for more than 15 years. The life spans of juvenile individuals for the 3 species are shown in Table 6.1.

Table 6.1. Life span of juvenile stage. *e* – expectation of juvenile stage continuance (in years), *se* – standard error, *n* – number of individuals

species	*e*	*se*	*n*
Anemone speciosa	9	0.34	50
Campanula tridentata	4.6	0.04	81
Carum caucasicum	4.7	0.02	158

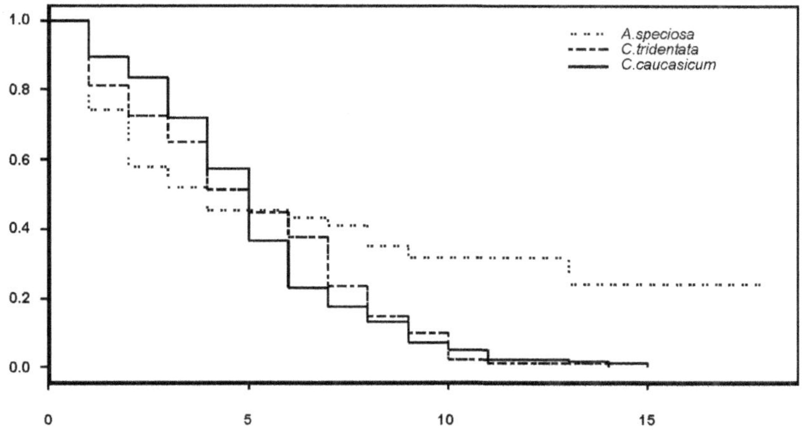

Figure 6.5. The curves of transition of juvenile individuals into the vegetative stage. Vertical axis – the proportion of individuals still staying in the juvenile state, horizontal axis – time (in years)

6.1.4.2. Vegetative stage

On reaching the vegetative stage an individual prepares for the production of generative shoots. The distribution of the length of the time period between the beginning of the vegetative state and first flowering is shown in Figure 6.6.

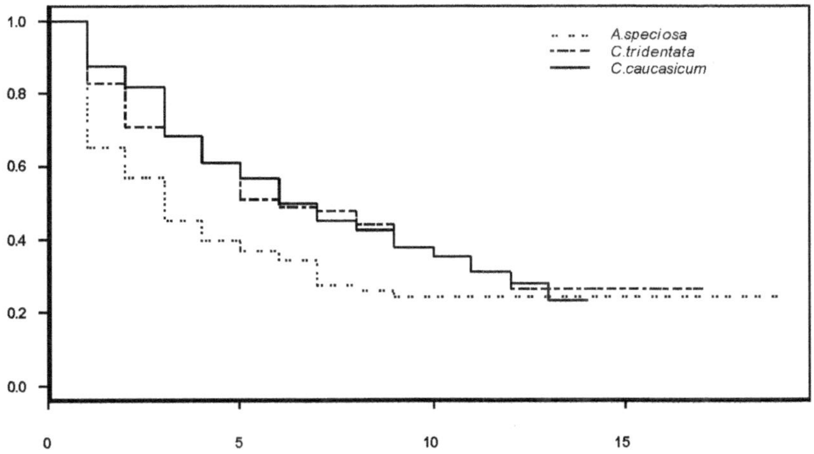

Figure 6.6. The curves of transition of vegetative individuals into the generative stage. Vertical axis – the proportion of vegetative individuals that have never been flowering yet, horizontal axis – time (in years).

The transition from the vegetative into the generative stage takes longer than the one between the juvenile and the vegetative stage. Whereas some individuals have already started to flower within the year after reaching the vegetative state, at least 20% of the vegetative individuals of the studied species did not flower for at least 10 years. Apparently, the start of flowering requires considerable energy, and differences in microhabitat conditions may also have a significant impact on the rate of the transition into the generative state.

The average length of the vegetative stage was approximately equal for all species, about 10 years (see Table 6.2.).

Table 6.2. Life span of the vegetative stage. *e* – expectation of vegetative stage continuance (in years), *se* – standard error, *n* – the number of individuals

species	*e*	*se*	*n*
Anemone speciosa	10	0.23	86
Campanula tridentata	11.3	0.02	103
Carum caucasicum	9.2	0.13	165

6.1.4.3. Generative stage

Increase of the life span is one of possible adaptations to the severe alpine environment. It allows to avoid the consequences of a low level of natural regeneration, which may be fatal for population. Our data suggest that the generative stage in *C. tridentata* and *C. caucasicum* lasts for at least a few decades and for *A. speciosa* – a few hundred years (Table 6.3). We assume that the generative stage ends sometime between the episode of the first flowering and the moment of death. For plants that survived until the end of the investigation the probability of death was taken as a constant. Using this assumption we can use Chiang's formulae (Sizov et al. 1999a).

Table 6.3. Life span of the generative stage. *e* – expectation of generative stage continuance (in years), *se* – standard error, *n* – the number of individuals

	e	*se*	*n*
Anemone speciosa	372	40.8	74
Campanula tridentata	46	2.2	62
Carum caucasicum	56	3.7	82

6.1.4.4. Life span of the studied species

Under the extreme conditions of the alpine zone a prolonged period of unfavourable conditions may easily prove fatal to a population. An increase in the life span in general and the generative period in particular is an adaptation to preclude such casual extinction. According to our data the expectation of the life duration is equal to 390 years for *A. speciosa*, 65 for *C. tridentata* and 70 for *C. caucasicum*. However, the generative period of individuals of the species can vary considerably. For example, if the final distribution of life spans belongs to the Gaussian type, it is easy to calculate that for each 1000 plants 10 of them will have generative periods lasting about 87, 123 and 1401 years for *C. caucasicum*, *C. tridentata* and *A. speciosa*, respectively.

6.1.5. Frequency of flowering

The breaks in flowering are frequent in the studied species, and is largely due to weather fluctuations. Generative plants of *A. speciosa*, *C. tridentata* and *C. caucasicum* are flowering on average every 2.3, 1.8 and 1.6 years. On the basis of the mean annual seed production (estimated at 5 seeds per individual for *A. speciosa*, 25 for *C. tridentata* and 37 for *C. caucasicum*) (Onipchenko & Komarov 1997), we can determine that adult individuals of these species produce approximately equal numbers of seeds during their life time (740 for *A. speciosa*, 800 for *C. tridentata* and 1110 for *C. caucasicum*).

We have already mentioned that the duration of the vegetative and generative stages for all species are stretched in time. It is reasonable to assume that the life span of the generative stage also varies. We can divide all generative individuals into two groups: one with individuals that produce generative shoots on average every 1-2 years and another with rarely blossoming individuals, producing generative shoots on average less frequently than once in every two years. The comparison of survival curves for these groups shows considerable differences for *A. speciosa* and C. *caucasicum* (Table 6.4), with groups of seldom-flowering individuals having a higher mortality level. Evidently, there are two possible explanations of this fact. Individuals of these species either have different viabilities and, consequently, more tolerance to the unfavourable environment, or there are many senile individuals.

Table 6.4. Comparison of survival curves of seldom and frequently flowering individuals of the studied species on the base of the Mantel-Haenszel test (Fleming & Harrington 1981). Symbols: χ^2 – value of chi-square statistics, μ – degrees of freedom, P – p-value (significance level). The calculations were made with using the statistical package S-Plus 4.5.

A. speciosa			*C. caucasicum*			*C. tridentata*		
χ^2	μ	P	χ^2	μ	P	χ^2	μ	P
3.4	1	0.065	10.5	1	0.001	0.1	1	0.73

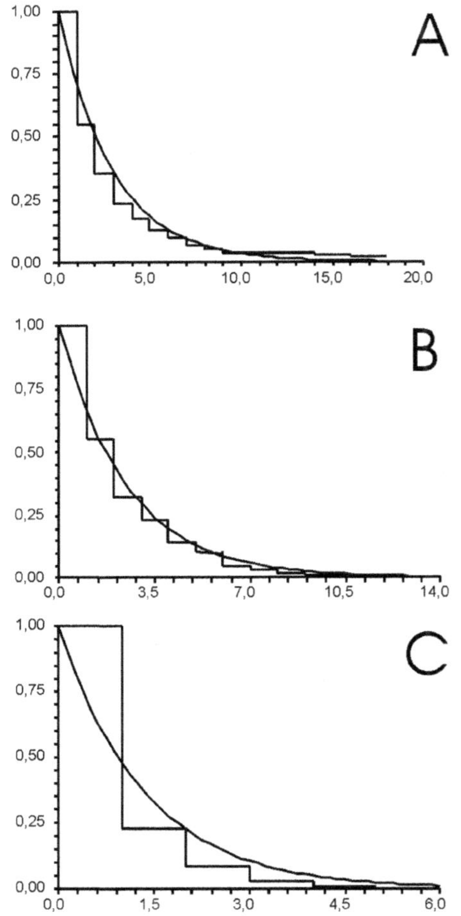

Figure 6.7. The distribution of uninterrupted flowering periods of single plants during the study. Horizontal axis – time (in years), vertical axis – the proportion of periods in which individuals keep on flowering annually. The stepped line was constructed on the base of observation data, the smooth one corresponds to a theoretical exponential distribution. A – *A.speciosa*, B – *C.tridentata*, C. – *C.caucasicum*.

The breaks in flowering are observed because the conditions in which these species are found seem to be subject to permanent fluctuations. The distribution of continuous intervals of flowering of single plants in the population for all species may be approximately represented as an exponential distribution (Fig. 6.7).

According to reliability theory (Cox & Oakes 1984) this exponential distribution appears when we have a lot of independent time variables, with the random variable being equal to the smallest of them. Apparently the flowering individuals keep producing inflorescences as long as all the required favourable conditions are met: e.g., good weather conditions, relatively weak competition of nearest neighbours, absence of disturbances, auspicious microhabitat, etc. As soon as any condition is no longer met, the plant interrupts blossoming until all needed conditions are met again.

We do not observe a good fit of the exponential distribution against the empirical data, apparently because the adults of the studied species, as mentioned above, can consist of two groups of individuals: seldom and frequently blossoming ones. We could expect that the conformity will be better in each separate group. Nevertheless, this was not the case, even when populations were split into given groups based on an arbitrary threshold in the characteristic. Perhaps, for different species we need to take different values. Moreover, there may exist more than two groups. On the other hand, one of the consequences of the exponential distribution must be the independence of the lengths of the flowering periods from one another or, in other words, we can not predict the length of the nearest continuous interval of flowering for an individual despite the fact that we know how it has blossomed in the past and present. If this is correct then the given property should be kept in all groups of individuals and thus the problem of sample heterogeneity may be solved. We have calculated the Kendall's correlation coefficient for lengths of two adjacent flowering intervals of individuals. The results, which are given in the Table 6.5, confirm our assumption on their independence.

Table 6.5. Rank-based correlation coefficient of Kendall (τ) for lengths of two adjacent flowering periods of individuals. Symbols: P – p-value (significance level) under the one-sided alternative hypothesis, N – sample size (the number of compared pairs of values).

Species	τ	P	N
Anemone speciosa	-0.097	0.058	79
Campanula tridentata	0.038	0.294	68
Carum caucasicum	-0.104	0.059	67

To summarize, the analysis has revealed the following properties of dynamics of adult plants in the studied species:

1. the amount of seeds produced by one individual during its life time is approximately equal for all studied species;

2. the adult individuals of *A. speciosa* and *C. caucasicum* fit into two groups: seldom and frequently flowering plants, with seldom flowering individuals having higher mortality levels;

3. the distribution of continuous intervals of flowering can be approximated by the exponential distribution, and this result suggests that the lengths of flowering intervals are independent from one another.

6.1.6. Analysis of the populations' dynamics using matrix models

Population dynamics can be simulated by using matrix models. In such models, the total population's size is a discrete value considered at discreet moments in time. Such a procedure corresponds to the actual population's census, which is carried out, as a rule, not continuously. Use of matrix models for plant population analysis seems to be a rather popular method, and many papers reflect different peculiarities of this modelling approach (Caswell 1982a,b, 1986, Law 1983, Manders 1987, Sarukhan & Gadgil 1974, etc.). We tried to apply this method for analyzing plant populations being split into different age-state groups. We used the approach of Lefkovitch (1965), who has used the extended variant of matrices of Lewis (1942) and Leslie (1945), where instead of the population's division into equal age groups division into present stage groups was used.

We calculated the transition matrix using experimental data for three main age-state groups: juvenile, vegetative and generative individuals. For each year we knew the proportion of plants in each age-stage. A set of each transition probabilities a_{ij} for these stages measured for each year was treated as a sample from the Gaussian random variable. Its means and standard deviations are shown in Table 6.6.

Table 6.6. Means and standard deviations for transition probabilities for all species (Stages: first column – transition **from** the stage, row – transition **to** the stage: sd – seedling, j – juvenile, v – vegetative, g – generative stages).

	Mean probability				Standard deviations			
Stages	sd	j	v	g	sd	j	v	g
				Anemone speciosa				
sd	0	0.667	0	0	0	0.25	0	0
j	0	0.786	0.127	0.006	0	0.021	0.017	0
v	0	0.031	0.733	0.234	0	0.001	0.008	0.008
g	0	0	0.304	0.696	0	0	0.013	0.013
				Campanula tridentata				
sd	0	0.411	0	0	0	0.24	0	0
j	0	0.8	0.095	0.003	0	0.019	0.012	0
v	0	0.026	0.668	0.291	0	0.001	0.017	0.015
g	0	0	0.361	0.63	0	0	0.016	0.016
				Carum caucasicum				
sd	0	0.447	0	0	0	0.197	0	0
j	0	0.813	0.09	0.003	0	0.004	0.003	0
v	0	0.103	0.562	0.303	0	0.004	0.037	0.032
g	0	0.016	0.29	0.67	0	0.001	0.026	0.022

It should be noted that almost all transition probabilities have small variances. The exception is only for the transition from the seedling to the juvenile stage. Evidently, this fact reflects the significant dependency on the microhabitat conditions for young small plants.

We fixed all coefficients in the model at their average values except one coefficient whose sensitivity we tested using its simulation as a random variable chosen from the Gauss distribution with the parameters from Table 6.5. We repeated this procedure 10 times to evaluate the variability of possible dynamic trajectories of our matrix model. It should be especially noted that we were not simulating strict population dynamics because for strict simulation it is necessary to take into account the limitation of available space for a plant. This cannot be done in the matrix model. Therefore, we can analyze the parameters' sensitivity only. Our results of the computer experiment are shown in Table 6.7.

Table 6.7. The results of simulation of the numbers of individuals on three species: *A. speciosa, C. tridentata* and *C. caucasicum;* the numbers estimated from the analysis of sensitivity of the transition's coefficients are used in the simulations as random variables chosen from the Gauss distribution. Standard deviations on the 100th steps of simulation in 10 runs are shown for different transitions (sd – seedling, j – juvenile, v – vegetative, g – generative stages).

		Standard deviation		
	transition	sd → j	j → v	v → g
Anemone speciosa		391	461	361
Campanula tridentata		12280	8301	27416
Carum caucasicum		96568	33210	119907

On the basis of the experiment we can determine that the transition from the juvenile to the vegetative state is the most critical for *C. caucasicum* and *C. tridentata* populations. As follows from the Table 6.6, the transition *j→v* of these species has the smallest variance, i.e. the most "rigid" one. While the occurrence of generative shoots (transitions *v→g, g→v*), seemingly, is determined by annual environment fluctuations, the proportion of individuals turned from the juvenile into the vegetative state is a stable variable and, most likely, depends on structural characteristics of a community. Numbers of these species have also the smallest variation as a result of casual fluctuations of the transition *j→v* (Table 6.7). At the same time, all transitions of *A. speciosa* have a rather small variance. This is quite natural, since one should expect a low rate of population renewal in this long-lived species, and consequently insignificant variability of its demographic parameters.

6.1.7. Conclusions

According to Fig. 6.2-6.4., populations of the 3 studied species, apparently, experience periodic oscillations in numbers in different age stages. However, the combined proportion of generative to vegetative plants remains approximately constant. In spite of well known fluctuations and periodical phenomena in plant population dynamics (Uranov 1975, 1977, Rabotnov 1975a), it seems doubtless that the mechanisms of these relatively complex phenomena result from the interaction of various processes.

The results obtained permit us to make some conclusions about the population dynamics of the studied species.

1. The total amount of seeds produced by an adult plant during its life time is approximately the same for all studied species.

2. Transitions from one age stage into another represent asynchronous processes extended in time in even-aged cohorts.

3. All three species have long life expectancies (390 years for *A. speciosa*, 65 for *C. tridentata* and 70 for *C. caucasicum*, respectively). Consequently, the severe conditions in the alpine zone affect mainly the numbers of young plants in the populations as well as the success of flowering of adult plants.

4. Length of a continuous period of flowering in all three species is a random variable that has an approximately exponential distribution. We believe that a temporary break in flowering is a consequence of exceeding a certain environmental threshold. One should expect that the occurrence of such an extreme situation is a stochastic process. The average quantity of seeds produced by one normally developed individual of any species during the life cycle is a relatively stable variable. If the life span of the species is greater, then variance of this variable is smaller.

5. The analysis of transition matrices has demonstrated that increase in the life span apparently leads to a decrease in the amplitude of random fluctuation of age-stage transitions.

6.2. Soil seed banks

6.2.1. Introduction

It has been recognised for a long time that dormant seeds in the soil have a great importance for the plant population maintenance, community functioning and succession after disturbance (Rabotnov 1958, 1982, 1986, Thompson & Grime 1979, Fenner 1985, Simpson et al. 1989, Thompson et al. 1993, 1996, McDonald et al. 1996, and many others). It is essential to know more about soil seed banks in high-mountain ecosystems, which develop under severe conditions and are very sensitive to human impact and therefore vulnerable. A thorough knowledge of alpine seed banks is necessary for successful planning of restoration of disturbed alpine habitats.

In our study we examined seed banks in soils under 8 alpine community types: alpine lichen heaths (ALH), *Festuca varia* grasslands (FVG), *Geranium-Hedysarum* meadows (GHM), snowbed communities (SBC), *Rhododendron caucasicum* shrublands (*Lerchenfeldio-Rhododendretum caucasici,* RC), alpine eutrophic fens (*Caro caucasici-Caricetum nigrae,* FEN) and open communities on dry and moist screes (*Campanulo ciliatae-Chamaesciadietum acaulis,* DSC and *Saxifragetum sibiricae,* MSC respectively) (see Chapter 4). Detailed descriptions of the study are given in Onipchenko (1987a), and Semenova & Onipchenko (1990, 1991, 1994, 1996). The main goals of this work were:

1) to determine the size and species composition of the seed banks in soils under these plant communities;

2) to compare species diversity of viable seeds in soils and species diversity of above-ground vegetation;

3) to study spatial and vertical seed distribution in soils;

4) to follow dynamics of seed germination from the seed banks over a period of a few years;

5) to compare two different methods of seed bank assessment under high-mountain conditions;

6) to estimate the role of viable seeds in populations of several common species in the studied alpine communities and try to understand how important is the presence and the size of the seed bank for characterisation of the plant life-history strategy.

6.2.2. Methods

In all studied plant communities except communities on dry and moist screes (DSC and MSC respectively) two sites were chosen in order to represent different variants of the corresponding plant community.

Ten soil samples of 10 cm x 10 cm x 10 cm were taken from each site, then the samples were divided into 2 cm-thick layers, air dried, packed and transported to the greenhouse of Moscow State University (Moscow). The samples from dry and moist screes were divided into 5 cm-thick layers.

Samples were collected in late July – early August, before new seed release but after spring seed germination. Thus, the results reflect only the persistent seed bank (Thompson & Grime 1979). Samples were taken in ALH in 1982 (Onipchenko 1987a), in FVG, GHM and SBC in 1986 and in RC, FEN, MSC and DSC in 1989 (Semenova & Onipchenko 1990, 1994).

Soil of FVG was collected for the second time in 1989 to test seed accumulation within *Festuca varia* tussocks and between them. Ten blocks, 10 cm x 10 cm, were cut to the depth of 5 cm from *Festuca varia* tussocks (5 samples from each site), and ten blocks were similarly sampled between the tussocks (Semenova & Onipchenko 1994).

Additional soil sampling was also made in ALH in 1990 in order to estimate a necessary size of samples and total sampling area which allows to reveal the actual species composition of viable seeds in soils (Semenova & Onipchenko 1996). A plot of 70 cm x 70 cm was marked, divided into 49 quadrates of 10 cm x 10 cm size, and soil was sampled to the depth of 2 cm (Fig. 6.8). One additional sample was taken alongside in order to simplify estimation. The total sampling area was therefore 5000 cm².

1	2	3	4	5	6	7
8	9	10	11	12	13	14
15	16	17	18	19	20	21
22	23	24	25	26	27	28
29	30	31	32	33	34	35
36	37	38	39	40	41	42
43	44	45	46	47	48	49

Figure 6.8. Sampling scheme with plots from 0.01 to 0.49 m² (from Semenova & Onipchenko 1996 with permission of *Oecologia Montana*)

All samples were exposed for seed germination in a greenhouse. The germination process was allowed during 3-4 months in spring and autumn and was discontinued during summer and winter for three years of the germination experiment. The samples were subject to common winter temperatures in order to stimulate seed germination. Emerged seedlings were counted and removed after identification; those that could not be identified were transplanted to separate pots and grown until identifiable.

Soil seed banks in four of the studied communities - ALH, FVG, GHM and SBC - were also examined in a field experiment (Semenova & Onipchenko 1991, 1994). For this purpose, 5 small plots of 500 cm² each were chosen in each study site (10 for each community type). The plots were fenced in by metallic hoops set 3-5 cm into the soil. The soil within the fence was ploughed and cleared of plants. To prevent seed loss, large roots, rhizomes and bulbs were washed in water, which was then used to water a corresponding plot. The plots were watered once or twice per week. To decrease seed invasion, we removed all reproductive shoots closer than 0.5 m and covered the plots with a fine net. The net also protected the seedlings from the strong light and prevented them being eaten by grasshoppers. Seedlings were counted every week and were removed when identifiable. The field experiment was established in early August 1988. Observations were made until September and were continued throughout the next growing season.

6.2.3. Results and discussion

6.2.3.1. Size of the seed banks

Soil seed bank composition of the studied plant communities is presented in Table 6.8 (for details see Onipchenko 1987a, Semenova & Onipchenko 1990, 1991, 1994).

The highest amount of viable seeds per m^2 in the greenhouse experiment was found in the soils of FEN (10970 seeds) and MSC (8080 seeds), and lowest in soils of FVG (1190 seeds) and ALH (350 seeds). Intermediate values were found for soils of GHM (3850 seeds), SBC (2810 seeds), DSC (2760 seeds) and RC (1615 seeds). Thus, the number of germinable seeds in soils increases from dry to moist alpine communities. These numbers of viable seeds are close to those reported for tundras, polar deserts and alpine meadows (Archibold 1984, Hodachek 1985, Ziroyan 1988, Morin & Payette 1988, Ebersole 1989, Hatt 1991, Diemer & Prock 1993).

6.2.3.2. Species composition of the seed banks

We found seeds of a considerable number of vascular plant species in the soils of the alpine communities (Table 6.8). Most of them were ordinary species of the corresponding above-ground vegetation, with a few species from neighbouring plant communities, for example, *Pulsatilla albana* in the soil of ALH, *Saxifraga exarata* and *Veronica gentianoides* in SBC, *Sagina saginoides* in FEN, etc. However, we found no germinable seeds of some abundant species, such as *Geranium gymnocaulon, Hedysarum caucasicum, Anthemis cretica, Leontodon hispidus, Deschampsia flexuosa, Antennaria dioica*, etc. *Festuca varia*, the dominant of the corresponding grassland, had very few viable seeds in the soil.

Table 6.8. Species composition and numbers of viable seeds per m^2 in soils of eight alpine plant communities: ALH = Alpine lichen heaths, FVG = *Festuca varia* grasslands, GHM = *Geranium-Hedysarum* meadows, SBC = snowbed communities, FEN = alpine eutrophic fens, RC = *Rhododendron* shrub community, DSC = open communities on dry screes, MSC = open communities on moist screes. Sample area for ALH – 2750 cm^2, MSC and DSC – 1000 cm^2, other communities – 2000 cm^2

Species	ALH	FVG	GHM	SBC	RC	FEN	MSC	DSC
Agrostis vinealis					5			
Ajuga orientalis		5						
Alchemilla caucasica	4	5		5				20
Alchemilla vulgaris					15	1065		
Anemone narcissiflora					5			
Anemone speciosa	7							10
Anthoxanthum odoratum		20	155	5	35	95		
Anthyllis vulneraria					10			70
Arenaria lychnidea	4							
Briza marcowiczii					50	285		
Campanula ciliata								10
Campanula collina					5			
Campanula tridentata	11	5	5		30			90
Cardamine uliginosa						1010		
Carex atrata		170	75	45	100	10	150	
Carex nigra						160		
Carex oreophila				35				
Carex pyrenaica				15	85	5	960	
Carex sempervirens	7							20
Carex spp.						5	20	
Carex umbrosa	18	5	10		5			
Carum caucasicum	18					55	50	10
Carum meifolium					5			
Catabrosella variegata				20			10	
Cerastium purpurascens		100						
Cerastium cerastoides						4040	670	
Chamaenerion angustifolium	18							
Cirsium esculentum					5			
Cirsium simplex						40		
Draba hispida					5			30
Empetrum nigrum					15			
Epilobium anagallidifolium						5		
Ericaceae (undeterm.)			10					
Erigeron sp.				5				
Eritrichium caucasicum	4							
Euphrasia ossica		100	10				10	230
Festuca brunnescens		20	30	5				
Festuca ovina	11	75	20	10				10
Festuca varia		5	5					
Gentiana aquatica	4							80
Gentiana pyrenaica	29	10				300		80
Gentiana septemfida	14							30

Table 6.8.(cont.)

Species	ALH	FVG	GHM	SBC	RC	FEN	MSC	DSC
Gentiana sp.		120						
Gentiana verna					15		10	170
Gnaphalium supinum				1475	20			
Helictotrichon versicolor	7							
Hyalopoa pontica							40	
Luzula multiflora		20	1600	90	775	2865	810	
Luzula spicata	4				5		60	710
Matricaria caucasica			1190	25	5			
Minuartia aizoides			5	5		35		
Minuartia circassica								10
Minuartia recurva		5	5					10
Minuartia sp.		5						
Murbeckiella huetii							720	
Myosotis alpestris		15						
Nardus stricta		455	210	20		15		
Oxytropis kubanensis	22				5	5		
Pedicularis comosa						5		20
Phleum alpinum			175	10	115	20		
Potentilla gelida								120
Primula algida	32				15			730
Primula auriculata						20		
Primula amoena					5		70	
Primula ruprechtii								20
Pulsatilla albana	11							
Ranunculus oreophilus	4	5						
Rhododendron caucasicum					20			
Rumex alpestris						5		
Sagina saginoides						785		
Saxifraga exarata				5	10	5		
Saxifraga sibirica						25	4030	
Sedum tenellum			35		145	5	220	10
Sempervivum caucasicum					5			
Senecio aurantiacus		5						
Senecio sp.	4							
Senecio taraxacifolius		5						
Sibbaldia procumbens			110	765	30	55	10	
Taraxacum sp.							10	
Taraxacum stevenii				245				
Trifolium polyphyllum	4							
Veronica gentianoides	32	25	150	5			20	130
Veronica minuta							140	
Viola altaica		5						
Other dicotyledons	80		20	5	65	50	50	140
Other monocotyledons		5	30	15	5		20	
Total number of seeds	349	1190	3850	2810	1615	10970	8080	2760
Fritillaria collina (bulbs)	4							
Gagea fistulosa (bulbs)		510	2840	990				
Polygonum viviparum(bulbs)						40		
Saxifraga sibirica (bulbs)						10	160	

Comparison of the seed bank size and mean seed production in four of the studied communities showed no clear relations (Fig. 6.9). For example, some species with high seed production were absent from the seed bank (*Geranium gymnocaulon*, *Hedysarum caucasicum*, *Leontodon hispidus*). Likewise, some species with a low seed production had a large seed bank, which apparently accumulates during many years (*Carex atrata*, *Luzula multiflora*, *Sibbaldia procumbens*, *Gnaphalium supinum*). Only in ALH the seed production of plant species is positively correlated with the size of their seed bank ($r=0.60$, $p<0.01$).

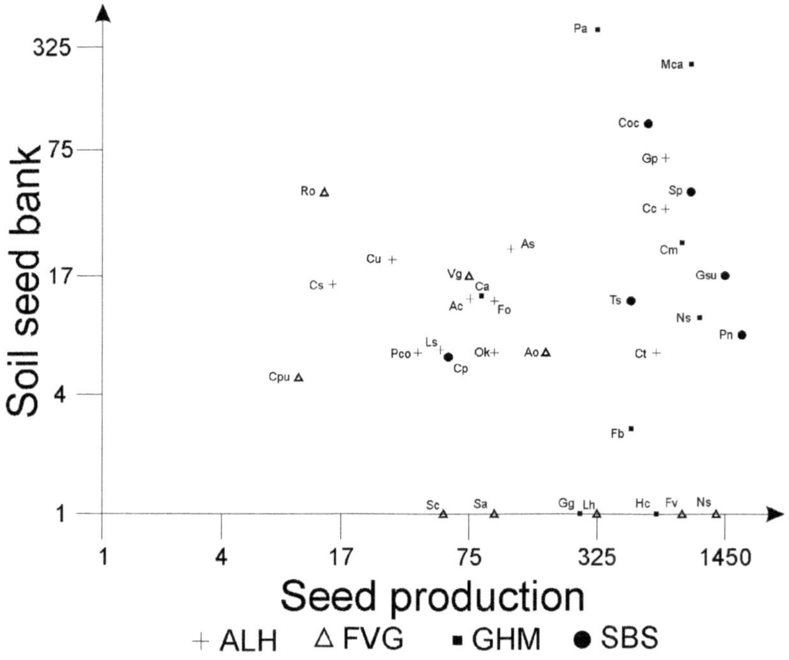

Figure 6.9. Correspondence between soil seed bank (seeds/m^2, y-axis) and mean annual seed production (seeds/m^2, x-axis) in four studied alpine communities. ALH = Alpine lichen heaths, FVG = *Festuca varia* grasslands, GHM = *Geranium-Hedysarum* meadows, SBC = snowbed communities. Species: Ac - *Alchemilla caucasica,* As - *Anemone speciosa,* Ao - *Anthoxanthum odoratum,* Ct - *Campanula tridentata,* Ca - *Carex atrata,* Cp - *Carex pyrenaica,* Cs - *Carex sempervirens,* Cc - *Carum caucasicum,* Cm - *Carum meifolium,* Cpu - *Cerastium purpurascens,* Coc - *Corydalis conorhiza,* Cu - *Carex umbrosa,* Fb - *Festuca brunnescens,* Fo - *Festuca ovina,* Fv - *Festuca varia,* Gg - *Geranium gymnocaulon,* Gsu - *Gnaphalium supinum,* Hc - *Hedysarum caucasicum,* Lh - *Leontodon hispidus,* Ls - *Luzula spicata,* Mca - *Matricaria caucasica,* Ns - *Nardus stricta,* Ok - *Oxytropis kubanensis,* Pco - *Pedicularis comosa,* Pn - *Pedicularis nordmanniana,* Pa - *Phleum alpinum,* Ro - *Ranunculus oreophilus,* Sc - *Scorzonera cana,* Sa - *Senecio aurantiacus,* Sp - *Sibbaldia procumbens,* Ts - *Taraxacum stevenii,* Vg - *Veronica gentianoides.*

Comparison of the seed weight with the size of the seed bank showed that large seeds weighing more than 4 mg were not present in the persistent seed bank (*Hedysarum caucasicum*, *Geranium gymnocaulon*, *Polygonum bistorta*, *Helictotrichon versicolor*), while small seeds weighing less than 0.7 mg accumulate in the soil in a significant amount – more than 100 per m^2 (*Nardus stricta*, *Luzula multiflora*, *Sibbaldia procumbens*, etc.).

6.2.3.3. Species diversity

The species diversity of seed banks is usually expected to be much smaller than the diversity of the corresponding plant communities. However, little is known about the necessary area for determination of the seed bank composition (Rabotnov 1958, Roberts 1981, Simpson et al. 1989). The sample areas used to estimate these parameters are usually very different, i.e. less than 1 m^2 for a seed bank and about 100 m^2 for aboveground vegetation (Rabotnov 1982).

In our study we compared the mean number of plant species on 10 cm x 10 cm plots with the mean number of species in the seed bank in the same area (Fig. 6.10). According to the t-test, these values do not differ significantly in FVG, GHM, SBC and RC. The species diversity of the soil seed banks in MSC, DSC and FEN was approximately twice as high as in the aboveground vegetation. So, what will happen if we carefully examine much greater numbers of samples from a larger area? It is now a well-known phenomenon that quite a lot of plant species do not possess a persistent seed bank, e.g. dominants of many plant communities (Chambers 1993, Diemer & Prock 1993, Semenova & Onipchenko 1994), therefore on the whole we could expect the seed pool to be poorer. On the other hand, some mechanisms limiting the number of species in the aboveground vegetation (e.g. competition) hardly play a role in a buried seed community. A dormant seed pool can also include species dispersed from the surrounding vegetation. These factors may all increase the species diversity, especially on a local scale.

How many samples do we need to evaluate the total species composition of viable seeds in soils of the alpine communities? How does the species diversity in the seed bank and aboveground vegetation change with sample size? To address these questions, we made the additional sampling in ALH according to the scheme described above (Fig. 6.8). Eighteen vascular plant species were identified in the seed bank from the samples of 1990, an additional two species germinated from small bulbs (Table 6.9, see also Semenova & Onipchenko 1996). Seventeen species germinated from 10 samples with a total area of 1375 cm^2, which were taken from the same study site as in 1982 (Onipchenko 1987a). The species lists slightly differ from each other: seedlings of *Arenaria lychnidea*, *Gentiana aquatica*,

Pulsatilla albana, *Trifolium polyphyllum* emerged only from the samples of 1982, whereas seedlings of *Erigeron alpinus*, *Euphrasia ossica*, *Minuartia circassica*, *Pedicularis comosa* and *Ranunculus oreophilus* were found in the samples of 1990. All these species but *Euphrasia ossica* were rare in the samples.

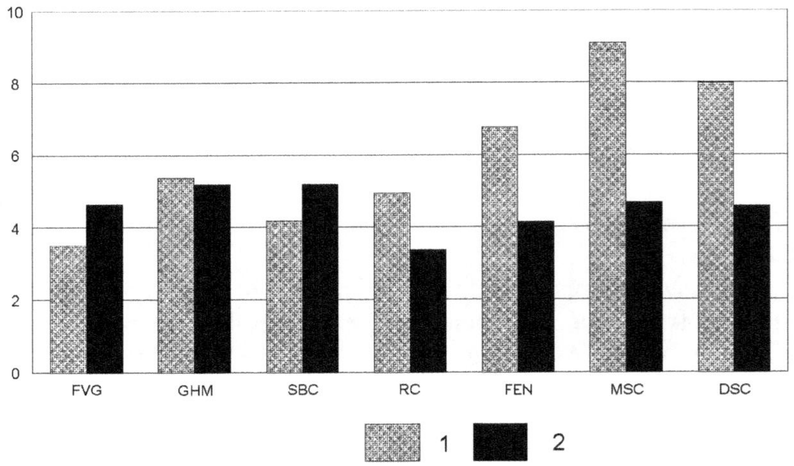

Figure 6.10. Average number of species (floristic richness) per 10 x 10 cm plots for soil seed banks (1) and recent plant communities (2). FVG = *Festuca varia* grasslands, GHM = *Geranium-Hedysarum* meadows, SBC = snowbed communities, FEN = alpine eutrophic fens, RC = *Rhododendron* shrub community, DSC = open communities on dry screes, MSC = open communities on moist screes (from Semenova & Onipchenko 1994 with permission of Geobotanical Institute SFIT)

The cumulative number of species depending on the amount of samples is shown in Fig. 6.11. Both curves are similar and level off at approximately ten samples. This simple cumulative curve shows that ten to twenty samples is enough to reveal most of the species pool persisting in the seed bank, which corresponds to the findings of Forcella (1984) for a pasture in south-eastern Australia and Gross (1990) for an annually ploughed field in Michigan, USA. However, to determine the total species composition of the seed bank we should sample a larger area. The sampling technique can be improved by subsampling several large units with very small subunits, as proposed by Bigwood & Inouye (1988).

Species-area dependence was analysed in a series of square plots with sides of 10, 20, 30, 40, 50, 60 (four replicates - from each corner of the sampling plot) and 70 cm (one replicate) (Fig. 6.8). Mean number of species in the seed bank per plot was compared with the mean number of species in

Table 6.9. The composition of the soil seed bank in an alpine lichen heath in additional experiment started in 1990 (number of seeds per m^2, mean and standard error, n=50).

Species	mean	s.e.
Anemone speciosa	16	6
Campanula tridentata	8	4
Carex sempervirens	10	5
Carex umbrosa	10	4
Carum caucasicum	24	10
Erigeron alpinus	6	3
Eritrichium caucasicum	28	9
Euphrasia ossica	344	59
Festuca ovina	18	6
Gentiana pyrenaica	62	16
Gentiana septemfida	22	8
Helictotrichon versicolor	20	7
Luzula spicata	54	10
Minuartia circassica	2	2
Pedicularis comosa	20	7
Primula algida	198	23
Ranunculus oreophilus	16	6
Veronica gentianoides	142	26
Dicotyledons (undeterm.)	74	18
Monocotyledons (undeterm.)	10	5
Total number of seeds	1084	75
Fritillaria collina (bulbs)	40	9
Lloydia serotina (bulbs)	10	5

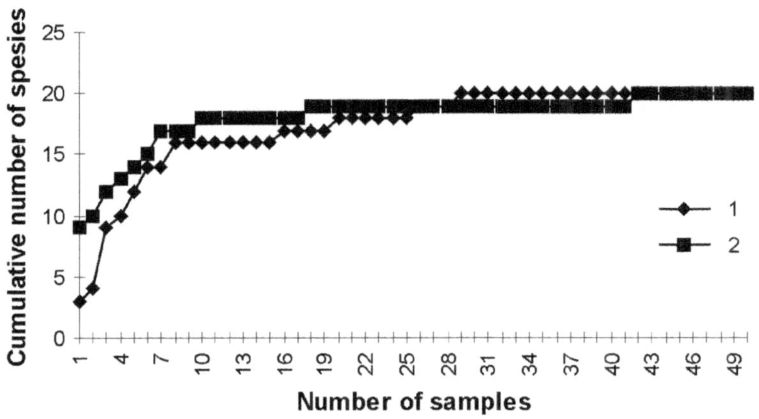

Figure 6.11. Cumulative number of species identified in soil seed bank samples from an alpine lichen heath based on regular (1) and random (2) sampling (from Semenova & Onipchenko 1996 with permission of *Oecologia Montana*)

the aboveground vegetation at the same plot size (Table 6.10). No significant differences were found for the numbers in the seed bank vs. aboveground vegetation at either 10 x 10 cm or 50 x 50 cm plots. The following equations were fitted (see Chapter 4.3.3):

$S = c \cdot A^z$ (1)

$S = a + b \cdot lnA$ (2)

where S = number of species, A = area, a, b, c, and z = regression parameters. Regression coefficients c and a of Equations (1) and (2) represent the number of species per initial m², coefficients z and b show the rate of increase in species richness as the area increases (Table 6.11). Equation (1) is from Arrhenius (1921) and Equation (2) is from Gleason (1922). The correlation coefficient (r^2) is higher for Equation (2) than for Equation (1). The same was found for aboveground vegetation of alpine lichen heaths (Onipchenko & Semenova 1995).

Table 6.10. The number of species in the seed bank and aboveground vegetation at different plot sizes. m = mean number of vascular plant species; SE = standard error. No significant differences were found in species richness of the seed bank and seeds+bulbs bank vs. the aboveground vegetation according to the t-test ($p>0.05$) (from Semenova & Onipchenko 1996 with permission of *Oecologia Montana*)

		Plot size, m²						
		0.01	0.04	0.09	0.16	0.25	0.36	0.49
Seeds	m	5.0	11.3	14.5	17.0	18.0	19.5	20
	SE	0.8	0.5	0.8	0.4	0.5	0.3	-
Seeds+bulbs	m	5.5	12.5	16.3	18.8	19.8	21.5	22
	SE	0.8	0.6	1.0	0.4	0.6	0.3	-
Aboveground vegetation	m	6.2	-	-	-	20.2	-	-
	SE	0.4	-	-	-	0.7	-	-

Table 6.11. Coefficients (c, z, a, b) of regression and correlation coefficient (r^2) for species-area regressions according to equations (1) and (2) (from Semenova & Onipchenko 1996 with permission of *Oecologia Montana*)

	Equation (1)			Equation (2)		
	c	z	r^2	a	b	r^2
Seeds	29.30	0.346	0.960	23.49	3.90	0.978
Seeds+bulbs	32.29	0.345	0.957	25.88	4.27	0.973
Aboveground vegetation	19.56	0.233	0.925	24.44	3.61	0.997

6.2.3.4. Vertical distribution

Most of the viable seeds contributing to the seed banks were found in the upper 2 cm layer of the soil: ALH 88%, FVG 93%, GHM 70%, SBC 95%, RC 67%, FEN 39%. However, their vertical distribution differed in the studied plant communities (Fig. 6.12). Soils of DSC and MSC were divided only into two layers, with 95% of the seed bank in DSC and 81% in MSC found in the upper 0-5 cm layer of soil.

Viable seeds were discovered to the depth of 10 cm in GHM, FEN and RC soils, 8 cm in FVG soils and 6 cm in ALH and SBC soils. The deeper penetration of the seeds into the soils can be explained by the intensive burrowing activity of voles (*Pitymys majori*) in GHM and submersion in peat in FEN. In RB it can be possible due to the large *Rhododendron* stems and more loose soil there. FVG is less influenced by voles than GHM, and the soils of ALH and SBC are not subject to much zoogenic disturbance.

Our data confirm that small seeds (in our case weighing less than 0.7 mg) can sink deeper (or stay longer viable!) into the soils than the larger ones. The lower soil layers contain mainly seeds of *Matricaria caucasica, Luzula multiflora, Nardus stricta, Carex atrata, Anthoxanthum odoratum, Cerastium cerastoides, Carex pyrenaica, Saxifraga sibirica.*

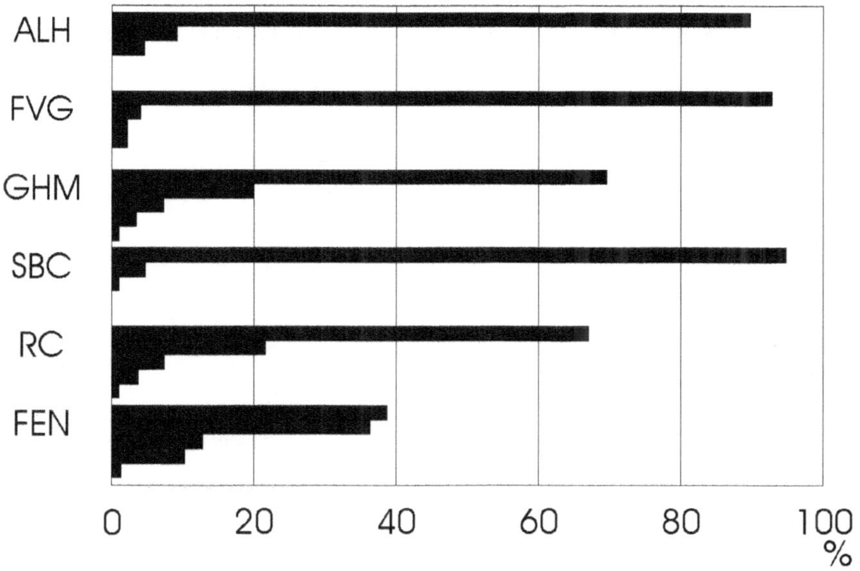

Figure 6.12. Vertical distribution of viable seeds in 2 cm-soil layers (%) in different communities. FVG = *Festuca varia* grasslands, GHM = *Geranium-Hedysarum* meadows, SBC = snowbed communities, FEN = alpine eutrophic fens, RC = *Rhododendron* shrub community, DSC = open communities on dry screes, MSC = open communities on moist screes (from Semenova & Onipchenko 1994 with permission of Geobotanical Institute SFIT)

6.2.3.5. Spatial distribution

The horizontal distribution of buried seeds in the soils is not homogeneous. The rate of heterogeneity of spatial distribution was estimated according to the coefficient of variation (CV = SD·100% / mean). The following values were obtained for the seed banks of the most numerous species: *Marticaria caucasica* 74%, *Gnaphalium supinum* 101%, *Sibbaldia procumbens* 109%, *Luzula multiflora* 113%, *Carex atrata* 110 – 121%, *Veronica gentianoides* 210%, *Phleum alpinum* 210%, *Taraxacum stevenii* 210%. Hatt (1991) also found considerable soil seed bank heterogeneity in the alpine meadows near Davos.

Because of the peculiar "tussock" structure of FVG, a difference in seed accumulation within *Festuca varia* tussocks and between them could be expected (Rabotnov 1986). However, our observations did not confirm this expectation (Table 6.12). No significant differences were found in the soil seed bank composition in samples from the tussocks and between them. It seems that heavy accumulation of litter can even out the horizontal distribution of seeds in this vegetation type.

Table 6.12. Soil seed bank composition of *Festuca varia* tussocks and the space between tussocks in FVG (number of seeds per 10 plots with total area 0.1 m^2) (from Semenova & Onipchenko 1994 with permission of Geobotanical Institute SFIT)

Species	Tussocks	Between tussocks
Ajuga orientalis	0	1
Anthoxanthum odoratum	1	0
Campanula collina	0	1
Carex atrata	27	11
Cerastium purpurascens	3	7
Empetrum nigrum	0	1
Festuca sp.	1	0
Festuca varia	3	10
Gentiana pyrenaica	1	0
Gentiana sp.	0	1
Luzula multiflora	1	0
Luzula spicata	1	0
Nardus stricta	1	6
Primula ruprechtii	1	0
Ranunculus oreophilus	2	1
Saxifraga exarata	1	0
Trifolium spadiceum	1	0
Valeriana alpestris	1	0
Veronica gentianoides	1	1
Viola altaica	1	2
Poaceae (undeterm.)	4	0
Total number of seeds	52	42
Gagea fistulosa (bulbs)	12	45

6.2.3.6. Dynamics of seed germination

The three-year period of observation has allowed the estimation of the seasonal and annual dynamics of seed germination in the soil samples (Table 6.13). Seeds of *Nardus stricta*, *Euphrasia ossica*, *Taraxacum stevenii*, *Veronica minuta* germinated only in spring. Most other seeds germinated

Table 6.13. Time of seed germination of some alpine species from soil samples in greenhouse experiments (number of seedlings for seasons and years) (from Semenova & Onipchenko 1994 with permission of Geobotanical Institute SFIT)

Communities and species	seasons		years		
	spring	autumn	first	second	third
FVG					
Carex atrata	34	0	23	3	8
Cerastium purpurascens	19	1	20	0	0
Euphrasia ossica	20	0	10	6	4
Festuca ovina	2	13	13	2	0
Nardus stricta	91	0	91	0	0
Total	219	19	193	30	15
GHM					
Anthoxanthum odoratum	29	2	27	4	0
Carex atrata	14	1	10	3	2
Luzula multiflora	213	107	198	81	41
Matricaria caucasica	38	100	179	49	10
Nardus stricta	42	0	42	0	0
Phleum alpinum	16	19	34	1	0
Sibbaldia procumbens	21	1	19	3	0
Veronica gentianoides	24	6	23	7	0
Total	519	251	564	151	55
SBC					
Gnaphalium supinum	272	23	155	97	43
Luzula multiflora	14	4	11	5	2
Sibbaldia procumbens	123	30	85	54	14
Taraxacum stevenii	49	0	49	0	0
Total	495	67	337	165	60
RC					
Carex atrata	20	0	18	2	0
Carex pyrenaica	13	4	11	6	0
Luzula multiflora	92	63	136	17	2
Phleum alpinum	1	22	23	0	0
Sedum tenellum	29	0	29	0	0
Total	227	96	284	34	5
FEN					
Alchemilla vulgaris	203	8	172	37	2
Anthoxanthum odoratum	14	5	18	1	0
Briza marcowiczii	44	13	44	11	2
Cardamine uliginosa	172	26	196	4	2
Carex nigra	31	1	23	7	2
Cerastium cerastoides	267	541	787	20	1

Table 6.13. (Cont.)

Communities and species	seasons			years	
Gentiana pyrenaica	60	0	60	0	0
Luzula multiflora	383	134	371	138	8
Sagina saginoides	139	18	133	24	0
Total	1370	765	1862	256	17
MSC					
Carex atrata	15	0	11	3	1
Carex pyrenaica	28	68	91	4	1
Cerastium cerastoides	25	42	65	1	1
Luzula multiflora	22	59	75	5	1
Murbeckiella huetii	27	45	69	2	1
Saxifraga sibirica	44	359	398	5	0
Sedum tenellum	22	0	22	0	0
Veronica minuta	14	0	14	0	0
Total	229	579	778	23	7
DSC					
Euphrasia ossica	23	0	16	7	0
Gentiana verna	17	0	14	3	0
Luzula spicata	70	1	52	13	6
Potentilla gelida	12	0	8	4	0
Primula algida	71	2	37	23	13
Veronica gentianoides	13	0	7	6	0
Total	270	9	170	66	43

mainly in spring. Seedlings of *Matricaria caucasica*, *Luzula multiflora*, *Carex pyrenaica*, *Murbeckiella huetii*, *Cerastium cerastoides*, *Saxifraga sibirica* and *Phleum alpinum* appeared both in spring and autumn, while seeds of *Festuca ovina* and *Festuca varia* germinated for the most part in the autumn of the first year.

Seedlings of *Nardus stricta*, *Taraxacum stevenii*, *Sedum tenellum*, *Veronica minuta* and *Festuca varia* emerged only in the first year. Seeds of most alpine species germinated mainly in the first year; in general, 60-96% seedlings appeared in different communities during the first year of the experiment. Seeds of *Gnaphalium supinum*, *Luzula multiflora*, *Carex atrata*, *Alchemilla vulgaris*, *Briza marcowiczii*, *Cardamine uliginosa*, *Carex nigra*, *Cerastium cerastoides*, *Carex pyrenaica*, *Murbeckiella huetii*, *Luzula spicata*, *Primula algida* and *Euphrasia ossica* germinated during all three years of observations. The highest proportion of seeds with delayed germination was found in SBC and DSC soils.

6.2.3.7. Field experiment

The following data were obtained by means of the field sampling method (Table 6.14): ALH 272 seeds/m^2, FVG 108 seeds/m^2, GHM 846 seeds/m^2 and SBC 256 seeds/m^2. These figures were a lot smaller than corresponding values in the greenhouse experiment (Table 6.8, see also Semenova & Onipchenko 1991), while the number of germinated small bulbs was

Table 6.14. Number of viable seeds (per m^2) and bulbs in soils of four alpine communities revealed during the field germination experiment. The sample area comprises 0.5 m^2 in each community type. * - species which were not found in the greenhouse experiment

	ALH	FVG	GHM	SBC
Alchemilla caucasica	12			
Anemone speciosa	22			
Anthoxanthum odoratum		6	72	
*Campanula collina**	2			
Campanula sp.*				2
Campanula tridentata	6			
Carex atrata		2	24	12
Carex pyrenaica				6
Carex sempervirens	14			
Carex umbrosa	20	8		
Carum caucasicum	36			
Carum meifolium			24	
Catabrosella variegata				16
Cerastium purpurascens		4		
*Corydalis conorhiza**				106
*Erigeron alpinus**	6			
Euphrasia ossica	6		20	
Festuca brunnescens	2		2	
Festuca ovina	12	10	4	4
Gentiana pyrenaica	66	4		
Gentiana sp.	20			
Gnaphalium supinum				16
*Juniperus communis**		2		
Luzula multiflora			20	
Luzula spicata	6			
Matricaria caucasica			200	
Minuartia recurva			2	
Myosotis alpestris		2		
Nardus stricta			10	2
Oxytropis kubanensis	6			
*Pedicularis comosa**	6			
*Pedicularis nordmanniana**				8
Phleum alpinum			328	18
Ranunculus oreophilus	10	48	2	
Sedum tenellum			80	
Sibbaldia procumbens			34	48
Taraxacum stevenii	4			12
*Trifolium spadiceum**		2		
*Vaccinium vitis-idaea**	2			
Veronica gentianoides	8	16	16	
Viola altaica		2		
Other dicotyledons	10	2	6	6
Other monocotyledons			2	
Total number of seeds	276	108	846	256
Fritillaria collina (bulbs)	24	10		
Gagea fistulosa (bulbs)		276	1198	404

comparable under field and greenhouse conditions (except small bulbs of *Fritillaria collina*, which germinated more in the field). These differences could result from year-to-year variation in seed production and seed viability during the years before sampling. However, we tried to diminish the influence of the year-to-year variation in the seed bank by choosing the sampling time and the beginning of the field experiment in the middle of the growing season (after the spring seed germination and before the new seed release). Other researchers who compared both methods also reported much fewer seeds found in the field method (Pfadenhauer & Maas 1987). Seed germination from the seed bank in the field is likely to depend on the microclimate conditions associated with the plot and the extent of its disturbance (Miles 1974, Pfadenhauer & Maas 1987).

There were also differences in the seed bank composition revealed by these two methods (Tables 6.8 and 6.14). Most of the differences were in the species that were rare in the samples, i.e., they could be absent from samples by chance. Note, however, that the total area of soil samples for the field experiment was 2.5 times larger than that for the greenhouse experiment.

Nevertheless, seedlings of *Nardus stricta* (FVG), *Primula algida* (ALH) and *Carex oreophila* (SBC) were numerous under greenhouse conditions, but absent in the field experiment. On the other hand, there were many seedlings of *Corydalis conorhiza* and *Pedicularis nordmanniana* (SBC) in the field experiment, which did not emerge in the greenhouse. This might mean that these species require specific conditions for germination, which were not met in the greenhouse experiment.

6.2.3.8. Fraction of dormant seeds in plant populations

In 1985 we studied the population composition, density and stage distribution of the 17 most abundant species in four alpine communities: ALH, FVG, GHM and SBC (Semenova & Onipchenko 1989). The analysis of the seed banks allows us to estimate the role of viable seeds in the populations of the studied species.

First of all, plant species can be divided into three categories according to the presence and the size of their seed banks: with a large seed bank, with a small seed bank and without a seed bank (Table 6.15), which corresponds to long-term persistent (permanent), short-term persistent and transient seed bank *sensu* K. Thompson (1992, 1993). The main dominants of alpine meadows as well as some other species with high biomass scores hardly possess a persistent seed bank. On the other hand, most of the alpine species do possess at least a small seed bank, as they grow in a quite stressful and unpredictable environment where the seed reserve may be of great importance for the maintenance of sexually reproducing species (Urbanska

& Schütz 1986). Furthermore, some of these seeds retain their germinating ability for a long time, accumulating in the soil in great numbers.

Besides this, the impact of the seed bank depends on the population density, the plant size and growth form and the relative importance of the juvenile stage. Thus, the proportion of the dormant stage is very high in populations of tussock grasses and sedges, where the tussock size is relatively large, and their juveniles are normally rare (*Festuca ovina, Nardus stricta, Carex atrata)*. Examples of plant population composition are shown on Fig. 6.13. The proportion of dormant seeds in populations of *Campanula tridentata*, *Anemone speciosa* and *Carum caucasicum* (ALH), which are relatively small non-clonal plants with a more pronounced juvenile stage, is not greater than 10%.

Table 6.15. Three categories of alpine plant species according to the presence and the size of their seed banks

No seed bank	Small seed bank	Large seed bank
Antennaria dioica	*Alchemilla caucasica*	*Anthoxanthum odoratum*
Anthemis cretica	*Anemone speciosa*	*Briza marcowiczii*
Calamagrostis arundinacea	*Anthyllis vulneraria*	*Cardamine uliginosa*
Deschampsia flexuosa	*Campanula tridentata*	*Carex atrata*
Geranium gymnocaulon	*Carex sempervirens*	*Cerastium purpurascens*
Hedysarum caucasicum	*Carex umbrosa*	*Cerastium cerastoides*
Leontodon hispidus	*Carum caucasicum*	*Gentiana pyrenaica*
Plantago atrata	*Catabrosella variegata*	*Gnaphalium supinum*
Polygonum bistorta	*Festuca brunnescens*	*Luzula multiflora*
Scorzonera cana	*Festuca ovina*	*Luzula spicata*
	Festuca varia	*Matricaria caucasica*
	Minuartia aizoides	*Nardus stricta*
	Minuartia recurva	*Phleum alpinum*
	Oxytropis kubanensis	*Primula algida*
	Pulsatilla albana	*Sagina saginoides*
	Ranunculus oreophilus	*Saxifraga sibirica*
		Sedum tenellum
		Sibbaldia procumbens
		Taraxacum stevenii
		Veronica gentianoides

Further considerations give an opportunity to understand how important the type of the seed bank is for characterisation of the plant population strategy (Chapter 6.3).

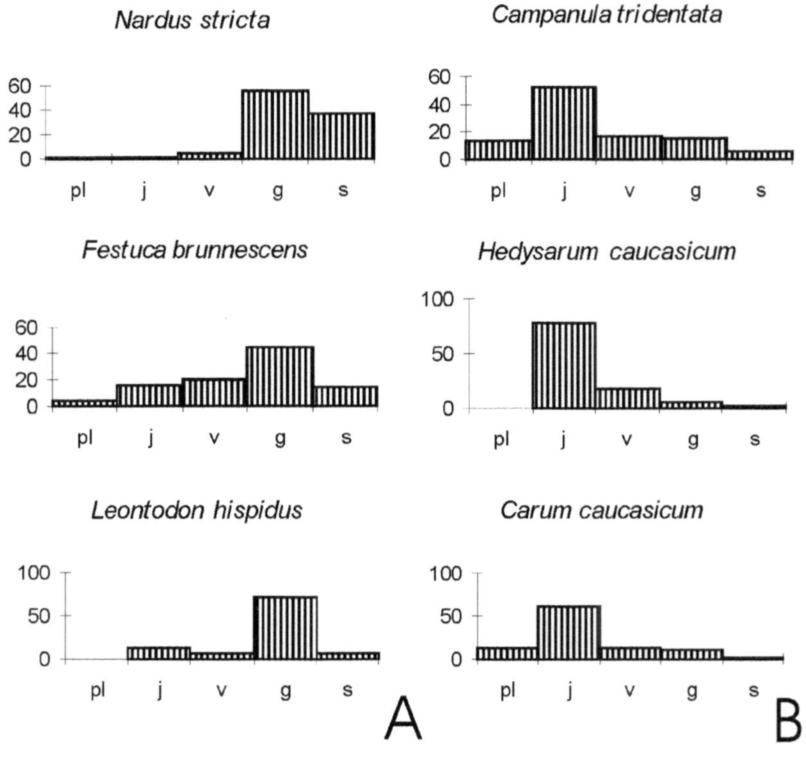

Figure 6.13. Examples of population stage distribution in: A) species with a reproductive stage prevalence, B) species with a juvenile stage prevalence. Pl – seedlings, j – juvenile plants, v – vegetative plants, g – generative plants, s – seeds in soils. Ordinate – percentage of plants in studied populations.

6.3. Plant population strategies

6.3.1. Introduction

The notion of life history strategy has been included in the works of many ecologists, both Anglo-American (MacArthur & Wilson 1967, Pianka 1970, 1983, Grime 1977, 1979, Whittaker & Goodman 1979, Solbrig 1980) and Russian (Ramenskii 1938, Rabotnov 1975b, 1985, 1993, Mirkin 1983,

1994, Zaugol`nova et al. 1988). Grime (1977, 1979) originally based his typology of plant strategies on a classification of three habitat types with regard to stress and disturbance in which the plants evolved. He then described the characteristics of plants adapted to these environments. Important parameters were the maximum relative growth rate and an index of competitiveness (Grime 1979; Grime et al. 1988). However, he considered juvenile and adult phases of the life cycle separately.

The population approach was further developed by the limnologist Romanovsky (1985, 1989a, 1989b, 1992), working mainly with *Cladocera*. He based his concept of primary life history strategies on species responses to peculiarities of resource dynamics. According to this concept, there are two ways for a population to succeed in the competition for limiting resources:

(1) reducing the equilibrium resource requirement, R* (sensu Tilman 1982), i.e. the resource level at which population growth and loss rates are equal;

(2) developing a high resource capture capacity and high population growth rate whenever the resource is available.

In the first case we deal with a 'patient' (i.e., stress-tolerant) strategy. Romanovsky pointed out that resistance of a population as a whole is determined by the tolerance of the most susceptible age group, i. e. the juveniles. A high population resistance to resource shortage is beneficial when the resource availability is steady (see also Abrosov et al. 1982; Tilman 1982). The ability to survive in resource-poor environments is associated with a low rate of resource consumption (Chapin 1980) and a low rate of population growth. As a consequence, 'patients' show a lower competitive ability when resource availability fluctuates considerably. In other words, there is a trade-off between tolerance to resource scarcity and growth rate or offspring production.

In the second case, plants have higher rates of resource uptake when resources are available and survive unfavourable periods (e.g. periods of resource depletion) either in a dormant (latent) stage – the 'explerent' (i.e., ruderal) strategy – or, as adults, in a state of low activity – the 'violent' (i.e., competitor) strategy. Species of the latter category have an advantage in relatively undisturbed habitats. They do not need much time for germination and growth, and can rapidly decrease resource availability for other species during periods of enrichment. These species have a strong impact on the resource regime within a community. 'Explerents', in their turn, are adapted to a less predictable resource regime. This strategy is associated with a short life span, high seed production and avoidance of resource limitation via dormancy or migrations.

The 'Grime' typology and the 'Ramenskii/Rabotnov/Romanovsky' typology are similar, but not identical (Onipchenko et al. 1998a). The main difference is concerned with the definition of 'competitors' ('violents'). According to Grime, a competitor is a fast-growing perennial in a fertile environment. Such a viewpoint does not take into account the long-term survival of these species. Real long-term competitors must be able to tolerate low resource levels they themselves contribute to create; however, they usually are not so fast-growing. Thus, we attribute dominants of the later stages of succession to real competitors, or 'violents' in Ramenskii's (1938) terminology, whereas the fast-growing species which can rapidly occupy free space (e.g. *Chamaenerion angustifolium*, *Urtica dioica*, etc.) are considered as 'explerents' (Onipchenko et al. 1998a).

Plant strategy analysis may be used to explore community structure and species coexistence under severe high-mountain conditions (Marchand & Roach 1980, del Moral 1983a, b, 1985, del Moral et al. 1985, Schimpf & Bann 1983, Urbanska 1984, Grabherr et al. 1987). This Chapter presents an attempt to identify different strategy types among plant species of four alpine communities in the northwestern Caucasus. For this purpose we examined four characteristics, as presented in Table 6.16. Three are concerned with seed characteristics, the forth is biomass production. The data were partially included in earlier papers (Onipchenko et al. 1991, 1998a). In addition we estimated the two main parameters used by Grime: Relative Growth Rate and Morphology Index.

Table 6.16. Three primary life-history strategies: definition and some population characteristics (from Onipchenko et al. 1998a, with permission of Opulus Press)

P-strategy ('patients')	**V-strategy ('violents')**	**E-strategy ('explerents')**
Competitive ability due to low equilibrium resource requirement	An advantage due to high resource capture capacity when resource is available. Survival of resource depletion in:	
tolerance of juvenile stage	adult tolerant stage	dormant stage
1. Occurrence under unfavourable conditions (abiotic and biotic stress)	1. Large biomass under more productive conditions	1. Occurrence under more productive conditions
2. Negative response to biomass removal and slow reaction to resource availability, negative response to disturbance as a whole	2. Negative response to biomass removal and positive reaction to resource availability, the result depends on the intensity of disturbance	2. Rapid reaction to resource availability and weak response to biomass removal, positive response to disturbance as a whole
3. High relative energy investment into an offspring	3. High seed production with relatively large seeds	3. High relative energy investment into reproduction
4. Small persistent seed bank	4. Lack of persistent seed bank	4. Large persistent seed bank or wide seed dispersal

6.3.2. Methods

To analyse strategies, we studied population characteristics of plants from four alpine communities (ALH, FVG, GHM, SBC) (Onipchenko et al., 1998a). The following parameters were used to analyse population strategies: above-ground biomass (see Chapter 5.2), seed production parameters (seed weight, annual seed yield, soil seed bank (Chapter 6.2), relative growth rate, morphology index (Grime 1979).

One hundred well-developed seeds (or one-seeded fruits, e.g. of *Asteraceae*, *Anemone*, etc.) were sampled from each species and weighed. Seed number produced per m^2 was calculated from reproductive shoot density and the mean seed productivity of a shoot. The latter was estimated in the following way: 100 reproductive shoots were cut, the number of fruits and seeds was counted for each shoot, and a mean fruit and seed number per shoot was calculated. We counted almost ripe seeds; these values, therefore, can characterize real seed productivity. Seed yield is the product of seed number and seed weight. Since real seed productivity and reproductive shoot number of alpine species vary significantly from year to year, we use mean values from three to four years of observations.

Values for *RGR* were obtained from a recent study (Semenova, in prep.) for 30 species, half of which are included in the present data set. For those species for which we did not yet have experimental data, RGR was estimated on the basis of field observations and comparisons with data from Grime & Hunt (1975), Grime et al. (1988) and van der Maarel (1988). By arranging the data according to a six-grade scale we obtained sufficient differentiation:

Grade:	1	2	3	4	5	6
RGR ($week^{-1}$)	0.41- 0.60	0.61 - 0.80	0.81 - 1.00	1.01 - 1.20;	1.21 - 1.40	1.41 - 1.60

Values for the morphology index were obtained according to the procedure described by Grime (1979: 74), i.e. calculated from the formula $M = (a + b + c) / 2$ where a is the estimated maximum height of the leaf canopy, b is the lateral spread, and c is the estimated maximum accumulation of persistent litter (see Onipchenko et al. 1998a for details).

The data matrix contained 42 species for which the following features were included: above-ground biomass, seed weight, seed production per m^2 – further called seed density, size of the seed bank, relative growth rate, and morphology index. For species found in two or three communities we took the values, which corresponded to the largest biomass value in any of the communities. Two multivariate analyses were applied to the resulting 42 x 6 matrix: 1. Agglomerated clustering with the Weighted Pair-Groups Method with Arithmetic Averages (WPGMA) using the Similarity Ratio (see van der

Maarel et al. 1978, Gauch 1982). All data were transformed dividing the values in every column by the maximum value of this column. 2. Principal Components Analysis as available in the program package CANOCO (ter Braak 1987). All data except RGR and M were log-transformed.

6.3.3. Results and discussion

6.3.3.1. Classification

The dendrogram obtained by the cluster analysis is presented in Fig. 6.14. The 42 species from all four communities could be grouped into six clusters, at the similarity level of ca. 30%. At the first fusion level we obtained two big groups, with the second group (clusters 4, 5 and 6) comprising species from the lichen heath, and the first group (clusters 1, 2 and 3) including species from all studied communities. The dominants of the *Festuca* grassland and *Geranium* meadow are contained in cluster 1. Cluster 3 includes species with moderate biomass, low seed weight, high seed production and a large seed bank: *Gnaphalium supinum*, *Luzula multiflora*, *Matricaria caucasica*, *Nardus stricta* and *Sibbaldia procumbens*.

6.3.3.2. Ordination

Fig. 6.15 shows the results of the PCA on the 42 x 6 matrix. The two factors from Grime's scheme – namely RGR and Morphology Index - are slightly negatively correlated, and the Morphology Index is positively correlated with Biomass. Seed weight is negatively correlated with both seed bank and RGR. In fact, there are four axes of variation explaining 35 %, 33 %, 13 % and 9 % of the variance, respectively. The species are differentiated in three main directions. The seed bank species are mainly found in the lower part of the diagram, but they are more distributed along axis 1; the species with higher *RGR* are in the lower right corner, and the species with slower growth and the large seeds are in the upper left corner of the PCA diagram.

6.3.3.3. Discussion and conclusions

Our results confirm that species vary in their combinations of traits, which means that they show a degree of 'violence' ('competitiveness'), 'explerence ('ruderalness') and/or 'patience' ('stress-tolerance'). Only a few species situated towards the periphery of the diagram may be indicated as 'pure types'. The occurrence of transitional types has been recognised from the beginning, when Grime (1979) introduced C-S, C-R, S-R and C-S-R strategists.

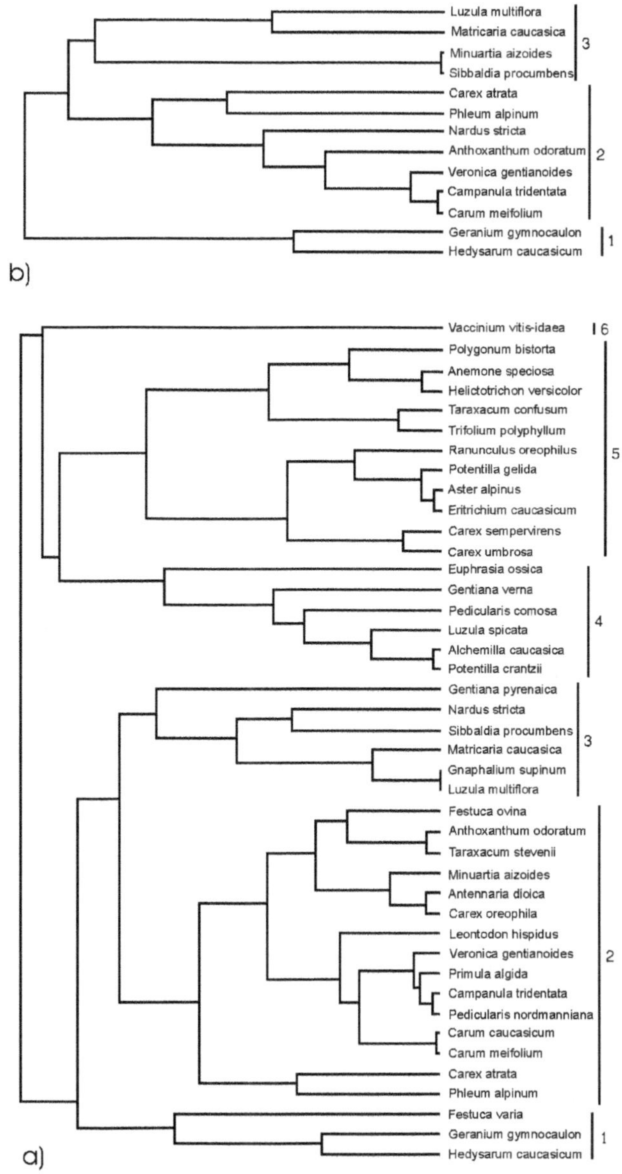

Figure 6.14. Dendrogram of the hierarchical agglomerative clustering (TABORD) of 42 species of 4 alpine communities. Suggested cluster structure as indicated below the dendrogram (from Onipchenko et al. 1998a with permission of Opulus Press)

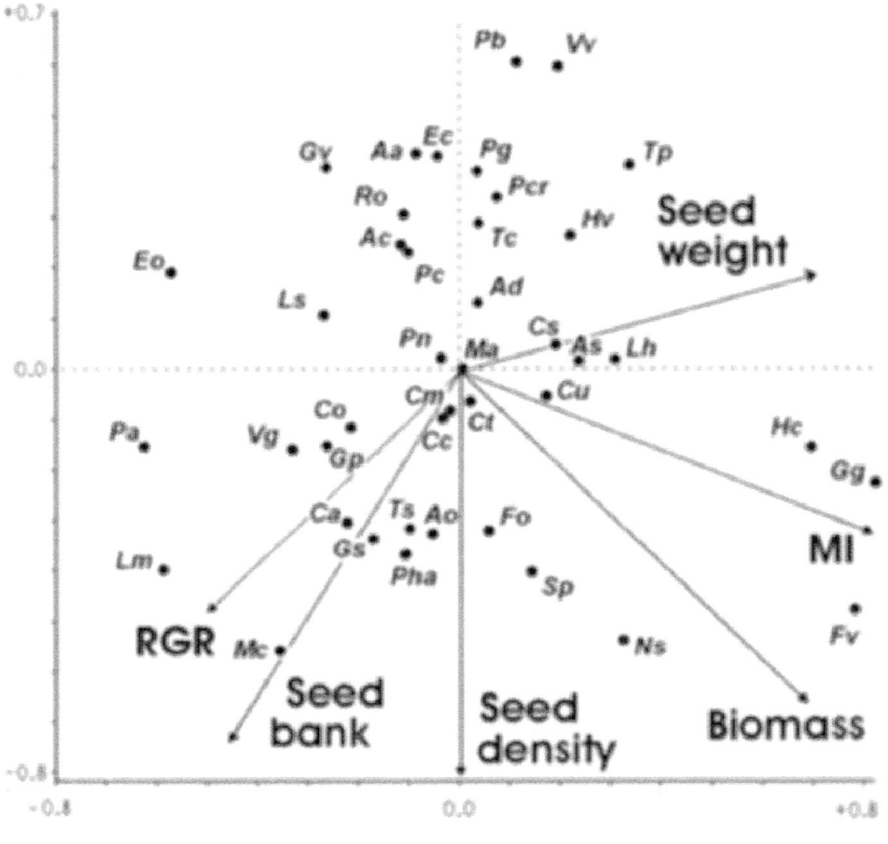

Figure 6.15. Plots of the first two axes of a PCA ordination of 42 species and their characteristics in four alpine communities. MI = morphology index; RGR = relative growth rate. Key to species: Aa - *Aster alpinus*; Ac - *Alchemilla caucasica*; Ad - *Antennaria dioica*; Ao - *Anthoxanthum odoratum*; As - *Anemone speciosa*; Ca - *Carex atrata*; Cc - *Carum caucasicum*; Cm - *Carum meifolium*; Co - *Carex oreophila*; Cs - *Carex sempervirens*; Ct - *Campanula tridentata*; Cu - *Carex umbrosa*; Ec - *Eritrichium caucasicum*; Eo - *Euphrasia ossica*; Fo - *Festuca ovina*; Fv - *Festuca varia*; Gg - *Geranium gymnocaulon*; Gp - *Gentiana pyrenaica*; Gs - *Gnaphalium supinum*; Gv - *Gentiana verna*; Hc - *Hedysarum caucasicum*; Hv - *Helictotrichon versicolor*; Lh - *Leontodon hispidus*; Lm - *Luzula multiflora*; Ls - *Luzula spicata*; Ma - *Minuartia aizoides*; Mc - *Matricaria caucasica*; Ns - *Nardus stricta*; Pa - *Primula algida*; Pb - *Polygonum bistorta*; Pc - *Pedicularis comosa*; Pcr - *Potentilla crantzii*; Pg - *Potentilla gelida*; Pha - *Phleum alpinum*; Pn - *Pedicularis nordmanniana*; Ro - *Ranunculus oreophilus*; Sp - *Sibbaldia procumbens*; Tc - *Taraxacum confusum*; Tp - *Trifolium polyphyllum*; Ts - *Taraxacum stevenii*; Vg - *Veronica gentianoides*; Vv - *Vaccinium vitis-idaea*.

The diagram obtained by PCA should be interpreted in terms of traits and trait combinations. The following groups arise from the diagram:

- Species with a large seed bank, small seeds, and relatively low biomass;
- Species with a high biomass, large seeds, and relatively high seed production;
- Species with large seeds, relatively low seed production, and low biomass.

By definition, 'violents' develop a large biomass under more productive conditions (Table 6.16). In our case this corresponds to the dominants of *Festuca* grasslands and *Geranium* meadows. The lichen heath and snowbed communities, which occur in more severe habitats, do not have real dominants. Species growing there can be considered as 'patients', but in their turn they may differ in their population characteristics.

Our data on the soil seed bank (see Chapter 6.2) confirm a tendency (Rabotnov 1975b, 1985) that 'violents' – *Festuca varia*, *Geranium gymnocaulon*, *Hedysarum caucasicum* in our case – usually do not accumulate viable seeds in the soil since they survive unfavourable periods in the adult stage. On the contrary, 'explerents' may have a large persistent seed bank (e. g. *Matricaria caucasica*, *Luzula multiflora*, *Gnaphalium supinum*, *Sibbaldia procumbens*).

On the basis of these results, we consider *Geranium gymnocaulon*, *Hedysarum caucasicum*, *Festuca varia* as 'violents'; *Viola altaica*, *Vaccinium vitis-idaea*, *Helictotrichon versicolor*, *Eritrichium caucasicum*, *Deschampsia flexuosa* as 'patients'; *Gagea fistulosa*, *Anthoxanthum odoratum*, *Matricaria caucasica*, *Veronica gentianoides*, *Gnaphalium supinum*, *Sibbaldia procumbens* as 'explerents'. The groups obtained by the clustering procedure are in agreement with such a scheme. For 42 species of all studied communities (Fig. 6.14), we consider the species of the second big group (clusters 4, 5 and 6) 'patients', whereas cluster 4 consists of species with some 'explerent' features; cluster 1 includes well-defined 'violents'; cluster 3 includes 'explerents'; and cluster 2 consists of species with many features of the 'explerent' strategy.

Fig. 6.16 shows the triangle ordination of the 42 alpine species based on *RGR* and *M* according to Grime's scheme (Grime 1979). Naturally, most of the alpine species are situated in the S-corner.

Response to disturbance differs in different groups of plants. Any disturbance event has two consequences: (1) it limits the plant biomass by its direct destruction (Grime 1979) - thus it has a negative effect; (2) it creates a source of available resources - thus it has a positive effect. 'Patients' have low biomass in grasslands and meadows, so their limitation by biomass removal is slight; but their response to resource availability is slow. Hence, on the whole 'patients' show a negative response to disturbance. 'Violents' lose more biomass because they are large. Their reaction to resource availability may be positive, but this depends on the intensity of the disturbance. When disturbance is small, adults of such species remain in undisturbed vegetation and can rapidly capture free resources. Moreover, adult plants represent a source of new propagules which can occupy such gaps, and the gaps are the necessary condition for recruitment of 'violents'. 'Explerents' hardly suffer from biomass removal, but greatly increase their abundance when resources become available, especially after severe disturbance. Thus, we conclude that species with a positive response to intense disturbance (Chapter 5.3.), such as *Anthoxanthum odoratum*, *Gnaphalium supinum*, *Luzula spicata*, *Sibbaldia procumbens*, *Veronica gentianoides*, have largely an 'explerent' strategy.

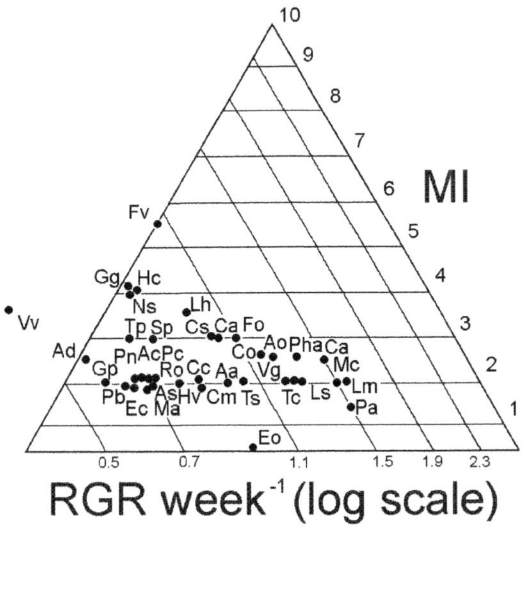

Figure 6.16. A triangular ordination of 42 alpine species according to Grime (1979). MI = morphology index; RGR = relative growth rate. Species as in Fig. 6.15 (from Onipchenko et al. 1998a with permission of Opulus Press)

In this study we considered only major soil surface disturbance by bears and wild boars which must favour the 'explerent' (or 'ruderal') strategy. We did not take into account the influence of natural grazing, a form of high-frequency and low-intensity disturbance which undoubtedly has a heavy impact on vegetation (Huntly 1987, Oksanen & Oksanen 1989, Oksanen &

Ranta 1992). However, the influence of ungulates is weak in our study area (Chapter 10), because the only ungulate - *Rupicapra rupicapra* L. - prefers steeper slopes in the summer and lower altitudes (subalpine belt) in the winter. Grasshoppers (Acrididae) and voles (*Pitymys majori*) have a higher impact, but their pressure was considered as an environmental factor so far (except the burrowing activity of voles). Nevertheless, the species which are known to tolerate grazing (e.g. *Festuca ovina*, *Gnaphalium supinum*, *Sibbaldia procumbens*) responded positively to large-scale soil surface disturbance.

The results of our investigation show that plants of lichen heaths have the most pronounced features of the 'patient' strategy, while plants of the more productive grasslands and meadows fit into either the 'violent' category - dominants of these communities - or into the 'explerent' category, which would be charcaterised by low biomass, large seed bank and a positive response to zoogenic disturbance. Species with clearly defined features of the 'explerent' strategy are abundant in snowbed communities; it is likely related to the short growing season in these habitats, where plants must possess an ability to develop rapidly.

Del Moral (1983b) reached a similar conclusion. He showed for subalpine meadows in the state of Washington that maximum differentiation into strategies occurred in intermediate habitats, namely in moderately stressed and disturbed sites with a productivity of ca. 300 g/m^2. Such conditions correspond to our *Festuca* grasslands and *Geranium* meadows, which are the most differentiated among the studied communities. Species of the lichen heaths and snowbed communities developing under significant abiotic stress are poorly differentiated. Our observations also confirm his conclusion that "adaptive strategies may contrast sharply within a narrow range of environmental conditions".

Our attempt to distinguish different strategies among alpine plants is not, of course, complete and definitive. For example, we did not consider possibilities of seedling establishment and growth to maturity under vegetation canopy and in gaps, etc. Furthermore, we did not take into account very important properties of clonal propagation, which could considerably complicate the analysis. Nevertheless we conclude that differentiation into strategy types was more obvious in the productive alpine grasslands than in the lichen heaths and snowbed communities. The grassland dominants (*Festuca varia, Geranium gymnocaulon, Hedysarum caucasicum)* showed the strategy of 'violents'. Species of the 'explerent' strategy were also well represented in these communities (*Matricaria caucasica, Sibbaldia procumbens, Gnaphalium supinum)*, but they especially predominated in the snowbed communities. Species of the 'patient' strategy prevailed in the lichen heaths. We conclude that plant strategy types are well differentiated, even in plant communities occurring under severe high-mountain conditions (Onipchenko et al. 1998a).

Chapter 7

Experimental study of plant relationships

7.1. Reciprocal transplantations in alpine plant communities

7.1.1. Introduction

Gigon (1987) suggested that in the alpine zone, environmental factors may play a leading role in excluding certain plant species from the less productive habitats (e.g., alpine heaths), while interspecific competition may play a more important role in the more productive habitats, e.g., in the alpine grasslands and meadows. Despite some recent contributions to the assessment of the relative importance of biotic and abiotic factors along productivity gradients (Grace 1993, Keddy et al. 1994, Wilson & Tilman 1995, Goldberg & Novoplatsky 1997), Gigon's hypothesis has not been tested with direct experiments.

There is a growing body of research employing direct field experiments in the arctic (Chapin et al. 1995, Chapin & Shaver 1996, Arft et al. 1999) and alpine ecosystems (Wielgolaski & Benedict 1993, Zhang & Welker 1996) to assess the possible range of biotic responses to future environmental change. The ultimate goal is to discover community assembly and response rules (sensu Keddy 1992) and to predict future changes in vegetation composition and abundance. One direct method is experimental reciprocal transplantations, when patches of intact vegetation are permanently removed from one community and placed into another and vice versa, and then the effects on both species' and community levels are monitored for a certain period of time. Because of the extensive efforts

V.G. Onipchenko (ed.), Alpine Ecosystems in the Northwest Caucasus, 223–262.

required, such experiments are rarely conducted, although they have a long history (Clements & Weaver 1924, Shennikov 1942, Gigon 1971, May et al. 1982, Del Moral 1983b, Wielgolaski & Benedict 1986, 1993, Partridge & Wilson 1988, Maranon & Bartolome 1993).

In this chapter we present results of a ten-year-long study of alpine communities' response to translocations. We analyzed responses of four alpine communities, arranged along a snow depth-soil moisture gradient, to the change in environmental conditions caused by reciprocal transplants. Following Gigon's hypothesis (1987), we expected to observe an overall positive response (e.g., increase in species numbers, biomass and/or number of shoots) with the transplantations into the more productive environments, and an overall negative response to the transplantations into the less productive ones.

7.1.2. Methods

Fragments of four typical alpine communities were reciprocally transplanted: alpine lichen heath (ALH), *Festuca varia* - dominated grassland (FVG), *Geranium gymnocaulon-Hedysarum caucasicum* - dominated meadow (GHM), and snow bed community (SBC).

The reciprocal transplantations of sod patches took place in 1989. The results were monitored for 9 years thereafter, until 1998. In each of the studied communities the experimental area was subsequently divided into 40 experimental squares 50 x 50 cm in size (Volkova et al. 1999). The resulting 160 sod squares (40 per community type) were dug around to the depth of about 15-25 cm, removed, and transplanted among the other communities. Ten squares from each community were used as control and were transplanted within the same community. The remaining 30 were reciprocally transplanted into 3 other communities (10 per each) into randomly assigned spaces. Three parameters were measured to estimate the effect of the transplanting on the communities: changes in floristic composition, changes in biomass of selected species during the sixth year, and shoot counts done for all species every year.

Biomass of aboveground shoots of the dominant 20 species was studied during the sixth year (1994) of the experiment. Of these, 16 species were found in only a single community, while four species were obtained from two or three communities for the total of 25 species' samples. Five to ten shoots from each sample were randomly chosen and cut from each experimental square from outside the counting plot, whenever the studied species occurred. The samples were then dried at 105°C until constant weight was reached.

All shoot counts were done inside transplanted square fragments on 25 x 25-cm counting plots. To estimate species response to transplantation we calculated a response coefficient K (Onipchenko et al. 1994) that compares the number of shoots under treatment with that in no-treatment (control) plots in the same year, and with the initial shoot numbers both under treatment and no-treatment. K was calculated according to the formula:

$$K_n = (N_n \times C_1) / (N_1 \times C_n),$$

where N is the number of shoots per experimental (transplant) plot in the first (*1*) and *n*-th year of observation, C – the number of shoots per control plot in the first (*1*) and *n*-th year. If $K_n > 1$, the species responded positively (increased its presence) in year *n*, if $K_n < 1$, it responded negatively (presence decreased). We used K only to compare the first two and the last two years of the experiment (1988-89 and 1997-98 respectively) by calculating mean shoot numbers of each pair of years. Cases in which denominator or numerator equaled zero, as well as cases when a species was not simultaneously present in control and in experimental plots, were excluded from the analysis.

7.1.3. Results and discussion

7.1.3.1. Changes in floristic richness

An earlier paper (Sennov & Onipchenko 1994) discussed preliminary results obtained after the first three years of the experiment. Moderate changes in floristic composition could be observed already by the 3rd year after the transplantations and became most apparent by the 5th-6th year of the experiment (Fig. 7.1). Most species survived for 10 years after the transplantations in the adult form, while some species disappeared and some new ones appeared. The number of newly invaded species was usually greater than the number of disappeared species after 10 years (Table 7.1.). In almost all cases, floristic richness reached its peak during the 5th year (1993) after the beginning of the experiment (Fig. 7.1.). Most invading species did not show dramatic increases in numbers, once established. Frequently, they were encountered for a few years only as seedlings or juvenile forms. For example, after transplanting fragments of *Festuca varia*-dominated grasslands into *Geranium-Hedysarum*-dominated meadows, juvenile *Geranium gymnocaulon* and *Hedysarum caucasicum* plants appeared during the year immediately following the transplantation (1989) and then continued to reappear every year. However, no adults have been established during the 10 experimental years.

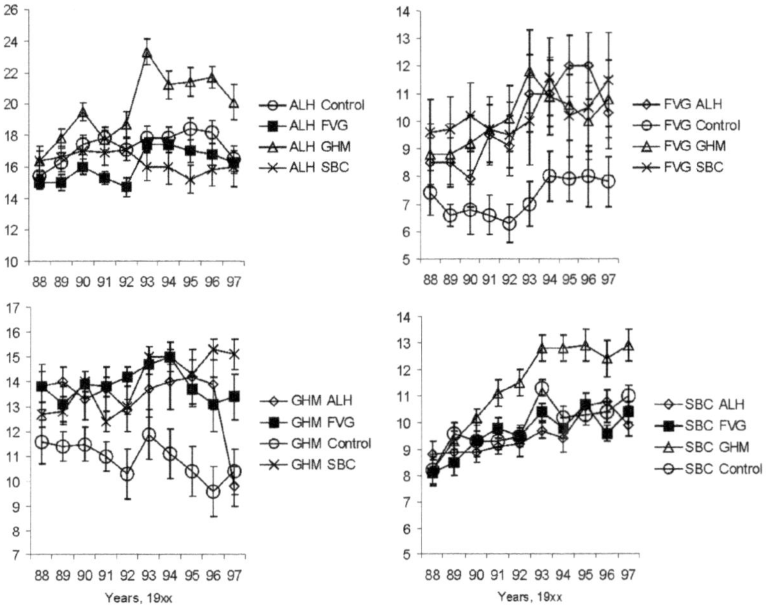

Figure 7.1. Changes in mean species richness per experimental counting plot (25 x 25 cm) in 16 transplantation treatments. Error bars represent one S.D. ALH – alpine lichen heath, FVG – *Festuca varia* - dominated grassland, GHM – *Geranium-Hedysarum* - dominated meadows, SBC – snow bed community. First abbreviations mark donor communities, second are recipients (or control).

When ALH fragments were transplanted, the greatest increase in mean floristic richness (i.e., number of species per 25 x 25 cm experimental counting plot) was observed in the GHM variant (four species added per plot as compared with 2 species added in control after 10 years). In this variant, the invading species included *Anthoxanthum odoratum, Carum meifolium, Festuca brunnescens, Matricaria caucasica* – all common species of the recipient GHM community. In comparison, in the ALH→FVG transplantation variant, only one species was added on average per plot for the duration of the experiment, while in the ALH→SBC variant the mean floristic richness did not change at all (*Anthoxanthum odoratum, Carum meifolium, Minuartia aizoides, M. recurva* were among added species, while, *Gentiana aquatica, Eritrichium caucasicum,* were among the disappearing ones).

Table 7.1. Changes in floristic composition on transplant plots. Year numbers since transplanting when a species appeared or disappeared are shown in brackets. Asterisk (*) marked species with more than 10 shoots at the last (1998) year of observation.

Communities		Species	
from	to	Disappeared	Appeared
ALH	ALH		*Anthoxanthum odoratum* (5), *Fritillaria collina* (1)*, *Gentiana aquatica* (2), *Luzula spicata* (1), *Polygonum bistorta* (4), *Scorzonera cana* (1)
ALH	FVG		*Anthemis cretica* (4), *Anthoxanthum odoratum* (4), *Carum meifolium* (3), *Cerastium purpurascens* (5), *Fritillaria collina* (1), *Leontodon hispidus* (4), *Scorzonera cana* (4)*, *Viola altaica* (3)
ALH	GHM	*Viola altaica* (2)	*Anthoxanthum odoratum* (4)*, *Carum meifolium* (1)*, *Cerastium purpurascens* (4), *Festuca brunnescens* (3)*, *Fritillaria collina* (1), *Geranium gymnocaulon* (1), *Hedysarum caucasicum* (1)*, *Leontodon hispidus* (2), *Matricaria caucasica* (2)*
ALH	SBC	*Eritrichium caucasicum* (5), *Gentiana aquatica* (2)	*Anthoxanthum odoratum* (1)*, *Carum meifolium* (3)*, *Catabrosella variegata* (3)*, *Fritillaria collina* (1), *Geranium gymnocaulon* (1), *Gnaphalium supinum* (4), *Minuartia aizoides* (5)*, *Minuartia recurva* (5)*, *Pedicularis nordmanniana* (5), *Phleum alpinum* (4), *Scorzonera cana* (1), *Sibbaldia procumbens* (2)*
FVG	ALH	*Sibbaldia procumbens* (6)	*Agrostis vinealis* (3), *Cerastium purpurascens* (4), *Campanula tridentata* (2), *Carum caucasicum* (2), *Euphrasia ossica* (1)*, *Helictotrichon versicolor* (5), *Polygonum bistorta* (5), *Taraxacum stevenii* (2), *Vaccinium vitis-idaea* (4)*
FVG	FVG		*Carum meifolium* (1), *Cerastium purpurascens* (4), *Fritillaria collina* (1), *Hedysarum caucasicum* (5), *Leontodon hispidus* (5), *Senecio aurantiacus* (5), *Taraxacum stevenii* (3)
FVG	GHM	*Anthyllis vulneraria* (1), *Viola altaica* (3)	*Geranium gymnocaulon* (1), *Hedysarum caucasicum* (1), *Veronica gentianoides* (1)*
FVG	SBC		*Anemone speciosa* (3), *Carum caucasicum* (2), *Cerastium purpurascens* (1), *Gnaphalium supinum* (5), *Taraxacum stevenii* (1)
GHM	ALH	*Erigeron caucasicus* (7), *Gentiana biebersteiniana* (6), *Luzula multiflora* (3),	*Anemone speciosa* (4), *Eritrichium caucasicum* (4), *Sibbaldia procumbens* (2)*, *Taraxacum stevenii* (4), *Vaccinium vitis-idaea* (3)*
GHM	FVG	*Luzula multiflora* (3), *Rumex alpestris* (5)	*Anemone spesiosa* (6), *Hieracium laevigatum* (2)

Table 7.1.(Cont.)

Communities		Species	
from	to	Disappeared	Appeared
GHM	GHM	*Campanula tridentata* (6), *Erigeron caucasicus* (6), *Gnaphalium supinum* (2)	*Festuca ovina* (5), *Sibbaldia procumbens* (2)
GHM	SBC	*Carum caucasicum* (5), *Festuca ovina* (2)	*Catabrosella variegata* (4), *Gagea fistulosa* (4)*, *Gentiana pyrenaica* (2)*, *Pedicularis nordmanniana* (4)*, *Sibbaldia procumbens* (3)
SBC	ALH	*Carum meifolium* (3), *Pedicularis nordmanniana* (5)	*Campanula tridentata* (4), *Euphrasia ossica* (1), *Festuca ovina* (3)*, *Gentiana pyrenaica* (6)*, *Ranunculus oreophilus* (6), *Scorzonera cana* (6), *Vaccinium vitis-idaea* (3)*
SBC	FVG	*Potentilla crantzii* (5), *Pedicularis nordmanniana* (5)	*Anthemis cretica* (4), *Festuca brunnescens* (1)*, *Festuca ovina* (2)*, *Festuca varia* (4)*, *Hieracium lactucella* (5)
SBC	GHM	*Pedicularis nordmanniana* (6)	*Festuca brunnescens* (1), *Geranium gymnocaulon* (1), *Hedysarum caucasicum* (2), *Matricaria caucasica* (2)*, *Scorzonera cana* (1), *Veronica gentianoides* (4)
SBC	SBC		*Carum caucasicum* (1)

When FVG fragments were transplanted, the fastest and greatest increase in richness was observed in the FVG→GHM variant prior to the 6th year, and in ALH after the 6th year of the experiment. The overall result was 1.4 species added on average per plot in the FVG→GHM, 1.6 species added in the FVG→ALH variant, 0.7 species added in the FVG→SBC variant, while the number of species in control remained unchanged.

When GHM fragments were transplanted, the only clear increase in floristic richness was observed in the SBC variant (1.4 species mean increase per plot).

Finally, when SBC fragments were transplanted, the greatest increase in floristic richness was observed in the SBC→GHM variant (3.6 species were added on average). At the same time, 1.6 species were added on average in control plots perhaps due to disturbance associated with transplanting fragments within the same community.

The most successful invaders were:

In ALH - *Campanula tridentata, Carum caucasicum, Euphrasia ossica, Festuca ovina, Gentiana pyrenaica, Helictotrichon versicolor, Vaccinium vitis-idea.*

In FVG - *Anthoxanthum odoratum, Deschampsia flexuosa, Festuca brunnescens, F. ovina, F. varia, Scorzonera cana.*

In GHM – *A. odoratum, Carum meifolium, F. brunnescens, Hedysarum caucasicum, Matricaria caucasica, S. cana, Veronica gentianoides.*

In SBC - *A. odoratum, C. meifolium, Gentiana pyrenaica, Gnaphalium supinum, Minuartia aizoides, Pedicularis nordmanniana, Sibbaldia procumbens.*

There are apparently three reasons for change in floristic composition: disappearance of some original species from the transplanted fragment, emergence of species that had been present in the transplanted patch in latent form, and invasion of new species from the surrounding recipient community. In a few cases, invading species were not present either in the transplanted, or in the recipient community, and apparently came from the general vicinity of the sites. For example, *Sibbaldia procumbens* appeared during the 3rd year in the GHM→ALH variant. It is unclear whether this species had already been present in latent form in soils, or was indeed brought in from other communities in the research area.

The alpine communities proved resilient to transplantations, because only a moderate change in floristic composition was observed. The total number of species per 25 x 25 cm experimental plot increased on average by 2–4 species, and in some variants remained the same (although the composition changed somewhat). This result could reflect the overall long-lived slow-growing nature of alpine species. The fact that the majority of variants showed an overall increase in the total number of species during 10 years signifies that the complete species turnover may take a longer time, and we are still observing the intermediate conditions, when some new species have already arrived, but the old species are still on the way out.

In all variants, *Pedicularis nordmanniana* disappeared from the transplanted plots by the fifth or sixth year, while *Euphrasia ossica* severely decreased in numbers. Both species are semi-parasites apparently suffering from disruption of connections to their hosts after transplantation.

7.1.3.2. Shoot biomass

Response of shoot biomass to transplantations varied across the studied species (Fig. 7.2.). In many cases a species responded negatively to transplantation into any non-native community by decreasing its biomass. In some cases the species biomass declined only after transplantation into drier habitats (ALH or FVG), but not into moister habitats (GHM and SBC). This group included *Geranium gymnocaulon* from GHM, *Scorzonera cana* from FVG, and, not surprisingly, five species from SBC.

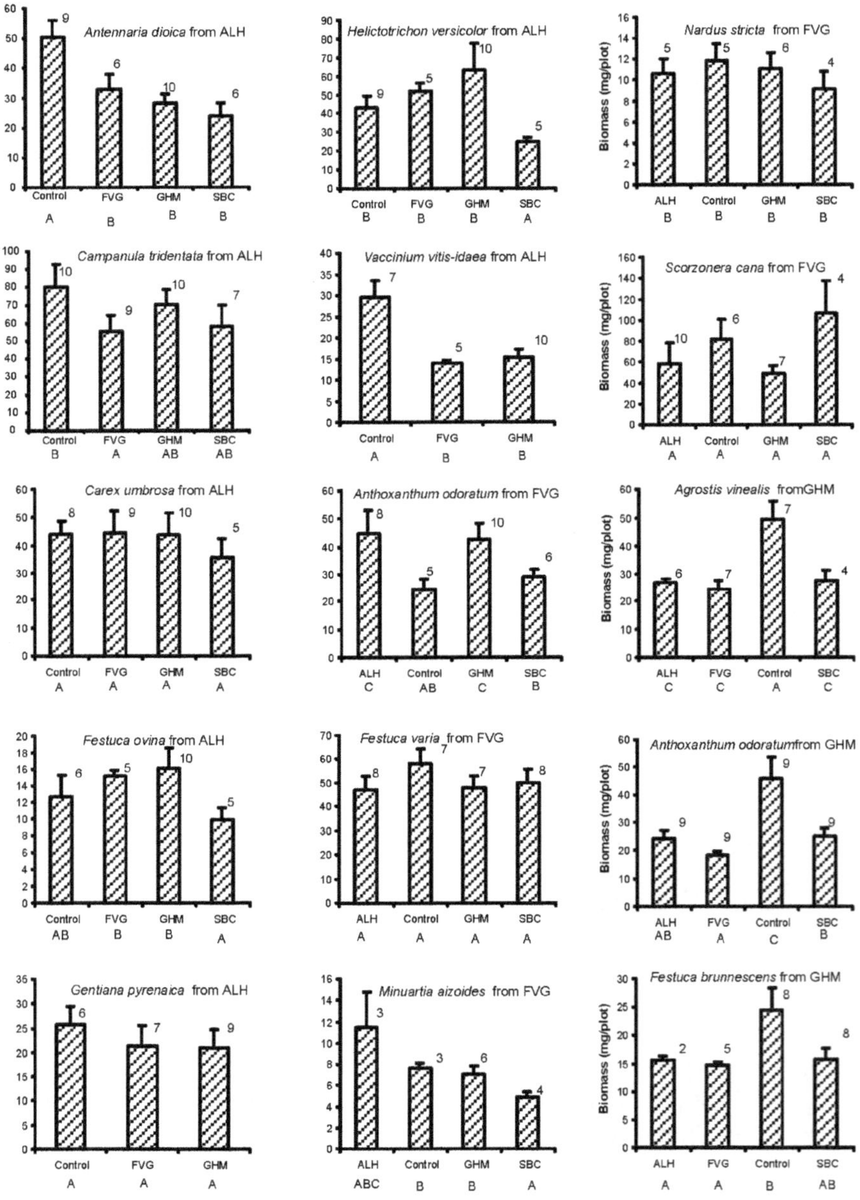

Figure 7.2a. Shoot biomass (mg per shoot or tiller) of selected species from the experimental plots five years after the transplantation (1995). Vertical bars represent 0.5 S.D. Numbers show number of plots from which plant shoots were obtained. Letters indicate significance of difference in means as tested with t-test (P<0.05). Means that share the same letter are not significantly different, those with different letters – are significantly different

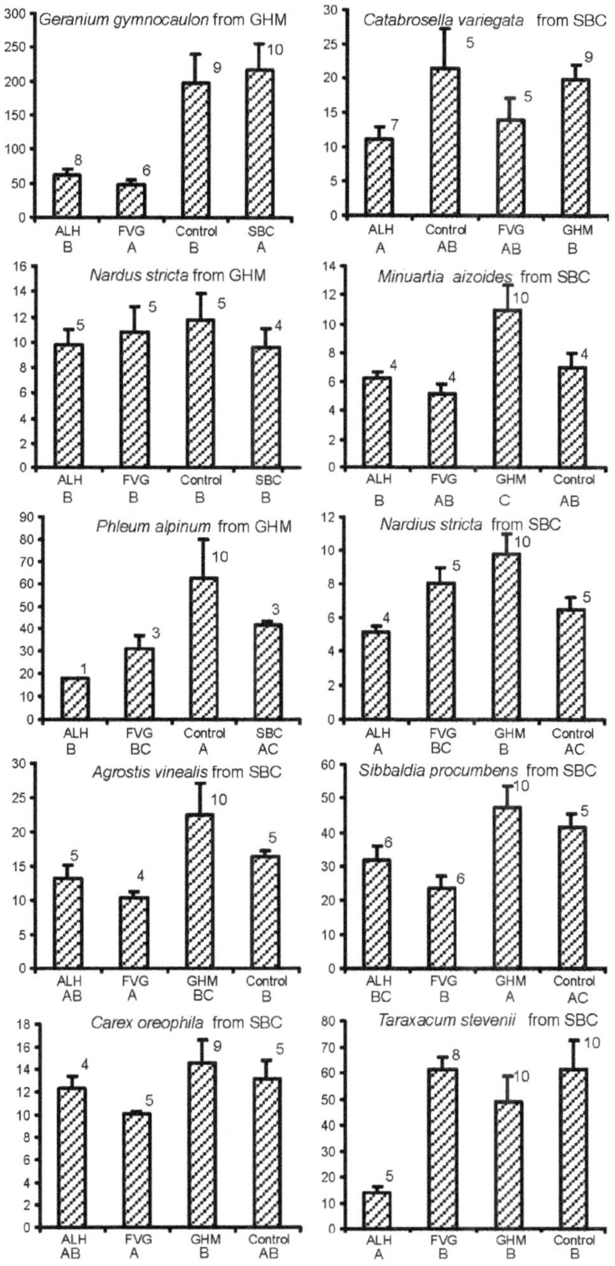

Figure 7.2b. Shoot biomass (mg per shoot or tiller) of selected species from the experimental plots five years after the transplantation (1995) (see *fig. 7.2a*).

In other cases the species biomass markedly increased after transplantation into alpine grassland and meadow (FVG and GHM) from the extreme conditions of ALH and SBC. This group includes *Nardus stricta* from SBC, and *Festuca ovina* and *Helictotrichon versicolor* from ALH.

Of the 20 species studied, most were dominant plants in terms of their biomass. Contrary to our initial expectations based on Gigon's hypothesis, there was a wide range of responses that can be classified into four different groups. The first includes species that occur in the apparently "physiologically optimal" habitat (i.e., for whom their realized and potential niches coincide). The fact that this group is large (13 samples) implies that many alpine species, at least among dominants, occur in their preferred habitat. One of the species in this group, *Vaccinium vitis-idaea* from alpine heath (ALH), was studied also by Wielgolaski & Benedict (1986, 1993), who found similar results in transplantation experiments in Norway. In their study, the species decreased its biomass when transplanted into deep-snow tundra from a snowless tundra. Based on the results of our study we conclude that in our area, *Vaccinium* and other ALH dominants (*Antennaria dioica, Campanula tridentata, Gentiana pyrenaica*, and, to lesser extent *Carex umbrosa*) apparently find their optimal habitat in the dry, relatively unproductive ALH community. Similarly, *Festuca varia* common in FVG indeed prefers this moderately dry and productive grassland, while *Phleum alpinum, Agrostis vinealis, Festuca brunnescens*, and *Nardus stricta* find their optimal habitat in the moderately wet and productive GHM community. Likewise, *Catabrosella variegata* and *Taraxacum stevenii* prefer the relatively unproductive and snowiest SBC community. Any future environmental change will likely have an adverse effect on such species, because they appear to be restricted to a rather narrow set of conditions.

The second group includes plants that did not perform well when transplanted into drier environments (i.e., ALH and FVG), perhaps due to a high level of environmental stress (desiccation and cold). However, this group showed a marked biomass increase under benign, moist conditions in GHM. It is probable that under natural conditions these species are excluded from this "optimal" habitat by increased competition with dominants in meadows, most notably *Geranium, Hedysarum*, and *Phleum* (as predicted by Gigon's hypothesis). The species in this group were all from the relatively unproductive SBC community, where they grow under enduring snowy conditions. Similar results were obtained in *Kobresia bellardii* transplantation experiments in the Olympic National Park (Bell & Bliss 1979). On the other hand, presence of competitive exclusion of mesic grassland species transplanted into dry *Kobresia* tundra was demonstrated by Theodose & Bowman (1997).

The third group includes species (*Festuca ovina, Helictotrichon versicolor, Nardus stricta*), that grow under severe conditions which are apparently not optimal for them (ALH and SBC), because they are excluded from moderate habitats (FVG and GHM) probably due to competition with dominants there. Such ALH and SBC species generally increased their biomass under the more benign conditions of meadows and grasslands (while competition was excluded), but fared poorly when transplanted from heath to snow beds, and vice versa. Thus, this group of species also conforms to the behavior predicted by Gigon's hypothesis.

Difference in response to the *same* conditions observed in the species from the fourth group – *Anthoxanthum odoratum* and *Minuartia aizoides* taken from *different* communities – may indicate ecotypical variation in local populations. In our area, these two species grow under a wide range of conditions and demonstrate elements of a ruderal ("explerent") strategy (see Chapter 6.3.). The biomass of *M. aizoides* from SBC and *A. odoratum* from GHM showed mildly negative responses to transplantation into ALH. *A. odoratum* from GHM showed a decrease in the mean biomass when transplanted into ALH (45 mg *versus* 20-25 mg). In contrast to that, individuals from the same species originating in FVG showed a marked increase in the mean biomass when put into ALH and GHM (increase from 25 mg to 45 mg). Transplantation experiments elsewhere showed evidence for the existence of local ecotypes in *Polygonum cuspidatum* in Japan based on elevation (Shiosaka & Shibata 1993), in *Dryas octopetala* in Alaska depending on snow bank accumulation (McGraw 1985), in *Carex curvula* in the Alps on different soils (Erschbamer 1996). However, Seliskar (1985) found that such ecotopic variation may disappear after transplanting halophytes into habitats with different salinity in Oregon tidal marshes.

7.1.3.3. Shoot dynamics

The studied species show quite varied responses to transplantations according their shoot number. Some examples are shown in Fig. 7.3. Such species as *Eritrichium caucasicum, Euphrasia ossica, Minuartia circassica,* and *Vaccinium vitis-idaea* decreased in shoot number when transplanted to all communities in comparison with the control (ALH). Other species increased in shoot number in non-native communities (Fig. 7.3.).

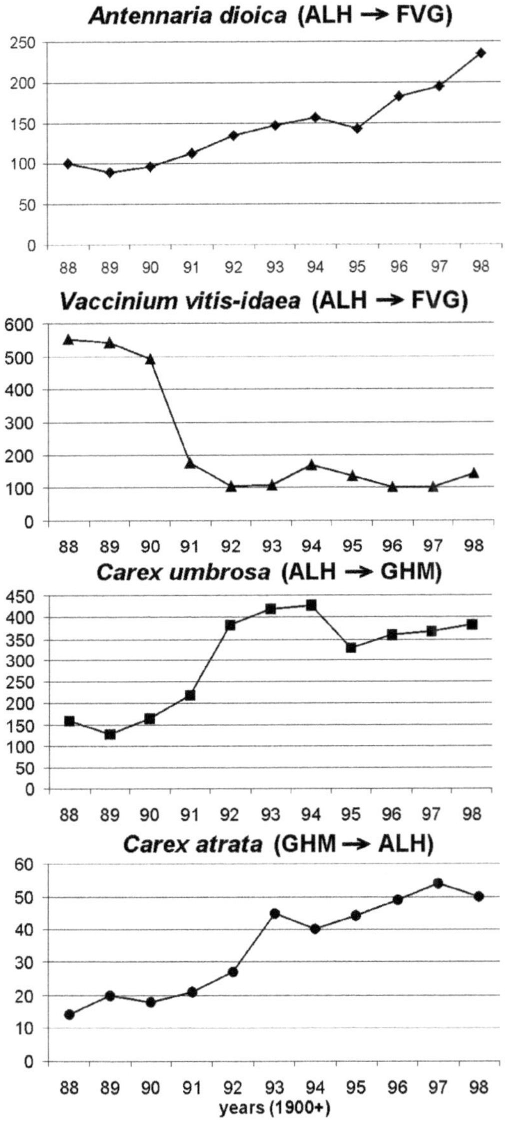

Figure 7.3. Examples of change in shoot number (ordinate) after transplantation during 10 years. Abbreviations: ALH – alpine lichen heath, FVG – *Festuca varia* grassland, GHM – *Geranium – Hedysarum* meadow.

Overall, we observed a wide range of responses among different species and communities. We calculated the number of species with different shoot dynamics according to K–values. Table 7.2. represents the number of species with positive and negative reaction to transplantation in comparison with the control.

Table 7.2. Number of species with different reaction to transplantation. "+" – increase in shoot number (K > 1,5), "-" – decrease in shoot number (K< 0,67). Communities: ALH – alpine lichen heath, FVG – *Festuca varia* grassland, GHM – *Geranium* – *Hedysarum* meadow, SBC – snow bed community (from Volkova & Onipchenko 1999).

Community: FROM	reaction	Community: TO ALH	FVG	GHM	SBC	Total number to other communities
ALH	+		4	4	4	12
	-		4	4	9	17
FVG	+	4		3	5	12
	-	5		3	6	14
GHM	+	3	4		5	12
	-	8	6		4	18
SBC	+	2	0	2		4
	-	5	7	4		16
Total number	+	9	8	9	14	
	-	18	17	11	19	

Especially important is the finding that every community contains species that behave in different ways in response to transplantations. Contrary to the expectations, some species, including species of the less productive habitats, actually performed *worse* when placed in the more productive habitats. On the other hand, some species favoured any non-native community over their native one. Another group of species largely conformed to the expectations, that is, increased numbers in the more productive habitats, but decreased in the less productive habitats.

In the study area, depth and duration of winter snow cover is the main factor determining community composition (Onipchenko 1994a). The least productive habitats are the least snowy (ALH) and the snowiest (SBC), while in the moderate and most productive habitats, are grasslands (FVG) and meadows (GHM) found in the middle of the toposequence. Since the root competition was suppressed for at least the first 6 years (Onipchenko 1994c), and there is little aboveground competition for light among short alpine plants (Onipchenko et al. 1994), we hypothesize that the change in abiotic conditions was the main factor affecting the plots in our experiment. If Gigon (1987) is correct, we should have observed an overall positive response to transplantations from ALH and SBC into more productive conditions of FVG and GHM, and an overall negative response to transplantations into less productive conditions of ALH and SBC.

Instead, we observed a wide range of responses among different species and communities. Table 7.2. illustrates the final results of the transplantation experiments at the community level. Very important is the finding that individual species from every community exhibited a wide variety of responses to transplantations. Contrary to the expectations, many species appear to prefer their own habitats, including the less productive ones, and actually perform worse in the more productive environments (e.g., *Vaccinium vitis-idaea* clearly favors ALH, *Catabrosella variegata, Gnaphalium supinum, Taraxacum stevenii* - SBC). Other species, in fact, favored any non-native community to their native one (e.g., *Carex umbrosa* when transplanted from ALH increased its shoot numbers in all other variants). Another group of species largely conformed to the expectations, that is, increased numbers in the more productive habitats, but decreased in the less productive habitats, as expected. Examples include *Antennaria dioica, Campanula collina, Gentiana pyrenaica, Luzula spicata* and *Plantago atrata* transplanted from ALH and *Phleum alpinum* and *Festuca brunnescens* from SBC, *F. varia* from FVG, *Geranium gymnocaulon, Phleum alpinum* from GHM.

7.2. Dominant removals

7.2.1. Alpine lichen heath

7.2.1.1. Introduction

Experimental studies of plant interactions in natural plant communities present a convenient way of exploring basic principles of community organization and enables us to establish a framework for making predictive statements about the structure and behavior of natural ecosystems (Tilman 1982, Keddy 1989). By removing dominant species, we can learn about different responses of the remaining taxa, and by analyzing the observed responses develop "assembly rules" for the community as a whole.

Competition is frequently believed to be the leading biotic factor influencing plant communities, even in cold, nutrient-poor and otherwise stressful environments such as arctic or alpine (del Moral 1983a, Chapin & Shaver 1985, Tilman 1988). An alternative hypothesis suggests that positive interactions among plants should be prominent in the stressful environments (Callaghan & Emanuelson 1985, Gigon & Ryser 1986, Callaghan 1987, Carlsson & Callaghan 1991, Callaway 1995). As the environment becomes more stressful, plants may facilitate survival of other species by providing shelter from desiccating winds or low temperatures, by increasing availability of water or nutrients, by providing protection from herbivores, etc. (Billings & Mooney 1968, Moen 1993, Callaway 1995, Shevtsova et al. 1995). Another important question is whether species compete with just a single main competitor, or each species experiences competitive pressure from a number of species, so called "diffuse competition" (MacArthur 1972). Field removal experiments have been carried out in many different plant communities in the last two decades (Abdul-Fatih & Bazzaz 1979, Fowler 1986, Aarsen & Epp 1990, Gurevitch et al. 1992, Gurevitch & Collins 1994), but the most common study areas were old-fields, not pristine ecosystems (Pinder 1975, Allen & Forman 1976, Hils & Vankat 1982). Also, most of these experiments did not exceed three years in length, therefore, no conclusions about long-term responses of species could be made. Only a handful of removal studies has been reported from alpine communities (del Moral 1983a, Jonasson 1992, Owen 1995).

In our work, we investigated responses of individual species of plants in alpine lichen heath (ALH) to the removal of five main vascular dominants. The following questions were examined: (1) is there evidence for competition between different vascular species of plants, and is there evidence that positive interactions between species are also important? (2) which species seem to experience pressure from competition? which are facilitated by positive interactions among plants? (3) is there a difference between short-term (two - three years) and long-term (ten - thirteen years) responses of species? (4) is there evidence for "diffuse competition", i.e., do a few species make a coordinated response to the removal of one of the dominants, or does only one species respond each time?

7.2.1.2. Objects and methods

The following dominant species of plants, each making up over 5% of the total plant biomass (Chapter 5.2), were selected for the removal experiments: *Anemone speciosa* - vegetatively sessile half-rosette polycarpic with racemous roots, *Antennaria dioica* - half-rosette polycarpic with a short rhizome, *Festuca ovina* - dense bunch grass, *Trifolium polyphyllum* - rosette polycarpic with rhizome and tap root, and a group of three species of sod-forming sedges, almost indistinguishable in vegetative state and belonging to the same life-form, and therefore treated here as a single unit *Carex* spp.(*Carex sempervirens*, *C. umbrosa*, *C. caryophyllea*).

The experiment started in 1983 and was continued until 1995 for the total of 13 years. There were six treatments: removals of the five above-mentioned species and no removals as control. Only one species was removed per treatment (Aksenova et al. 1998).

To estimate the significance of shoot number response for the remained species we use ANOVA as well as response coefficient K (see Chapter 7.1.2). These methods were described in Aksenova et al. (1998) in detail.

7.2.1.3. Results

It was shown that the number of statistically significant cases, when plants on average increased or decreased their number of shoots following the removal, is close (84 and 91 correspondingly) (Aksenova et al. 1998). This indicates that both negative (competitive) and positive (facilitative) relations between plants may be equally common in the studied community. Most species reacted either consistently negatively, or positively, in response to the removal of any of the dominant species, if a significant effect was at all detected. In a few cases, species responded differently depending on which dominant was eliminated. Thus, *Minuartia circassica* showed a positive response to the removal of *Festuca ovina*, but a negative one to the removal of all other species, while *Antennaria dioica* positively responded to the removal of *Anemone speciosa*, *Festuca ovina* and *Trifolium polyphyllum*, but slightly negatively to the removal of *Carex* spp. (Fig. 7.4.).

Among the five dominants, positive responses prevailed (Fig. 7.5.). We view this as evidence that all five dominants are strongly competing with each other in the studied community. Two subordinate species (*Helictotrichon versicolor* and *Luzula spicata*) also showed consistently positive responses, which indicates that they are also mostly experiencing competition from other plants in the community.

On the other hand, all other main subordinate species (*Eritrichium caucasicum, Carum caucasicum, Campanula tridentata, Gentiana pyrenaica, Minuartia circassica, Euphrasia ossica*) showed uniformly negative responses (relative decrease in the number of shoots) to the removal of the dominant plants (Fig. 7.6.). Apparently, presence of these species in the ALH is facilitated by the five dominating species.

As expected, in most cases the reactions between pairs of species were not symmetrical. If species A react positively to the removal of species B, species B did not necessarily reacted in the same way towards the removal of species A. The following pairs demonstrated symmetrical positive reactions to the removal of their counterparts - an evidence for competition between them (Fig. 7.5):

Antennaria dioica -Trifolium polyphyllum,
Antennaria dioica - Anemone speciosa,
Festuca ovina - Carex spp.

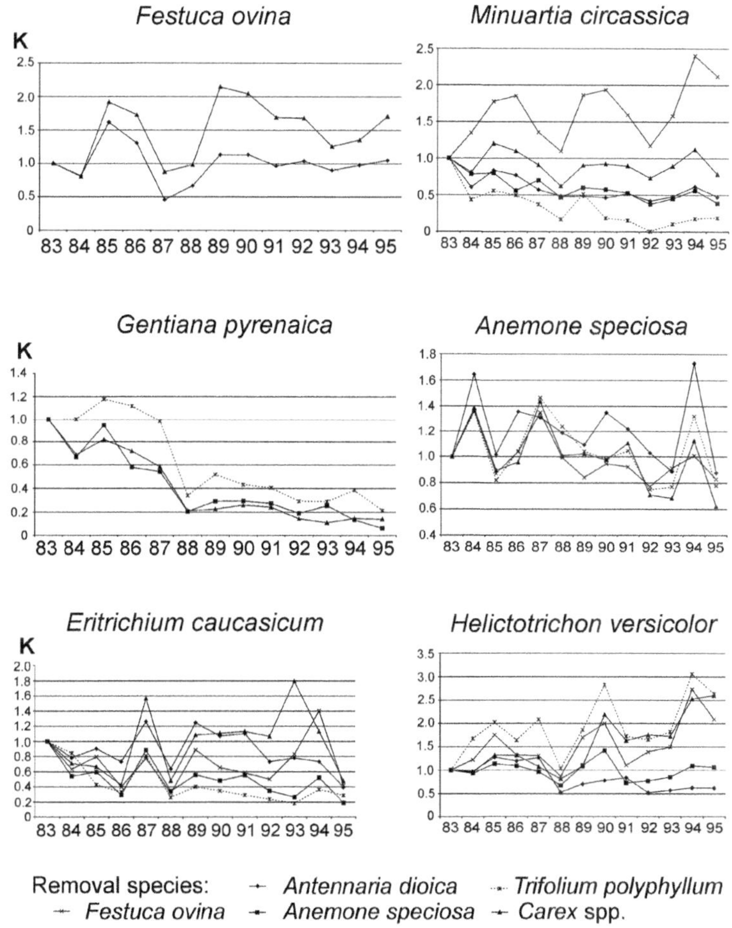

Figure 7.4. Annual dynamics of the response coefficient K for 6 species. The three-year smoothed average is shown (from Aksenova et al. 1998 with permission of *Ecoscience*)

7.2.1.4. Discussion and conclusions

Up to 40 vascular plant species occur in the Teberda Reserve alpine lichen heath per 100 m^2 (Chapter 4.3.3), while only about 15 are common. Responses of the rare species could not be statistically tested due to their low occurrence on the experimental plots. However, among those species no sudden increase in numbers was noted. Both the total number of species and floristic richness per plot remained fairly constant for the whole duration of the experiment.

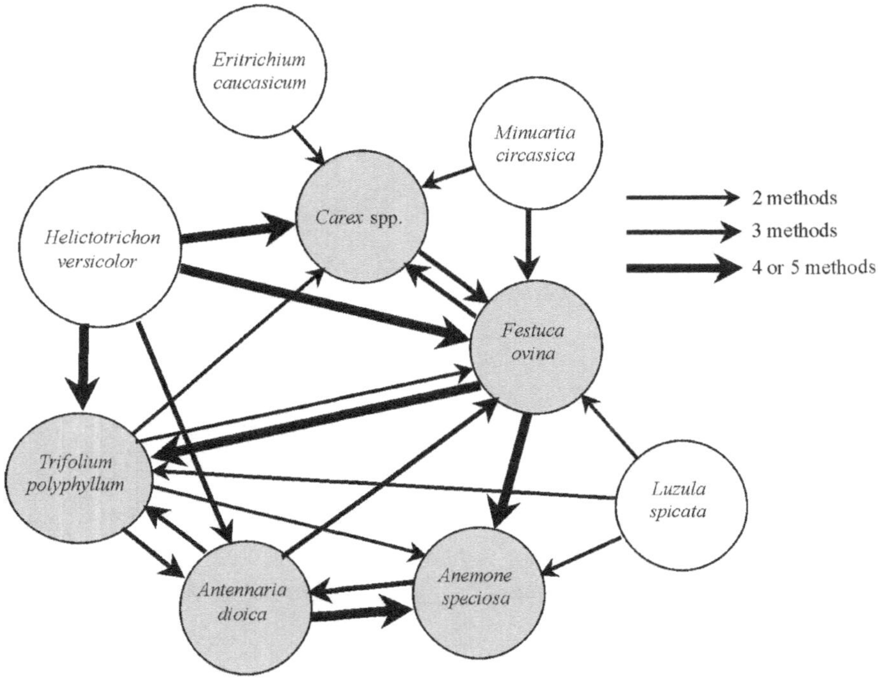

Figure 7.5. Negative relationships (positive responses to the removals) between the alpine species. Arrows are directed from the responding species to removed dominants. The dominants are shown as dark circles, other species as white circles. Width of arrows corresponds to numbers of statistical methods confirming significance of the response (from Aksenova et al. 1998 with permission of *Ecoscience*)

The overall community composition proved to be quite resistant to change. Overall species richness of the community did not change. Not a single species disappeared completely, nor did any new species appear despite the fifteen years of experiments. Analogous results were obtained in Estonian polydominant meadows, where removal of a number of species did not speed up establishment of new species (Zobel et al. 1994). Conversely, removal of the absolute dominant *Dactylis glomerata* on rich soils in an old-field community with much higher productivity and in a less stressful environment than the ALH led to a dramatic increase of species diversity. Such response was made even more obvious by adding fertilizer (Gurevitch & Unnasch 1989). In our case, lack of newcomers could be in part explained as being caused by the slow growth of the alpine plants, and in part by the lack of potential available newcomers.

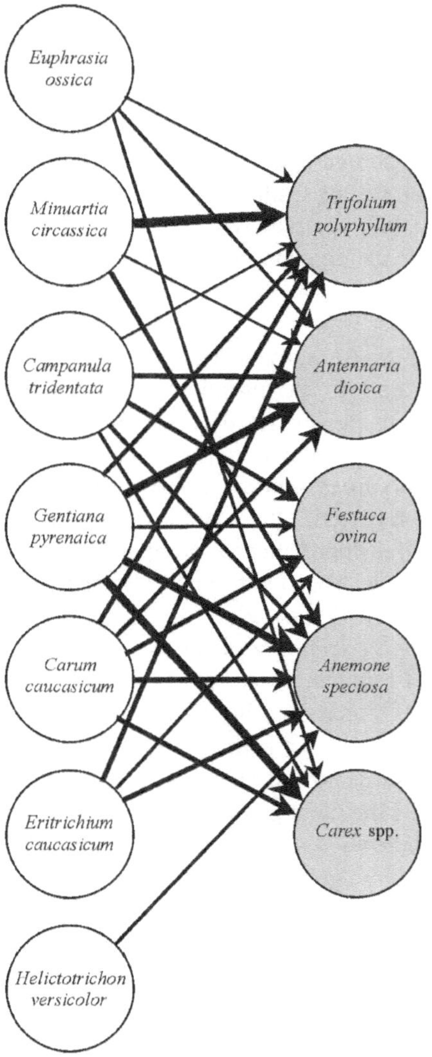

Figure 7.6. Positive relationships (negative responses to the removals) between the alpine species (from Aksenova et al. 1998 with permission of *Ecoscience*)

The presence of competition among vascular plants in the ALH was demonstrated earlier in the series of experiments with lichen removal and root cutting (Chapter 5.3). It was also shown that in the ALH lichens occupy vacant spots of bare soil in between the scattered clumps of vascular plants, and do not compete with the latter. Vascular plants do compete for resources with each other through the interaction of root systems below the ground. Our current results suggest the widespread presence of competition among vascular plants, especially among dominants. However, we also found evidence that subordinate species seem to experience decline after the removal of the dominant plants, not an increase in numbers as would be expected had the competition been present. We believe that our results support the hypothesis of Callaghan and others (Callaghan & Emanuelson 1985, Callaghan 1987, Carlsson & Callaghan 1991) that positive interactions among vascular plants are common in stressful habitats, such as in the alpine zone.

Such results differ from experiments in nutrient-rich habitats, such as salt marshes (Silander & Antonovics 1982) or wetlands (Keddy 1989), where competition appeared to be prominent, while no evidence for positive interactions between species was detected. Similarly, Moen (1993) found no evidence for positive interactions between *Oxyria digina* seedlings and resident *Ranunculus nivalis* in an alpine field. Instead, neutral to negative interactions were reported. Other studies from high-elevation communities also demonstrate the importance of competition (del Moral 1983a, Egerton & Wilson 1993). On the other hand, Shevtsova et al. (1995) obtained convincing evidence that some kind of positive relationship existed between *Empetrum nigrum, Vaccinium myrtillus* and *Vaccinium uliginosum* in northern forests of Finland. Likewise, proximity to dwarf shrubs, moss mats or artificial shelters apparently improved survival of *Carex bigelowii* in an alpine habitat (Carlsson & Callaghan 1991).

While most species showed the same responses year after year (Fig. 7.4.), many demonstrated the most pronounced reaction by the third or fourth year of the experiment (1986–1987) (Aksenova et al. 1998). This is probably due to the generally slow growth response typical for alpine plants. A few species, however, changed their behavior in the course of the 13 experimental years from positive to negative, or vice versa. For example, *Anemone speciosa* showed an initially positive response to the removal of all four other dominants, while after 1989 it changed its reaction to the removal of *Carex* spp., *Festuca ovina* and *Trifolium polyphyllum* to moderately negative (Fig. 7.4). This may be due to the different response time for different species. Some are quicker to respond favorably to reduced competition, but then they get outcompeted by the slower responding community members. We agree with Tilman (1988) that more long-term competition experiments are necessary to learn about the difference in short- and long-term competitive behaviour of plants.

The negative response to the removal of the dominants (Fowler 1981, Callaway 1995) may be generally attributed to the detrimental changes in the environment as a result of removal of large, protective dominant plants, increase in desiccation by wind, exposure to lower temperatures, drop in soil moisture, lack of protection from herbivores, etc.; increase in competition from other plants that were previously suppressed by the dominant species; removal or weakening of the main host plants. In our case, the decline of *Euphrasia*, a semiparasitic plant from the Scrophulariaceae family, can be attributed to the removal of its main hosts - sedges and grasses (Larin et al. 1950, Seel & Press 1994).

Overall, our results suggest that many subordinate species in the alpine tundra depend on the dominant species for survival. Many showed a steady decline for the whole period of the experiment (e.g., *Eritrichium caucasicum* after removal of *Trifolium polyphyllum* and *Anemone speciosa*, or *Gentiana pyrenaica* after removal of *Anemone speciosa, Carex* spp., and *Trifolium polyphyllum*). Not all subordinate plants responded negatively to the removal of the dominants, however. Thus, as was mentioned earlier, both *Helictotrichon versicolor* and *Luzula spicata*, plants from different families but with somewhat similar growth form, reacted positively to the removal of both *Festuca ovina* and *Carex* spp. This implies that as a group, sedges, rushes and grasses (i.e. graminoids) in the studied community make up a guild of tightly competing species.

In most cases, generative shoots increased or decreased in numbers along with the vegetative ones. In case of *Campanula tridentata* (following removal of *Festuca ovina* and *Anemone speciosa*) and *Carum caucasicum* (following removal of *Anemone speciosa*, *Carex* spp. and *Trifolium polyphyllum*) a general decline in the number of shoots was accompanied by a simultaneous increase (relative or absolute) in the number of flower-producing, generative shoots. Since population analysis of these species revealed the prevalence of juvenile individuals in populations of both species (Semenova & Onipchenko 1989, Onipchenko & Komarov 1997), it appears that young non-flowering plants reacted negatively to the removal, while older generative plants reacted positively (i.e., increased their numbers).

The responses of species with different growth forms were different. Rhizomatous, stolon-producing, or otherwise vegetatively mobile species on average responded more favorably to the removal of dominants than did the ones dispersing by seeds only. Examples of the former group include *Antennaria dioica*, *Trifolium polyphyllum*, *Helictotrichon versicolor*, and *Festuca ovina*. The second group includes plants lacking stolons (*Anemone speciosa, Eritrichium caucasicum*).

We also noticed that the reaction of species to the dominant removal was not connected with the depth of the bulk of absorbing roots. Investigations of the root systems of the ALH plants (Onipchenko 1987b) showed that the fine roots of *Festuca ovina*, *Antennaria dioica* and *Carex* spp. are located mostly in the A horizon down to the depth of 6 to 8 cm, while the roots of *Anemone speciosa* and, apparently, *Trifolium polyphyllum*, are located in the lower AC horizon at 6 to 16 cm depth. All of these species, however, responded in a similarly positive manner to the removal of dominants.

Finally, our experiment suggests the presence of "diffuse" (non-species specific) competition in the studied community (MacArthur 1972). Most species responses were dominant-insensitive. Also, removal of a given dominant made a few, not just a single species, increase their presence in the community. The same behaviour was observed in many other experimental studies of competition in plant communities (Fowler 1981, Silander & Antonovics 1982, Keddy 1989, Moen 1989).

To summarize our findings, results of this research demonstrate the following: (i) There is evidence that the dominant five species of flowering plants in the ALH compete strongly with each other. (ii) There is evidence that the subordinate species experience decline after removal of the dominant plants, which possibly indicates the presence of positive interactions between the two groups, so called the "Callaghan's effect" in the harsh environments of the ALH. (iii) There is not much difference between short-term (two-three years) and long-term (ten-thirteen years) responses of species, although in many cases the strongest response is observed during the third or fourth year of the experiment. Most species' responses were either consistently negative, or positive, towards the removal of a given dominant during the whole 13-year period of the study. A few (e.g., *Anemone speciosa* and *Carex* spp.) initially showed a positive response, which after six years changed to a mildly negative one. (iv) Responses of particular species are non-species specific, which indicates the presence of "diffuse competition" in the alpine lichen heath.

7.2.2. Removals in the *Geranium gymnocaulon – Hedysarum caucasicum* meadow

7.2.2.1. Aims and methods

This experiment was carried out in a fashion similar to the ALH dominant removals. Removals can help investigate interactions among species, both direct and indirect (Connell 1990, Gigon 1994). Removal of dominants is especially interesting from this point of view. How will the structure of a community change if the dominant species are excluded? *Geranium – Hedysarum* meadows are relatively productive communities of the alpine zone, so we expected the presence of stronger competition than in the above-mentioned ALH. The aim of this experiment was to reveal reactions of GHM vascular plants to the removal of GHM dominants: *Geranium gymnocaulon* (Geraniaceae) and *Hedysarum caucasicum*

(Fabaceae). The experiments were carried out during six years (beginning in 1996). There were 4 variants:

1. *Geranium gymnocaulon* removed
2. *Hedysarum caucasicum* removed
3. both *Geranium gymnocaulon* and *Hedysarum caucasicum* removed
4. no removals (control)

A total of 16 experimental plots 1.5 x 1 m were laid out (4 per variant). Three sample plots (25 x 25 cm) were marked within each quadrat for shoot counts. Thus 48 sample subplots were laid out in total. Removal took place from the whole area of the experimental plots, but shoots have been counted only on small counting subplots. Dominants were removed repeatedly during the first year (3-5 times during the vegetative season), and in the next years 1 – 2 times per season. The shoot counting was carried out annually in the first half of August.

For statistical processing we have applied the ANOVA/MANOVA analysis (Statistica 5.0 for Windows). As a basic parameter we used the total number of shoots for every species with high frequency (present on more than 50% of subplots). A separate analysis was carried out for reproductive shoots and juvenile plants of the most frequent species.

7.2.2.2. Dynamics of shoot numbers

The results of two-way MANOVA have shown that the influence of a year on species performance (shoot number) was more frequent than the influence of a variant. For some species with a significant reaction to the removal of the dominant the natural year dynamics of shoot and juvenile plant number were observed (Table 7.3). Total shoot number depends on the

Table 7.3. Shoot number dynamics in GHM control plots (per m^2). t – total shoot number; j – juvenile number; v – vegetative shoot number

Species	Shoot type	1996	1997	1998	1999	2000	2001
Agrostis vinealis	t	84	85	11	28	35	55
Geranium gymnocaulon	j	11	5	3	28	12	28
Hedysarum caucasicum	j	52	35	73	109	85	88
	v	88	40	49	93	132	96
Minuartia aizoides	t	19	11	0	0	0	0
Festuca brunnescens	t	107	51	49	93	132	96
Matricaria caucasica	j	1	15	24	48	11	28

year for 6 species (*Agrostis vinealis, Festuca brunnescens, Geranium gymnocaulon, Hedysarum caucasicum, Matricaria caucasica, Minuartia aizoides*). The maximum total shoot number was observed in 2001 for two of these species (*Agrostis vinealis* and *Festuca brunnescens*), and the highest number of juvenile plants for three species (*Geranium gymnocaulon, Hedysarum caucasicum, Matricaria caucasica*). The summer of 2001 was rather wet. The lowest total shoot number was observed in 1998 for *Agrostis vinealis, Festuca brunnescens, Minuartia aizoides*, and of *Geranium gymnocaulon* juveniles. The species changed their shoot numbers in correlation with each other. The summer of 1998 was very dry with August totaling a mere 4 mm of rain! Obviously, water stress caused the decrease in shoot numbers for most species.

7.2.2.3. The influence of dominants on the community structure

To compare variant influence on the shoot number of the GHM plants we have calculated the significance of difference between each variant and the control. The influence of a variant is estimated as the percentage of cases with a significant change against the possible number of cases for each variant.

The variant with *Geranium gymnocaulon* removed showed the most significant level of change. The removal of *Hedysarum caucasicum* had a much smaller impact on the community than that of *Geranium gymnocaulon* (Fig. 7.7).

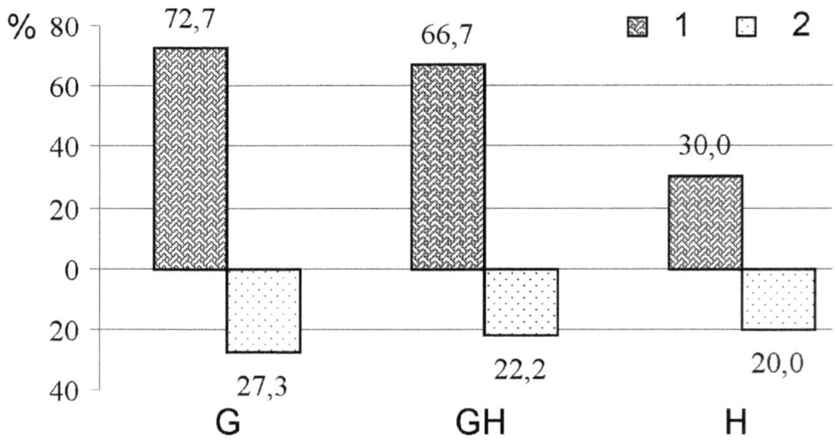

Figure 7.7. Influence of the removal of the dominants on the shoot number of GHM plants. **G** – *Geranium gymnocaulon* removal; **GH** – *Geranium gymnocaulon* and *Hedysarum caucasicum* removal; **H** – *Hedysarum caucasicum* removal. Ordinate – percentage of cases shoot number increase (1) and decrease (2)

More then half of the studied species (9 out of 16) have shown a significant response to the dominant removal, namely *Agrostis vinealis, Anthemis cretica, Carum meifolium, Festuca brunnescens, Geranium gymnocaulon* on the removal of the *Hedysarum caucasicum; Hedysarum caucasicum* on the removal of *Geranium gymnocaulon,* and *Leontodon hispidus, Matricaria caucasica,* and *Minuartia aizoides* on the removal of both dominants (Table 7.4). All these species (except for *Geranium gymnocaulon* and *Hedysarum caucasicum*) have changed their shoot number in response to the removal of *Geranium gymnocaulon*. The response to the removal of *Geranium gymnocaulon* and the removal of both dominants were similar for these plants (Fig. 7.7, Table 7.4).

Figure 7.8. shows the year-to-year changes of total shoot numbers for some species with significant response to the dominant's removals. These plants decreased their shoot numbers in 1998, when a strong drought took place. It is important to note that during the drought the shoot numbers were reduced more dramatically in the control plots, rather than in the removal plots. For example, *Festuca brunnescens* decreased its shoot numbers after the 1998 drought in the control plots, but actually increased its numbers in the removal plots after the drought (Fig. 7.8). The same was true for other

Table 7.4. Species responses to dominant removal in *Geranium-Hedysarum* meadows. Different letters show significant changes in shoot number by comparison with the control (arrow up – increase, arrows down – decrease in shoot number). The species and shoot types without significant changes of shoot number are excluded from the table. Variants: G – *Geranium gymnocaulon* removal, H – *Hedysarum caucasicum* removal, GH – both dominants removed; shoot type: g – generative shoots, v – vegetative shoots, j – juveniles, t – total shoot number

Species	Shoot type	Variant			
		G	H	GH	Control
Agrostis vinealis	v	↑A	B	↑C	B
	t	↑A	B	↑C	B
Anthemis cretica	t	↓A	↓A	↓A	B
Carum meifolium	g	↑A	BC	↑C	B
Festuca brunnescens	v	↑A	B	↑A	B
	t	↑A	B	↑A	B
Geranium gymnocaulon	j	-	↓A	-	B
Hedysarum caucasicum	v	↑A	-	-	B
	j	↓A	-	-	B
Leontodon hispidus	t	↑A	B	B	B
Matricaria caucasica	j	↓A	↑C	↓A	B
Minuartia aizoides	v	↓A	↓C	↓AC	B
	t	↓A	↓C	↓AC	B

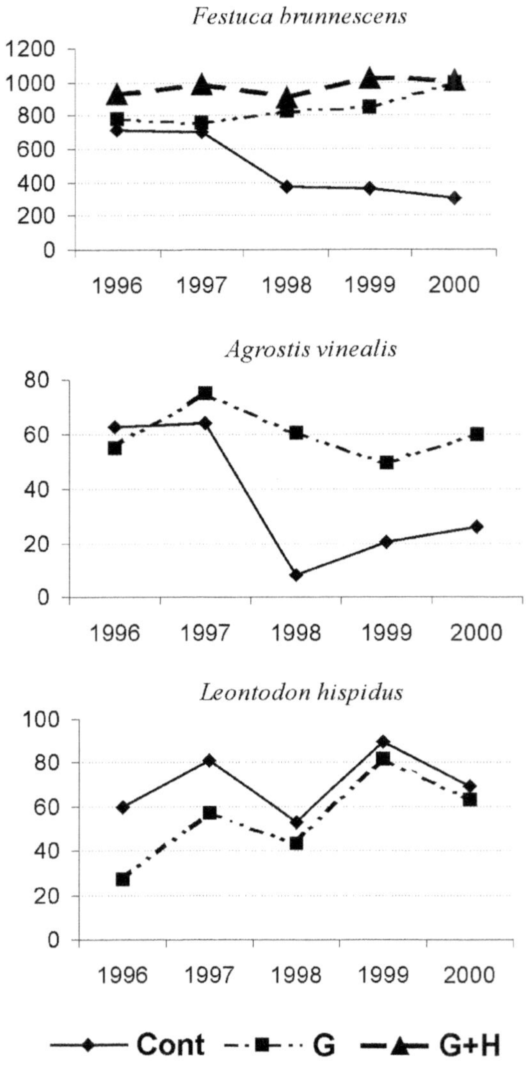

Figure 7.8. Dynamics of total shoot number (ordinate) in experimental plots. The graphics are constructed only for the control and variants with significant differences. Cont – no removal; G – *Geranium gymnocaulon* removal; G+H – both dominants removed

species. *Minuartia aizoides* (Tab. 7.4.) has disappeared from the control plots after the drought of 1998. Its shoot numbers in the experimental plots at first greatly decreased, but later began to increase.

It seems that water stress during the 1998 drought had a stronger impact in the control plots compared to the experimental plots. Thus, competition for water may be the main reason for weaker response to the drought by species in the experimental plots in contrast with control ones. In removal variants the competition has decreased, and the plants have gone through the summer drought with greater ease. Droughts in the alpine zone are common. However, our data show that the influence of competition in GHM can be stronger than the influence of the climatic factors, such as water stress.

The number of *Matricaria caucasica* juveniles decreased in plots when *Geranium gymnocaulon* and both dominants were removed, and has increased in the *Hedysarum caucasicum* removal variant. Juvenile numbers of *Geranium gymnocaulon* and *Hedysarum caucasicum* have decreased in response to removal of each other (Tab. 7.4.).

The decrease in juvenile numbers in the removal plots can be explained by the increase in the adverse environmental stress caused by dominant removals (see also Fowler 1981, Callaway 1995). The abiotic stress is particularly detrimental for seedlings and juveniles, since plants at early stages of development are especially sensitive. It is obvious, that the removal of the *Geranium* canopy leads to a change of microclimate in the plant cover, e.g., drying of the topsoil, increase in insolation, and increase in wind speed. Such microclimatic changes can lead to reduction of water availability and to increase of transpiration losses, which cause high plant mortality from water stress.

Positive interactions between adult dominants and juvenile stages of *Hedysarum caucasicum*, *Geranium gymnocaulon* and *Matricaria caucasica* take place in GHM. But the relative strength of positive relationships among GHM plants is lower than among ALH plants. It seems that competition plays a more important role in the relatively productive alpine meadows than in the stressful, poor ALH communities.

7.2.2.4. Features of GHM dominants

It is interesting to note that removal of *H. caucasicum* had small impact on the community structure. *Hedysarum* has a very long taproot and consequently is less likely to compete for water with the majority of other GHM species. *H. caucasicum* is also capable of nitrogen fixing. This may decrease competition for soil nitrogen with other species in the community. Besides, *Hedysarum caucasicum* has open pinnate foliage that creates little shade.

Geranium gymnocaulon, in contrast to *H. caucasicum*, has a large system of fine roots stemming from the rhizome in the upper topsoil. Roots of most GHM species are likewise found in this layer. *G. gymnocaulon* is capable to create significant shading with its dense leaf cover. According to our field observations, few other plants ever occur under the *Geranium* canopy, while many species do well under the *Hedysarum* canopy. Therefore, the removal of *G. gymnocaulon* could reduce the intensity of competition in the community and enabled other species to increase their abundance.

The increase of some species' shoot number in response to the removal of both dominants can be explained mainly by the influence of *Geranium* removal. In this variant, the increasers are all the same species as in the *G. gymnocaulon* removal variant (Fig.7.8.). Based on the two-way MANOVA, the results of the experiment with these two variants do not differ significantly (Tab. 7.4.).

To summarize, the two main dominants of GHM diverge in terms of their life strategy: *H. caucasicum* occupies a separate ecological niche that allows it to avoid competition with other species of the community. On the contrary, *G. gymnocaulon* is the main strong competitor. The removal of this species from the community promotes the growth of several other species.

7.3. Shading experiments

7.3.1. Introduction

A large role in the co-existence of alpine species is played by temporal variability in environmental factors, enabling each species to have a "temporal window" of optimal growth at different times during the growing season (Körner 1999). Even when the photosynthetic apparatus is permanently deployed during the whole growing season, species were shown to have distinct periods of predominant photosynthetic activity (Goryshina 1979, Abdaladze & Voznesensky 1984). Because alpine plants react opportunistically to the early snow melt (Prock & Körner 1996), under global warming they are expected to demonstrate phenological acceleration. Lack of light during the early part of the growing season, for example, because of increase in shade from taller competitors, may be critical for alpine plant survival. Also, because alpine plants' light requirements are relatively high (Körner 1999), lack of light may have a severe negative impact on biomass production even when the vegetation is sparse. An experimental study of the carbon balance of *Carex curvula* in the Alps (Körner 1982) demonstrated that reciprocal shading of plants in dwarf (3-5 cm) alpine sedge meadows led to a 20% decrease in the actual productivity, as compared to the potential. Partial shading of communities was proven to cause a change in floristic composition (Chapin & Shaver 1985, Hester 1987, Jackson & Caldwell 1992, Maranon & Bartolome 1993, Huber 1994, Zhang & Welker 1996, Eek & Zobel 1997, Cui & Caldwell 1997,

Onipchenko et al. 2001a). For example, Zhang & Welker (1996) reported that after experimental shading of a *Kobresia* meadow in Tibet, the relative biomass of Cyperaceae increased, while the total biomass decreased to mere 23% of the control. Experiments with artificial shadings in the Alaskan tundra (Chapin & Shaver 1985) demonstrated a high sensitivity of the tallest alpine dominants (*Betula nana, Ledum palustre, Carex bigelowii, Eriophorum vaginatum*) to the reduction in light intensity. At the same time, high tolerance of shading was found among short alpine plants (*Vaccinium vitis-idaea, Rubus chamaemorus*).

In this section we compare reactions to experimental artificial shading of alpine species from four alpine communities located along a topographic/snowpack gradient (ALH, FVG, GHM, SBC). Three research questions were investigated: (i) whether there is a difference in overall community-level (floristic) response to shading among four communities with different lengths of the growing season (ranging from 2 to 6 months); (ii) what are individual species' responses to shading across all four communities? What are the most and the least tolerant-to-shading species (potential "winners" and "losers" in the competition against tall invaders from lower zones); (iii) are there phenologically divergent groups of species, specifically, if there are species that show a difference in response to shading during the early, or late part of the growing season.

7.3.2. Field experiments

Shading experiments were carried out from 1986 to 1998 (Onipchenko et al. 2001a). Each community was shaded for three years: ALH in 1986-1988, FVG in 1989-1991, GHM in 1993-1995, and SBC in 1995-1998. Experimental design was identical in all 4 communities. Four treatments were applied: (i) no shading (control); (ii) shading during the whole growing season (for ALH – from early June until early September, for FVG – from mid-June until early September, for GHM – from early July until early September, and for SBC – from mid-July until mid-September), (iii) shading for the first half of the growing season, (iv) shading for the second half of the growing season.

Two-layer cloth screens were used for shading. The screens easily allowed water and air through, but filtered about 95% of direct solar radiation. This is close to the CO_2–compensation point for many alpine plants (Körner 1982, Larcher 1995). Methods of the field experiment and data analysis are described more fully in Onipchenko et al. (2001a).

7.3.3. Influence of light deficiency on the floristic richness of the four alpine communities

The total number of species found in four different communities after 3 years of shading for the whole growing season showed a pronounced decrease (Figure 7.9). However, the share of the remaining species was

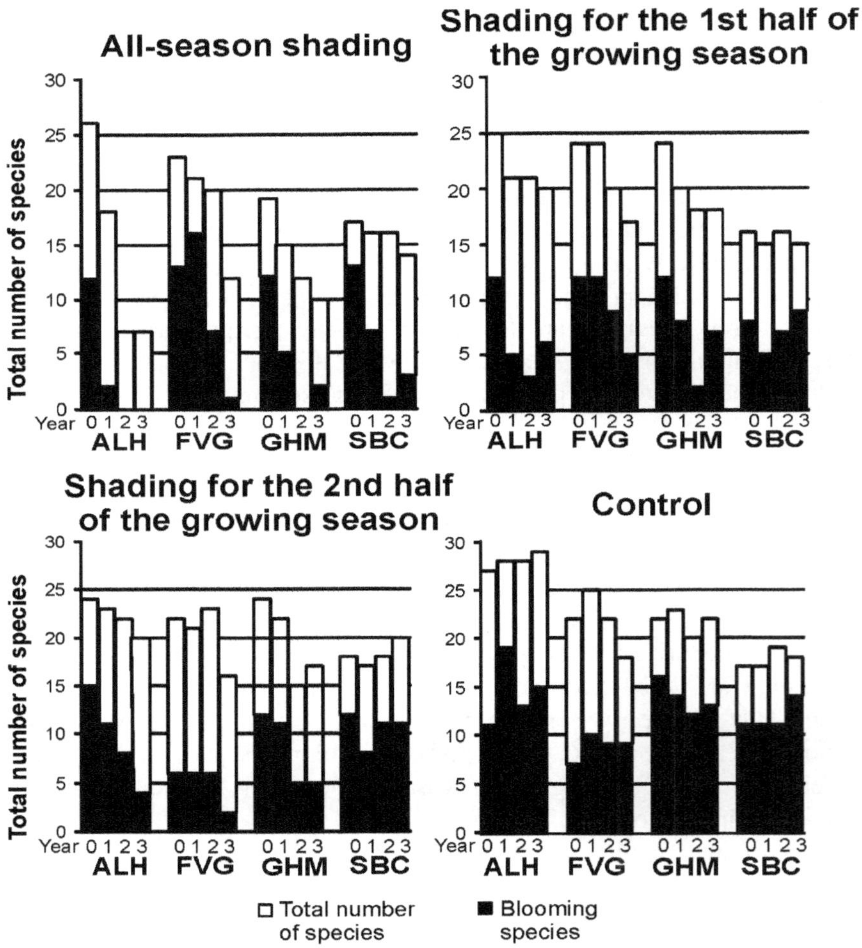

Figure 7.9. Annual changes in the total number of blooming and all species in four experimental treatments for 3 years. (from Onipchenko et al. 2001a with permission of *Arctic, Antarctic and Alpine Research*)

different in different communities. Thus, in alpine lichen heaths (ALH) the total species loss after 3 years was over 70%. In grasslands and meadows (FVG and GHM), about half of all species was lost. The snowbeds lost only about 18% of its species.

Interestingly, in the other two shading variants (first half of the season and second half of the season), new species emerged by the second and third experimental year. These species were either dormant in the soil seed bank, or encroached from the outside of the plots by vegetative propagation. In the all-season shading variant, no such species were observed, which evidently exacerbated the total species loss.

In all communities, decline in the number of species bearing generative shoots was greater than that in the total number of species (Fig. 7.9). For instance, in ALH no blooming species were found as early as the second year of the experiment. In the other three communities one to three species still were able to bloom by the third year of the experiment.

The original floristic richness (species per 0.25 x 0.25 m plot) differed among the four communities. In the all-season shading variant, the decline in floristic richness was proportional to the decline in the total number of species (Fig. 7.10). A significant decrease (66% fewer species) was found in ALH, in FVG – 75%, in GHM – 50%. Again, the smallest effect was noted in SBC (down from 10.7 to 7.1 species per plot, or 33%). Overall, the all-season shading had a more pronounced effect on the floristic composition than either early or late-season shading. In SBC, the total number of species remained unchanged in the early and late-season shading variants.

7.3.4. More and less shade-tolerant species in the four communities

Based on the relative decrease in shoot numbers in the all-season variant, we compiled lists of species ranked by their decreasing shading tolerance for each community (Table 7.5). The most tolerant species were *Gentiana pyrenaica* in ALH, *Viola altaica*, *Antennaria dioica*, *Gentiana pyrenaica* in FVG, *Luzula multiflora*, *Rumex alpestris*, *Carex atrata* in GHM, and *Corydalis conorhiza, Hyalopoa pontica*, *Potentilla crantzii*, *Carum meifolium*, *Catabrosella variegata*, and *Taraxacum stevenii* in SBC. The all-season shading had the most significant impact of the 3 shading variants (Table 7.6). For example, after three years of shading in ALH and FVG, no species there increased their shoot numbers. Two common species of alpine heath (ALH), *Gentiana pyrenaica* and *Vaccinium vitis-idaea*, although decreasing their numbers by half, still retained relatively high numbers of shoots until the end of the experiment. Both species also demonstrated similar reactions in FVG, where *Nardus stricta* and *Anthoxanthum odoratum* also preserved relatively high numbers of shoots after three years.

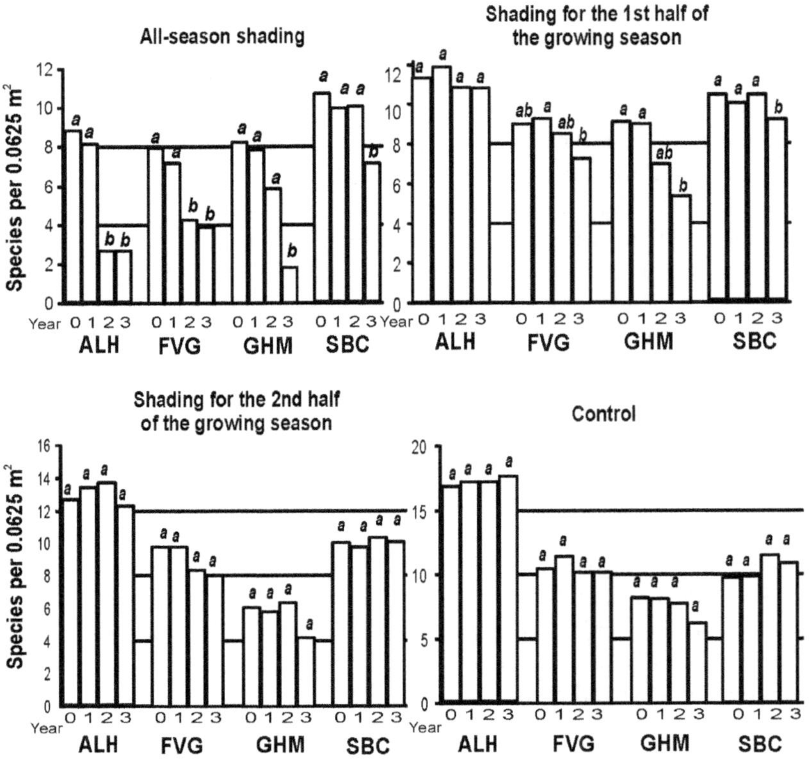

Figure 7.10. Annual changes in floristic richness (species per individual in 0.25 x 0.25 m counting subplot) (from Onipchenko et al. 2001a with permission of *Arctic, Antarctic and Alpine Research*)

Most species of *Geranium-Hedysarum* meadow (GHM) substantially decreased their shoot numbers after three years. The main dominant, *Geranium gymnocaulon*, showed a 40% decrease in shoot numbers, but still retained a substantial number of shoots by the end of the experiment. Two species, however, did not show a significant change (*Carex atrata, Luzula multiflora*). However, both of these species had initially low numbers and, therefore, the result may not be significant.

Species of the alpine snowbed community (SBC) were the most tolerant of shading. For example, SBC dominants *Taraxacum stevenii, Catabrosella variegata,* and *Carex oreophila,* retained a substantial number of shoots (they lost between 25 and 50%). Some species in this community even showed an increase in shoot numbers interpreted as a positive response to the all-season shading. In particular, *Corydalis conorhiza* increased its shoot numbers by a factor of three on shaded plots, while increasing by a factor of two on control (not shaded) plots (Table 7.6).

Table 7.5. Mean three-year value of the response coefficient K for selected abundant species in 4 alpine communities (see Chapter 7.1.2). K is calculated for the first (before shading) and 3rd year after shading of the whole summer variant. Species for each community are ranked in the order of decreasing K (i.e., from the most shade tolerant to the least shade tolerant)

Species	K	Species	K
ALH		GHM	
Gentiana pyrenaica	0.67	*Luzula multiflora*	4.69
Anemone speciosa	0.36	*Rumex alpestris*	0.89
Trifolium polyphyllum	0.30	*Carex atrata*	0.52
Vaccinium vitis-idaea	0.29	*Hedysarum caucasicum*	0.31
Antennaria dioica	0.26	*Carum meifolium*	0.30
Campanula collina	0.19	*Leontodon hispidus*	0.25
Carex umbrosa	0.17	*Agrostis vinealis*	0.21
Ranunculus oreophilus	0.05	*Phleum alpinum*	0.19
Carum caucasicum	0.04	*Potentilla crantzii*	0.19
Helictotrichon versicolor	0.03	*Catabrosella variegata*	0.15
Festuca ovina	0.03	*Matricaria caucasica*	0.13
Campanula tridentata	0.01	*Anthoxanthum odoratum*	0.05
FVG		*Nardus stricta*	0.03
Viola altaica	0.88	SBC	
Antennaria dioica	0.72	*Corydalis conorhiza*	2.20
Gentiana pyrenaica	0.67	*Hyalopoa pontica*	1.91
Anthoxanthum odoratum	0.46	*Potentilla crantzii*	0.89
Carex umbrosa	0.43	*Carum meifolium*	0.89
Festuca varia	0.27	*Catabrosella variegata*	0.82
Nardus stricta	0.22	*Taraxacum stevenii*	0.80
Festuca ovina	0.16	*Festuca brunnescens*	0.64
		Phleum alpinum	0.56
		Carex oreophila	0.43
		Minuartia aizoides	0.40
		Pedicularis nordmanniana	0.38
		Nardus stricta	0.37
		Gnaphalium supinum	0.29
		Agrostis vinealis	0.27
		Sibbaldia procumbens	0.14
		Festuca ovina	0.01

To make community-wide comparisons of shading tolerance, a mean three-year value of K for all species was calculated for each community (Fig. 7.11.). In the all-season variant the mean value of the response coefficient K changed from 0.2 in ALH to 0.34 in FVG, to 0.32 in GHM, to 0.69 in SBC (as elsewhere in this chapter, the communities are arranged from the least to the most snowy).

Table 7.6. Changes in shoot number of the most abundant species from four alpine communities in four experimental treatments of shading. B – before and A – after 3 years of shading. For statistical estimations see Onipchenko et al. (2001a)

Species	No shading (control)		All-season shading		Shading for the first half of the vegetative season		Shading for the second half of the vegetative season	
	B	A	B	A	B	A	B	A
Alpine lichen heath								
Anemone speciosa	55	50	38	3	28	21	30	23
Antennaria dioica	131	138	90	0	81	15	127	56
Campanula collina	25	42	20	7	38	20	60	32
Campanula tridentata	89	69	81	0	125	27	100	30
Carex umbrosa	149	154	142	0	211	215	170	203
Carum caucasicum	10	24	35	2	26	14	24	22
Festuca ovina	471	447	723	0	665	436	352	195
Gentiana pyrenaica	124	99	89	44	102	110	107	179
Helictotrichon versicolor	54	57	30	0	73	0	41	20
Trifolium polyphyllum	89	75	28	0	116	136	68	41
Ranunculus oreophilus	48	44	20	1	16	8	36	9
Vaccinium vitis-idaea	248	285	314	152	280	210	221	109
***Festuca varia* grassland**								
Anthoxanthum odoratum	80	141	33	11	56	70	29	18
Carex umbrosa	100	121	83	6	65	102	47	51
Festuca ovina	124	423	4	0	75	94	15	15
Festuca varia	727	1027	361	2	760	495	761	688
Gentiana pyrenaica	10	19	41	19	48	70	46	17
Nardus stricta	1306	2386	1323	45	1086	1086	938	728
Viola altaica	97	13	1	0	47	4	45	16
***Geranium – Hedysarum* meadow**								
Agrostis vinealis	47	30	33	0	34	10	27	22
Anthoxanthum odoratum	196	257	225	0	255	121	216	189
Carex atrata	31	46	27	21	12	9	16	35
Carum meifolium	498	208	297	18	499	113	418	72
Catabrosella variegata	47	108	55	0	124	186	68	102
Festuca brunnescens	279	212	17	0	106	55	204	103
Geranium gymnocaulon	168	181	253	157	170	214	181	182
Hedysarum caucasicum	21	16	28	1	17	5	32	6
Luzula multiflora	20	33	9	23	2	5	37	17
Matricaria caucasica	200	97	93	3	130	83	206	91
Nardus stricta	190	517	44	0	162	125	221	165
Phleum alpinum	383	375	316	47	395	185	289	192
Snow bed community								
Agrostis vinealis	156	147	186	5	198	5	337	180
Carex oreophila	88	140	127	86	182	218	83	174
Carum meifolium	10	11	67	26	34	28	27	34
Catabrosella variegata	173	206	290	145	218	193	256	735
Corydalis conorhiza	10	23	14	42	17	44	23	81

Table 7.6. (Cont.)

Species	No shading (control)		All-season shading		Shading for the first half of the vegetative season		Shading for the second half of the vegetative season	
	B	A	B	A	B	A	B	A
Gnaphalium supinum	609	786	275	0	230	56	149	70
Minuartia aizoides	788	1079	1072	43	1136	1193	1062	1709
Nardus stricta	440	1016	541	66	1011	1006	874	1159
Pedicularis nordmanniana	24	52	72	16	83	48	88	122
Phleum alpinum	55	96	57	17	38	40	116	116
Sibbaldia procumbens	1422	1850	1011	21	1493	811	1075	998
Taraxacum stevenii	350	393	426	397	343	324	410	423

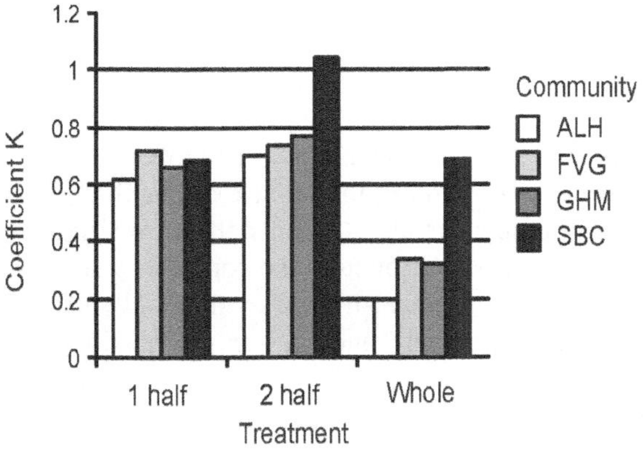

Figure 7.11. Mean 3-yr values of response coefficient K in the four alpine communities. (from Onipchenko et al. 2001a with permission of *Arctic, Antarctic and Alpine Research*)

7.3.5. Difference in species reaction to early *versus* late-season shading

For most species, no significant difference in reaction was found when the early and late-season shading variants were compared (Table 7.6). Only in a few instances did we find some differences. In the alpine heath (ALH), *Trifolium polyphyllum* showed a statistically significant reaction to the changes in shading regime. While the number of shoots in control fell a little over three years (from 89 to 75), in the first half-season shading variant, it showed a modest increase in shoot numbers (from 116 to 136) and a decrease in the second half-season shading (from 68 to 41). Perhaps, growth during the second half of the season, when the next-year buds are developed, is more important for this species.

Antennaria dioica and *Helictotrichon versicolor*, both having low tolerance of shading (both disappeared from the community by the second year under the all-season shading), showed tendency for less response to the shading during the second half of the season than to shading during the first half. Apparently, for these species growth during the first half of the summer is more important than during the second half.

In *Festuca varia* grassland (FVG), only *Gentiana pyrenaica* showed any difference in reaction to the timing of shading. When shaded for the second half of the season, it decreased its numbers by 66%, whereas when shaded for the first half of the season, it actually increased its numbers by 50%. Marginally significant was the difference in reaction in *Anthoxanthum odoratum*, which showed a little decrease in the second half-season shading, and a modest increase in the first half-season variant.

In the *Geranium-Hedysarum* meadow (GHM), the most interesting behavior was observed in *Gagea fistulosa*, an ephemeral species. Immediately after the end of the experiment the highest count of its shoots was recorded, three times that in the control (147 $\pm$ 11 mean and S.E. per 0.25 x 0.25 m plot, n=12), on the plots shaded for the second half of the season (Figure 7.12.). Somewhat lower numbers (127$\pm$15), but higher than in the control, were found in the all-season shading. Also higher than in the control were numbers in the second half-season shading variant (99 $\pm$ 7). The lowest numbers were found in the control (no shading) variant (50 $\pm$ 6).

Geranium gymnocaulon showed a marginally significant difference in reaction between the early and late-season shading. The species actually increased its mean number of shoots in the early-season treatment and showed little change in the late season treatment.

The only species of the SBC that significantly increased its shoot numbers in all 3 shading variants, but not in the control, was *Corydalis conorhiza.* Regrettably, in each variant it was only found on 4 out of 12 experimental plots, which was too small a number to make the result statistically significant (Fig. 7.12). Nevertheless, this species clearly increased its shoot numbers under shading. While it increased its shoot numbers by a factor of 2.3 in the control plots (apparently due to a natural fluctuation, since no manipulation was made there), it increased its numbers in the first half-season variant 2.6 times relative to the initial conditions, 3 times in the all-season variant, and 3.5 times in the second half-season variant. This species' reaction was thus numerically similar to that of *Gagea fistulosa* in the GHM.

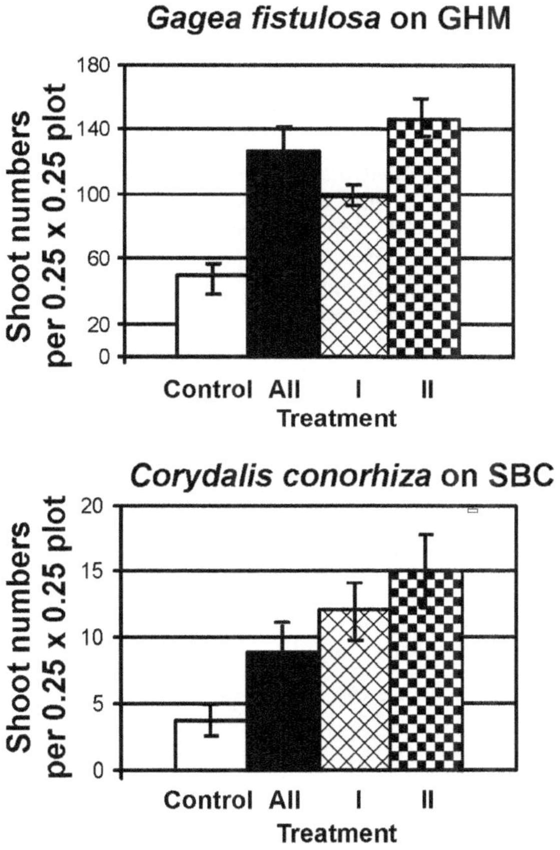

Figure 7.12. Mean shoot counts for two spring ephemeral species per 25 x 25 cm counting subplot (n=12) during the year after the end of shading (from Onipchenko et al. 2001a with permission of *Arctic, Antarctic and Alpine Research*)

In the snowbed community (SBC) *Agrostis vinealis*, *Catabrosella variegata,* and *Minuartia aizoides* showed some different reaction to shading during the first and second half of the growing season (Table 7.7.). For example, *Catabrosella variegata* somewhat decreased when shaded during the first half of the season, and showed a 3-fold increase when shaded during the second half of the season.

Table 7.7. Species of snowbed community (SBC) classified as based on their reaction to experimental shading (from Onipchenko et al. 2001a with permission of *Arctic, Antarctic and Alpine Research*)

Reaction to timing of shading (first or second half of the season)	Species tolerance to the all-season shading	
	Low	High
Present	*Catabrosella variegata, Agrostis vinealis, Sibbaldia procumbens*	*Carex oreophila*
Absent	*Nardus stricta, Phleum alpinum, Carum meifolium, Minuartia aizoides, Gnaphalium supinum*	*Taraxacum stevenii, Corydalis conorhiza*

Noteworthy also is the similar reaction to shading among a few specicies of sedges (*Carex* spp.) found in different communities. In the ALH three sedges (*Carex umbrosa, C. sempervirens,* and *C. caryophyllea*) were counted as one unit (listed under *C. umbrosa* in the tables) due to impossibility of distinguishing them in their vegetative state. Only *C. umbrosa* grows in the FVG, while *C. atrata* is common in the GHM and *C. oreophila* is common in the SBC. All sedges showed a high tolerance to light deficiency and did not reduce, or even relatively increase, their numbers when shaded during the second half of the growing season (Table 7.6).

In the alpine snowbed community (SBC) there were no species that were more tolerant of shading during the first half of the season than during the second half. Altogether, SBC species can be divided into four groups on the basis of their reaction to shading (Table 7.7.). The group comprising the least tolerant species (e.g., *Nardus stricta, Phleum alpinum, Carum meifolium*) are found in a wide range of habitats, e.g., different types of grasslands. The other three groups comprise species mostly restricted to the snowbed community (SBC) (Table 7.7).

Overall, the four studied communities can be divided into two groups based on the reaction to shading in general, and to the reaction to shading in the second half of the season in particular. The first group includes alpine lichen heath (ALH), *Festuca varia* grassland (FVG), and *Geranium-Hedysarum* meadow (GHM). The second group includes snowbed communities (SBC). Species from the communities in the first group have both species that responded more favourable to shading during the first half of the season relative to the second, as well as those that showed the opposite pattern. Therefore, for this group the mean values of K differ little between the two shading variants (Fig. 7.11). In contrast, in the SBC, there were no species that responded better to shading during the first half of the season. Thus, the mean values of K for the two shading variants show a great difference. Also, SBC species showed a much higher tolerance to the all-season shading (the mean K value was twice as high as in the other three communities). This is to be expected, because the SBC normally develop on sites with the greatest snow accumulation and therefore with the shortest growing season of the four communities.

7.3.6. Discussion

As expected, shading that removes 95% of the direct solar radiation had a strong negative impact on most alpine plants in all four communities. The all-season shading had the greatest negative effect both in terms of the number of species that disappeared after three years and in terms of shoot numbers. Shading during the first half of the growing season had a smaller, but also major, impact on all four communities. In contrast, shading during the second half of the growing season impacted on fewer species, which is consistent with the conclusions by Prock & Körner (1996) that early season development of cold climate plants is opportunistic, while late season phenology is deterministic. In other words, alpine plants start rapid growth just before or immediately after the snow melt, whenever that happens, which is the most critical period of growth during the year. Late-season growth is generally much less important because by that time enough biomass has been produced to ensure successful winter survival.

Our results demonstrate an increasing level of shading tolerance in the community series arranged along the gradient of increasing winter snow depth and duration (from alpine lichen heath to grasslands and meadows to snowbed communities). As documented in many other alpine areas (e.g., Billings & Bliss 1959, Komarkova 1993, Stanton et al. 1994) this topographically-determined gradient is actually a complex one, also being characterized by progressive shortening of the growing season, increasing nutrient availability, and increasing soil moisture. Relatively high tolerance to shading observed among plants in nutrient-rich habitats has been noted by many researchers (Smith & Huston 1989, Grubb et al. 1997, Graglia et al. 1997). The link between shading tolerance and length of growing season, on the other hand, is less clear. We could expect that plants from habitats with a short growing season would have a high dependence on light, since they only have a short period of time to utilize it. Apparently, in our area, soil fertility plays a greater role.

Our data are consistent with earlier findings (Hester 1987, Iason & Hester 1993) that generative organs respond more negatively to shading than the vegetative ones, because generative organs are more sensitive to the carbon deficiency caused by shading.

Members of the Cyperaceae showed a uniformly high degree of tolerance to all variants of shading. This can be explained by the presence of overwintering green shoots in these plants, which enabled photosynthesis before and after the tents were removed. In a shading experiment in Tibet (Zhang & Welker 1996), the relative abundance of Cyperaceae on the experimental plots similarly increased. In our study, other highly tolerant species included the evergreen *Gentiana pyrenaica* and *Vaccinium vitis-idaea.*

The ephemeroids *Gagea fistulosa* and *Corydalis conorhiza* demonstrated a uniform response: both increased their shoot numbers in the second half-season shading variant. Such reaction agrees well with these species' phenology: their period of active growth occurs in the early summer, and by the time the shading is over during the second half of the growing season, their main competitors weaken considerably, while the ephemeroids enjoy full illumination during the first half of the summer. Ephemeroids generally responded well to shading, which is likely to result from their overall ruderal (*sensu* Grime 1979) strategy. Ruderal species can quickly mobilize resources whenever competition with other species is reduced.

The somewhat unexpected reaction of the perennial grass *Catabrosella variegata*— a similar increase of shoot numbers, if shaded during the second half of the season — can be explained by the «hidden ephemeroidity» (*sensu* Goryshina 1979) of this species. Such species, while having well-developed leaves during most of the year, have a clear peak in photosynthetic ability in the early part of the growing season. A typical example from the temperate region is *Aegopodium podagraria* (Goryshina 1979).

Among the four communities, the snowbed community had the largest share of species with strong ruderal traits (Chapter 6.3, Onipchenko et al. 1998a). Similarly to ephemeroids, these species apparently have a peak in their photosynthetic activity immediately after the snowmelt in the first half of the growing season. On the SBC this happens in early July, when other conditions are most favourable to growth (longest daylenght period, warmest temperatures, highest soil moisture).

The overall conclusion is that while phenological divergence is weakly expressed in most alpine plants, species from the SBC have the weakest reaction to the timing of shading. SBC species are the most tolerant of shading and are thus more likely to endure pressure from taller competitors moving into the alpine zone from lower elevations.

Chapter 8

Soil algae and cyanobacteria in the alpine plant communities

8.1. Introduction and methods

Soil algae and cyanobacteria are important components of most terrestrial ecosystems (Shtina & Gollerbakh 1976). As photosynthetic organisms, they take part in the primary production of the plant communities and are important food resources for the soil invertebrates. Cyanobacteria also play an important role in the nitrogen balance.

The study of the algal flora of the Teberda alpine communities took place in 1982 (Leinsoo et al. 1987) and 1990-93 (Shtina et al. 1995). The floristic diversity, abundance, biomass and distribution of soil algae and cyanobacteria were examined in all four main alpine communities discussed elsewhere in this volume: alpine lichen heaths (ALH), *Festuca varia*-dominated grasslands (FVG), *Geranium gymnocaulon-Hedysarum caucasicum*-dominated meadows (GHM) and snow bed communities (SBC). Each community was represented by two plots: "upper" (U) with a lower snow accumulation, and "lower" (L) with a higher accumulation of snow, higher moisture and a later start of the vegetation season.

Mixed samples of soil, plant litter and the aboveground plant biomass (each consisting of 10-20 individual pinch samples) were taken from the four communities, both from the "upper" and "lower" plots. Soil samples represented depths of 0-2, 2-5, 5-10 cm, and in some cases deeper samples. All samples were taken under sterile conditions, dried slowly and kept in paper envelopes to prevent the activity of algophagans. Different types of culturing (water culture, moist glass culture, agar cultures) were used for studying the species diversity (Lund 1945). To analyse species composition,

V.G. Onipchenko (ed.), Alpine Ecosystems in the Northwest Caucasus, 263–270.

criteria like number of species, systematic relationship, cyanobacteria to green algae ratio (Cya/Chl, also known as the aridity index) were used. Cluster analysis was applied for the characterisation of the species composition and floristic similarity.

To determine numbers of algae cells and their biomass, 1-cm^3 soils samples were collected about two weeks after the snowmelt, when the maximal accumulation of algal biomass was expected. The samples were taken at the soil surface (0-1 cm) and from a 5-6 cm depth with 3-10 replications. All samples were fixed with formaldehyde. The quantity and biomass of algae and cyanobacteria was determined by the method of direct calculation and measuring of algal cells (Shtina 1984).

8.2. Floristic composition

A total of 126 algae and cyanobacteria species (sometimes genera or species group) were identified in soil and litter from the four studied plant communities. Table 8.1 contains the list of species listed by sample.

The main characteristic of the algal flora from the alpine soils is the prevalence of green algae with broad ecological preferences. Litophylic as well as hydrophylic forms are present. The depth of the snow cover and the length of the snow-free period are very important for the structure of algal communities. The peculiarity of the algal flora of the studied communities is the prevalence of green algal species, most of whom are widely distributed species of soil algae ("cosmopolites"). In most samples cyanobacteria were rare or absent, especially Nostocales. Only few species of diatoms were found, mainly *Pinnularia borealis* and *Navicula mutica*, but the first one formed most of the biomass due to high numbers.

The diversity of cyanobacteria decreases in the series from ALH to SBC. In particular, nitrogen fixing species of the order Nostocales disappear in the middle of the gradient. The presence of cyanobacteria species is rather low everywhere; most of them could be found on the soil surface and on the plant material. In comparison with the ALH, the number of species of green algae in the more moist alpine meadows and snow bed communities is smaller (Tables 8.1 – 8.2). E.g., species of Desmidiales and Mesotaeniales are completely absent in the latter communities. The same tendency was found for diatoms. Only two species of them were found in different communities and soil depths - *Pinnularia borealis* and *Navicula mutica.* The typical soil diatom *Hantzschia amphioxys* was found only in a few samples from the ALH and FVG.

The number of algal species in the relatively dry communities of the ALH and FVG (64 and 66 species) is higher than in the moister communities of the GHM and SBC (44 and 40 species). Generally, all studied

Table 8.1. Distribution of soil algae and cyanobacteria by soil depth in four alpine plant communities. Abbreviations: s - surface of a bare soil, p - on plants and litter, 1 = 0-2cm, 2 = 2-5 cm, 3 = 5-10 cm, 4 = 10-15 cm, 5 = 15-20 cm, 6 = 10-20 cm, 7 = 20-30 cm, 8 = 30-40 cm (from Shtina et al. 1995)

Species	Communities			
	ALH	FVG	GHM	SBC
Cyanobacteria				
Anabaena sp.	1			
Chlorogloea microcystoides Geitl.			p	
Gloeocapsa minor (Kutz.) Hollerb.	p, 1	2	2	
G. minuta (Kutz.) Hollerb.	p, 1		1	
G. turgida (Kutz.) Hollerb.	p			
Gloeothece palea (Kutz.) Rabenh.	p			
Lyngbya aerugineo-coerulea (Kutz.) Gom.	p, 1		p	
Microcystis muscicola (Menegh.) Elenk.		1		6
M. pulverea (Wood) Elenk.	2		2	p,1
Nostoc commune Vauch.	1			
N. punktiforme (Kutz.) Hariot		s, p		
N. sp.	1	4		
Oscillatoria amoena (Kutz.) Gom.				p
O. brevis (Kutz.) Gom.	1	p	1	
O. kuetzingiana Nag.		s, p, 1	p, 1	
O. splendida Grev.		s, p, 1		
O. sp.	1			p
Phormidium angustissimum W. et G.S.West	1,3			
Ph. autumnale (Ag.) Gom.	1	s, p, 1	p, 1	
Ph. foveolarum (Mont.) Gom.	1		p, 1, 2, 3	6
Ph. uncinatum (Ag.) Gom.	p, 1			
Ph. sp.				p
Plectonema edaphica (Elenk.) Vaul.		1		
P. sp.	p, 1			
Schizothrix sp.	p, 1			
Synechococcus aeruginosus Nag.	p, 1	p		p
S. cedrorum Sauv.	1			
Synechocystis aquatilis Sauv.			p	
S. pevalekii Erceg.		2		
Chlorophyta				
Actinotaenium cucurbitinum (Biss.) Teil	p, 1			
Borodinellopsis sp.	1			
Bracteacoccus aggregatus Tereg.		s,p,1,2,3,4,5,8		
B. grandis Bischoff et Bold		s, p, 2,3,4	p	
B. minor (Chodat) Petrova	1,2,3	s,p,1,2,3,4,5,7,8	p,1,6	1
Chaetophorales gen. sp.		s, 1,2		
Chlamydomonas chlorococcoides Ettl et Schwarz		s, p, 1,2,3,4,5	p,1	p, 1
Ch. gloeogama Korsch. f. *humicola* Hollerb.		s, p,1,2,3,4,5	1,2	p,1
Ch. macrostellata Lund		s, p,1,2,3		p,1,2
Ch. oblongella Lund		s, p, 1,2,3		p,1,2
Ch. proboscigera Korsch		s,p,1,2,4		

Table 8.1. (Cont.)

Species	Communities			
	ALH	FVG	GHM	SBC
Ch. pseudopulsatilla Gerloff		s, p, 2,4		
Ch. spp.	p, 1,2,3	s,p,1,3,4,5		
Chlorhormidium dissectum (Gay) Fareoqui	1,2,3	s, p, 1,2	1,3	p, 6
Ch. flaccidum (Kutz.) Fott	1,2,3	s, p, 1,2,3,4,5	p, 1,2	p, 1,2,6
Chlorella minutissima Fott et Novacova	p,1,2	4,5,p	1	p, 1,6
Ch. mirabilis V.Andr.	p, 1,3			p, 1
Ch. protothecoides Kruger			1	
Ch. vulgaris Beijerinck	p,1	s, p,1,5	p,1,2,3	1,2
Ch. sp.			2,6	
Chlorococcum sp.	p, 1,2,3	s,p,1,2,3,4,5,7,8	p,1,3,6	p,1
Chloromonas paradoxa Korsch.		s,p,1,3,5,8		
Chlorosarcinales gen.sp.			p	
Chlorosarcinopsis sp.	1	s,p		
Choricystis chodatii (Jaag) Fott		s, p, 1,2,7	p, 1	p, 1
Ch. sp.	1			
Coccomyxa solorinae Chodat	1,2,3			p
Cosmarium anceps Lund	p			
C. crenatum Ralfs	1			
Cylindrocystis brebissonii Menegh.	p, 1,3			
C. crassa DeBary	p, 1,2,3			
Dictyochloris fragrans Visch.		1		
Dactyococcus varians Germ. ex Starr		p, 1		
Elliptochloris bilobata Tsch.-Woess		s, p, 1,3,7,8		
Fernandinella alpina Chod. emend. Korsch.		p		
Gloeocystis sp.			p	
Gloeotila caldaria Kutz.				p
Kentrosphaera bristolae G.M.Smith	1			
Leptosira terricola (Bristol) Printz		p, 2		
Macrochioris dissecta Korsch.		s, p, 3	1	
Mesotaenium endlicherianum Nag.	p, 1,2,3			
Myrmecia biatorellae (Tsch.-Woess) Boye-Pet.		p,1,2,3,4,5,7	1,2	1
M. bisecta Reisigl	p,1,2,3	s,p,1,2,3,4,5,7,8	p,3,6,7	p,1
M. incisa Reisigl		2,3,4,5		
Neochloris bilobata Vinatzer	1,3			
N. wimmeri (Rabenh.) Archibald et Bold		s, p, 3,4,5	p,1	
Oocystis sp.	2			
Planktosphacria sp.		4	1	
Pleurastrum terrestre Printz et John		p		
Pseudochlorococcum sp.			3	
Pseudococcomyxa simplex (Mainx) Fott	s,p,1,2,3, 4,8	8	p,1,3,6,7	p, 1,6
P. sp.	1			
Radiosphaera sphaerica (Korsch.) Starr	3			p
Schizochlamydella delicatula (West) Korsch		s, p, 4		
Scotiella levicostata Hollerb.	p, 1,3			
Spongiochloris minor Chantanachat et Bold		s, p, 1,2		1
Stichococcus bacillaris Nag.	1	s, p, 1,2		2

Table 8.1. (Cont.)

Species	Communities			
	ALH	FVG	GHM	SBC
S. chodatii (Bial.) Heering		7	1	p,1
S. exiguus Gern.		s, p	p	1,6
S. fragilis Gay	3			
S. lacustris Chod.			p	
S. minor Nag.	p, 1	s, p, 1,2,3,5	p,1,2,3,6,7	1,2,3,6
S. palescens Chod.		s, 7		
S. variabilis G. et G.S.West			6	
Tetracystis aplanosporum Arce et Bold		s, p, 1,2		
T. sp.	1,3	s, p, 1,2,3,4	1	1
Ulothrix subtilissima Rabenh.	1,2,3	s, p, 1,2,4,5		
U. tenerrima Kutz.		1		
U. variabilis Kutz.		s, 1,3,4		
Xanthophyta				
Botrydiopsis arhiza Borzi		s,p,1,4		p
B. minor (Schmidle) Chod.	1			
Botryochloris minima Pasch.		s,p,1		
Chlorocloster terrestris Pasch.		s		
Chloropedia plana Pasch.		1		
Ellipsoidion solitare (Geitl.) Pasch.		s,p,1,2,7	p, 1,2	1,2,3
E. sp.	p, 1,3			
Heterococcus sp.	2			
Heterothrix exilis Pasch.	1			
Monodus dactylococcoides Pasch.		s,1		
Nephrodiella semilunaris Pasch.		s, p, 1,7		3
Pleurochloris commutata Pasch.				p,1
P. magna Boye-Pet.		s, p, 1,2,5	1,2,3,6	p,1,2,3
P. sp.	p,1,2,3		p,2	
Polyedriella helvetica Pasch.				1
P. irregularis Pasch.				1
P. sp.	2,3			
Bacillariophyta				
Hantzschia amphioxys (Ehr.) Grun.		s, p, 1		p
Navicula contenta Grun.	p			
N. mutica Kutz.	p,1	s, p, 1	p	
N. pelliculosa (Berb.) Hilse	1,2			
Nitzschia palea Kutz.	p			
N. sp.	p			
Pinnularia borealis Ehr.	p, 1,2	s, p, 1,2,3,4,5	p,1	p, 1,3
P. sp.	1			
Stauroneus parvula Lund	1			

communities had a dominance of green algae (by the number of species) and a lower diversity of diatom algae. Cyanobacteria represented 23%, green algae - 56%, yellow-green - 14% and diatoms - 7% of the total number of algal species.

Table 8.2. Total number of algal species in soils (from Shtina et al. 1995)

	ALH	FVG	GHM	SBC
Cyanobacteria	19	12	10	7
Clorophyta	31	43	29	24
Xanthophyta	6	8	3	7
Bacillariophyta	8	3	2	2
Total	**64**	**66**	**44**	**40**
Cya/Chl	0,61	0,28	0,34	0,29

Only 14 genera were represented by more than three species. Five genera are represented by five and more species: *Chlamydomonas* (9), *Stichococcus* (8), *Oscillatoria* (5), *Phormidium* (5), *Chlorella* (5). The most frequent species among the cyanobacteria were *Phormidium autumnale*, the green algae - *Chlorococcum* sp., *Stichococcus minor, Myrmecia bisecta, Bracteacoccus minor, Chlorhormidium flaccidum, Pseudococcomyxa simplex, Chlamydomonas magna* and *Ch. gloeogama,* and the yellow-green algae - *Pleurochloris magna* and *Ellipsoidion solitare.*

Cluster analysis shows a higher specificity of algae communities of the alpine heaths compared to the other three communities, especially the FVG (Fig. 8.1). The comparison showed floristic similarity inside the communities between the "upper" and "lower" plots, as well as similarity of algal species' composition of the GHM, SBC and in some cases the FVG to each other. The higher difference between the SBC (U) and the SBC (L) could be based on greater snow accumulation in the "lower" plots.

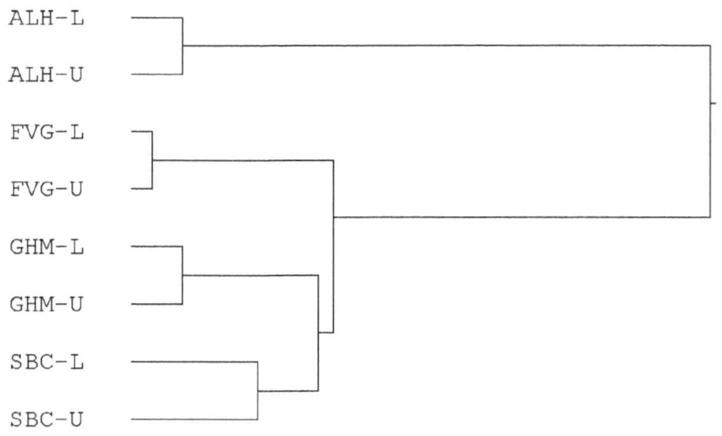

Figure 8.1. Cluster dendrogram for algal community composition in eight studied plots (U – upper plot and L – lower plot for each community) (from Shtina et al. 1995)

Algae, being phototrophic organisms, prefer the upper layers of the soil, especially bare places devoid of higher plants (Table 8.1). Algae can also grow as epiphytes on plants and litter, but could also be found in deeper soil layers, sometimes as a result of the digging activity of animals or plant root penetration. The number of cyanobacteria decreased most rapidly with depth. One can hardly find them deeper than 5 cm. Most of the algae found in the deeper layers belong to green algae genera *Bracteacoccus, Myrmecia, Chlorococcum* and *Chlormidium.*

8.3. Biomasss

The biomass of soil and epiphytic algae in alpine lichen heath was examined from the perspective of their seasonal dynamics (Leinsoo et al. 1987). The obtained values are low: 1.2-1.6 g of fresh biomass/m^2, or about 13,000–32,000 cells per gram of dry soil. Bacillariophyta was the most abundant group in terms of biomass.

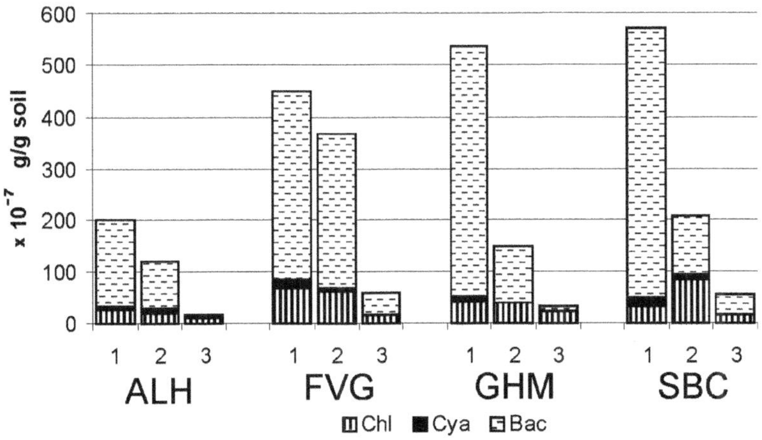

Figure 8.2. Algal fresh biomass in the soil of the 4 alpine communities. Cya – Cyanobacteria, Chl - Chlorophyta + Xanthophyta, Bac - Bacillariophyta. 1 – soil surface under lichens or bared, 2 – soil surface under vascular plants, 3 – soil at 5-6 cm depth (from Shtina et al. 1995)

The biomass of soil algae at the surface and at 5-6 cm depth was examined for all four alpine communities (Shtina et al. 1995). In most samples diatoms predominated. The whole biomass decreases rapidly with depth. Algal biomass in the upper soil layer was higher under fruticose lichens or inside bare soil patches than under vascular plants. A slight increase of algal biomass could be observed in the series from ALH to SBC. (Fig. 8.2). According to the data, the mean fresh algal biomass is about 1-3 g/m^2 in our communities.

As a whole, the biomass of soil algae in the alpine plant communities was of the same order of magnitude as the values for many woodland and meadow ecosystems (Shtina 1984) and it was much lower than recorded for most tundra, steppe or desert communities. In particular, the absence of large macroscopic colonies of *Nostoc* and the small role of cyanobacteria in general differentiate our communities from other tundra or steppe ecosystems.

In general, the algal communities of the high mountain and alpine ecosystems of the north-western Caucasus at 2700-2800 m elevation could be classified as primitive, as expected for moist acidic soils. The number of species is rather low in comparison with well developed woodland ecosystems (Feher 1933). The species composition was characterised by the predominance of green algae with broad ecological tolerance. A significant proportion of the species were epiphytes developing on litter and the aboveground parts of vascular plants. The diversity of species decreased along the gradient of increasing snow cover, soil moisture and reduction in the length of the growing season. Different plots inside the same plant community show high floristic similarity. The most distinct algal communities were found in the alpine lichen heaths.

The biomass of soil algae is low: hardly 1-3 g of fresh weight/m^2. The number of algal cells is about 13,000-32,000 per gram of dry soil.

Chapter 9

Fungi, mycorrhiza, and microbiological processes in the alpine ecosystems

9.1. Soil microfungal communities

9.1.1. Introduction

Numerous studies on soil microfungi have been carried out since Adametz in 1886 first isolated fungi from a soil. Despite this fact, our knowledge on the soil fungi of many regions remains incomplete.

Microfungi of high-altitude soils were first studied in the Alps (Mosca 1957, 1960, Gams 1959, Horak 1960). Then some studies were devoted to other mountain regions of Europe, Asia and North America (Mirchink 1963, 1971, Rall 1965, Dal Vesco 1974, Franz 1975, Janečková *et al.* 1977, Bissett & Parkinson 1979a, 1979b, 1979c). Only scant research on this topic has been done in the Caucasus (Leinsoo *et al.* 1991, 1999). Taking into account a high diversity of habitats characteristic for mountain regions, the picture obtained from these works can hardly be regarded as representative for all these habitats.

Besides the interest of such studies for the fundamental knowledge of biodiversity of alpine ecosystems, they are of a certain practical value because remote cold and wet habitats can be regarded as one of the underexplored sources of producers of new pharmaceuticals (Dreyfuss & Chapela 1994).

The distribution of soil microfungi usually is weakly affected by environmental parameters (*e.g.* soil type, vegetation, geographical location, *etc.*) and many species show a rather cosmopolitan distribution (Gams

V.G. Onipchenko (ed.), Alpine Ecosystems in the Northwest Caucasus, 271–302.

1992). A great part of the taxa listed in the numerous studies on microfungal communities of distant soils still overlap. According to Christensen (1989), apparent lack of habitat specificity presumably stems "from high species diversity and recognition of infrequent species over and above more common but unfamiliar forms". The comparisons are rendered more difficult by the fact that the taxonomy of many groups of soil microfungi has changed several times since the end of the 19th century. However, we believe that different soils have some specific features, at least in a proportion of certain fungal taxa and in the presence of some taxa that still show a strong correlation with their habitat.

In the above-mentioned papers about fungi in high-altitude soils, we see a large proportion of cosmopolitan species and a small proportion of typical arctic and alpine species. But lowland tundra soils generally have similar microfungal communities to alpine soils (Dowding & Widden 1974, Bissett & Parkinson 1979a, Christensen 1981, Leinsoo 1994). This fact is not surprising in view of the similarities in climate, soils and vascular plant communities.

9.1.2. Study sites

We studied fungal communities in seven alpine plant communities (alpine lichen heaths - ALH, *Festuca varia* grasslands - FVG, *Geranium gymnocaulon* – *Hedysarum caucasicum* meadows - GHM, snow bed communities - SBC, *Rhododendron caucasicum* shrubs – RC, and two fens – FEN_1 and FEN_2) and two subnival screes (relatively dry scree – SDS and moist scree – SMS). The vegetation and soils of the alpine communities are described in Chapters 1-4 of the present volume. The subnival sites are covered by primitive stony soils with very short mineral profiles (AC-C). They lack a closed vegetation cover. Some plants just form discrete patches on the otherwise bare soil surface. A few dominating species (in SDS *Minuartia imbricata* and *Primula amoena*, in SMS *Catabrosella variegata*) and a number of minor ones occur in these communities.

Because of the described discontinuity in plant cover on the one hand and homogeneity of the soil morphology and vegetation on the other (and our expectation of simple soil microfungal communities), the subnival screes represent good models for studying the horizontal distribution of soil microfungi within one site. The basic question is whether the soil microfungal community sampled below a patch of a certain plant species is different from that in the surrounding bare soil. Therefore, we investigated the mycobiota under the patches of each dominating plant species and that of bare soil separately. We analysed ten samples from every variable at the two sites.

9.1.3. Methods

The observed species composition of soil fungi is strongly affected by the methods applied (Frankland *et al.* 1990, Gams 1992). No method yields a complete picture of the fungi populating a certain soil. Any method is selective to a certain degree and recovers only a part of the species present. Therefore, fuller results are obtained if several methods of isolation are applied.

A basic method applied for isolation was the dilution plate technique in which we spread application of 0.1 ml of suspension on the surface of Czapek agar with 300 ppm streptomycin (Marfenina in Zvyagintsev 1991). This method favours fungi that grow fast on rich media.

Soil microfungal communities of FEN_1, FEN_2, SDS and SMS were also studied using a cellulose bait technique (Marfenina in Zvyagintsev 1991). A similar method, buried cellophane films, was used successfully by Gams (1960). However, our method is different and our results are hardly comparable with those obtained with cellophane films. The modification was applied as follows. Autoclaved pieces of filter papers (2×2 cm) were buried *in situ* at 3 cm depth in the soil for two weeks. After recuperation, the pieces were kept air-dry at 4°C for not more than six months, then placed unwashed on tap-water agar (one piece onto each Petri dish) with 300 ppm streptomycin and incubated at 24°C for 20 days. Fungi growing up were observed and transferred from sporulating structures for subsequent identification.

For the estimation of species abundance in certain sites we calculated the frequency of occurrence, *i.e.* the number of Petri dishes where a species was present in relation to the total number of dishes. Microfungal communities of different soils were ordinated using the procedure of correspondence analysis (Jongman *et al.* 1995). Species diversity was calculated using Shannon's index (Shannon & Weaver 1964). Sample series were compared using multidimensional scaling analysis based on the matrix of distances calculated as $\Delta = 1 - CC$, where CC is the Sørensen coefficient of community (Jongman *et al.* 1995). Statistical routine procedures were done using MVSP 2.2.

9.1.4. Results and discussion

9.1.4.1. Autecology

A total of 125 species of microfungi were isolated from the high-altitude soils of the Teberda Reserve. The most abundant ones are listed in Tables

9.1 and 9.2. In addition, numerous isolates of dark and hyaline sterile mycelia were obtained.

The list of the microfungal species can be divided into the following 3 groups:

1) species typical of high altitudes;

2) species with a rather ubiquitous and worldwide distribution;

3) species with a limited distribution, which is determined by other factors than altitude or latitude.

Pseudogymnoascus roseus, including isolates with only conidial reproduction belonging to the anamorph species *Geomyces pannorum* s.l. (van Oorschot 1980), was most abundant in dilution plate isolations made from all the alpine soils and the subnival SDS. The species was also quite common in dilution plate isolation from SMS and on cellulose bait isolations. This fungus has constantly been reported in all studies concerned with the mycobiota of mountain, tundra, arctic and antarctic habitats (Traaen 1914, Mosca 1957, 1960, Gams 1959, Horak 1960, Tubaki 1961, Heal *et al.* 1967, Latter & Heal 1971, Dowding & Widden 1974, Dal Vesco 1974, Franz 1975, 1981, Bissett & Parkinson 1979a, Widden & Parkinson 1979, Zabawski 1982, Bab'eva & Sizova 1983, Egorova 1986, Mortemartini Corte & Gestro 1992, Kirtsideli & Tomilin, 1997, Kirtsideli 1999a, 2001a, 2001b). The distribution of this species aggregate is by no means limited to the above-mentioned habitats and there are numerous reports of it from other habitats (for references see Domsch *et al.* 1980). However, taking into account that low-altitude temperate habitats have been studied much more widely, a lower percentage of papers reporting *Pseudogymnoascus roseus* and related anamorphs suggests that these fungi could be less common. *Geomyces pannorum* is known to be strongly stimulated by various toxic chemicals (Kuthubutheen & Pugh 1979) and its stress tolerance may also be a factor determining its high abundance in mountain soils.

Species of *Mortierella* subgen. *Mortierella* (especially *M. alpina, M. elongata, M. parvispora*) were often reported from wet alpine and tundra habitats (Gams 1959, Horak 1960, Mosca 1960, Heal *et al.* 1967, Dal Vesco 1974 *etc.*). The fungi of this subgenus were abundant in all soils we examined. *Mortierella parvispora* was found in all the soils studied, whereas each of the other four species (*M. alpina, M. elongata, M. gamsii, M. humilis*) was isolated from only some of them but their distribution can hardly be interpreted. We specifically emphasise the role of the subgenus *Mortierella* because the subgenus *Micromucor* is quite distinct morphologically (Gams 1977), biochemically (de Ruiter *et al.* 1993) and ecologically (Sewell 1959).

Table 9.1. Dominant species of soil microfungi in the high-altitude communities of Teberda Reserve. Results of dilution plate technique. The values presented are the frequencies of occurrence. The species list includes those isolated with a frequency of occurrence =10% from at least one of the communities. Frequencies <10% are indicated by "+".

Species	Communities							
	ALH	FVG	GHM	SBC	FEN	RC	SDS	SMS
Zygomycetes								
Mucorales								
Mortierella alpina Peyr.	+	-	15	90	+	+	-	-
M. elongata Linnem.	-	-	20	-	-	-	15	+
M. gamsii Milko	-	-	-	40	-	-	-	-
M. humilis Linnem. ex W. Gams	-	10	10	-	+	-	-	-
M. vinacea Dixon-St.	-	-	-	-	11	75	+	-
M. parvispora Linnem.	45	55	10	10	11	10	+	+
M. ramanniana var. *angulispora* (Naumov) Linnem.	-	-	-	-	-	60	-	-
Mucor hiemalis Wehmer	-	-	-	30	22	10	18	+
Ascomycetes								
Onigenales								
Pseudogymnoascus roseus Raillo, incl. strictly anamorphic isolates (*Geomyces pannorum* (Link) Sigler et Carmichael)	35	90	95	100	73	85	69	30
Sordariales								
Melanospora sp.	-	10	10	+	-	-	-	-
conidial ascomycetes								
Acremonium berkeleyanum (P.Karsten) W. Gams	+	-	-	25	58	-	+	13
Acremonium cf. *roseogriseum*	20	-	+	-	-	-	-	-
Beauveria bassiana (Balsamo) Vuillemin	-	-	20	-	+	-	+	-
Botrytis cinerea Pers.:Fr.	+	-	10	-	-	+	+	+
Cladosporium cladosporioides (Fres.) de Vries	-	10	-	-	+	+	+	+
C. herbarum (Pers.:Fr.) Link	+	-	10	10	+	23	13	+
Cylindrocarpon destructans (Zinssm.) Scholten	15	-	-	-	-	-	-	-
C. magnusianum (Sacc.) Wollenw.	+	-	-	-	77	-	26	+
Fusarium merismoides Corda	-	-	10	-	+	-	-	-
F. sambucinum Fuckel	100	70	95	20	-	-	29	23
Lecythophora spp.	-	-	-	-	+	28	+	-
Paecilomyces carneus (Duché & Heim) A. H. S. Brown et G. Sm.	20	10	+	-	-	-	-	-
Penicillium citrinum Thom	-	-	+	-	+	18	-	+
P. griseofulvum Dierckx	15	-	-	-	-	-	-	-

Table 9.1. (Cont.)

Species	Communities							
	A L H	F V G	G H M	S B C	F E N	R C	S D S	S M S
P. janthinellum Biourge	-	15	-	-	-	-	+	-
P. lanosum Westling	-	100	30	-	-	-	-	27
P. raistrickii G. Sm.	-	10	-	-	-	+	-	-
P. sacculum Dale	-	-	-	-	-	-	-	25
P. simplicissimum (Oudem.) Thom	+	65	85	80	+	+	+	51
Tolypocladium cylindrosporum W.Gams	-	-	-	-	18	13	-	-
T. geodes W.Gams	-	25	70	-	+	+	-	+
T. inflatum W.Gams	-	-	+	100	+	43	-	+
Trichoderma hamatum	-	10	-	-	-	-	-	-
T. koningii Oudem.	15	-	-	-	-	-	-	+
Verticillium lecanii (A. Zimmerm.) Viegas	20	10	-	10	21	-	+	-
V. psalliotae Treschow	-	-	-	-	18	-	-	-
A19	-	-	-	-	27	-	-	-
mycelia sterilia								
Dark	55	+	-	15	53	18	13	14
Hyaline	20	+	-	-	11	-	+	+

The phenomenon of dominance of *Pseudogymnoascus roseus* and *Mortierella* subgen. *Mortierella* spp. can presumably be ascribed to their psychrophilic nature. These fungi grow at 1°C with a rate exceeding 20% of their maximum, while many others grow much more slowly at such temperatures (Latter & Heal 1971). The experiments of Carreiro and Koske (1992) on the influence of temperature on the composition of microfungal communities in decomposing litter showed that *Mortierella* spp. dominated at 0 and 10°C. In contrast, fungi isolated by Carreiro at 20°C consisted almost entirely of deuteromycetes, with the genera *Trichoderma*, *Humicola*, and *Sporothrix* being most abundant.

Among the group of fungi with worldwide distribution, we see high frequencies of *Cladosporium cladosporioides*, *C. herbarum,* and *Mucor hiemalis*.

Some other species showed a strong preference of particular sites. Three species of the genus *Tolypocladium* (*T. inflatum, T. cylindrosporum* and *T. geodes*) are also known to be typical for alpine and tundra soils. In our study the distribution of these fungi was confined to the soils that are covered with snow in winter, *i.e.* all except the ones under ALH and SDS. Similar results were gained by Bissett & Parkinson (1979a). They studied soil microfungi of three different high-altitude sites on the southeast ridge of Mt. Allen in the Kananaskis valley. They isolated *T. cylindrosporum* and *T. geodes* from the only site characterized by little snow accumulation during winter whereas both others were highly exposed to winds and therefore lacked snow cover. Latter & Heal (1971) found that "proportional

Table 9.2. Dominant species of soil microfungi in the high-altitude communities of Teberda Reserve. Results of cellulose bait technique. The values presented are the frequencies of occurrence. The species list includes those isolated with a frequency of occurrence =10% from at least one of the communities. Frequencies <10% are indicated by "+".

Species	Communities		
	SDS	SMS	FEN
Zygomycetes			
Mucorales			
Mortierella elongata.	68	58	+
Mortierella parvispora	-	-	17
Mucor hiemalis	+	-	47
Ascomycetes			
Onigenales			
Pseudogymnoascus roseus, including strictly anamorphic isolates	30	-	10
conidial ascomycetes			
Acremonium berkeleyanum	-	-	27
Acremonium ser. *Luzulae* sp.	11	+	+
Alternaria alternata (Fr.) Keissler	91	95	20
Arthrinium arundinis (Corda) Dijko & B. Sutton	+	12	17
Bipolaris coicis (Nisikado) Shoemaker	11	+	-
Cladosporium cladosporioides	71	53	37
Cladosporium herbarum	28	32	33
Cladosporium macrocarpum Preuss	28	42	+
Curvularia clavata B.L. Jain	13	+	-
Cylindrocarpon magnusianum	-	+	20
Epicoccum nigrum Ehrenb.	49	42	+
Fusarium sambucinum	13	15	-
Penicillium simplicissimum	+	95	-
Phoma herbarum	-	-	10
Trichocladium opacum (Corda) Hughes	+	12	-
Trichoderma atroviride	+	-	10
Ulocladium atrum Preuss	39	23	-
Ulocladium botrytis Preuss	46	30	-
Ulocladium chartarum (Preuss) E.G.Simmons	16	13	-
Ulocladium consortiale (Thüm.) E.G.Simmons	29	10	+
Verticillium lecanii	-	-	50

growth at 1°C" (percentage of the maximal growth rate at the higher temperatures) of *T. inflatum* and *T. cylindrosporum* is significally lower than that of, *e.g.*, *Geomyces pannorum* and some species of *Mortierella* subgen. *Mortierella.* Bissett (1983) observed minimal or no growth of *Tolypocladium* spp. at 5°C. However, he also found that *T. geodes* and several other species of this genus, except for *T. inflatum,* are also sensitive to high temperatures showing no and minimal growth at 30°C respectively. Thus, species of *Tolypocladium* are obviously not strongly tolerant to low temperatures and there is no clear explanation of their relatively high frequency in reports from alpine and tundra soils.

In our research three species of *Tolypocladium* had the highest frequencies of occurrence in different habitats though sometimes two species were isolated from the same subsample. We tried to summarise which common properties in a range of soils could determine the presence of a certain species of *Tolypocladium* in them. However, we should immediately assume that our inferences might be doubtful. *T. inflatum* was isolated most frequently from the soils of GHM, SBC, RC, and one of the FENs. These soils are covered with snow for the longest period in summer. *T. cylindrosporum* was isolated from the soils of the fens and RC, *i.e.* the habitats where peat formation takes place. *T. geodes* was isolated with highest frequencies from GHM and FVG and occasionally from RC and FEN.

The genus *Tolypocladium* is well-known because of its role in applied mycology. *T. inflatum* is the best known producer of the immunosuppressive cyclosporins, which also possess antifungal (Dreyfuss *et al.* 1976) and insecticidal (Weiser & Matha 1988) properties. Cyclosporin A is one of the most valuable fungal drugs in medicine, commonly used to prevent rejection of transplanted organs (Dreyfuss *et al* 1976). Another important group of its metabolites, the efrapeptins, are potent inhibitors of mitochondrial and prokaryotic ATPases and have antifungal and insecticidal activity too (Krasnoff *et al.* 1991). Some isolates of *T. inflatum* were found to have entomopathogenic properties (Weiser 1987, Kal'vish 1990). Hodge *et al.* (1996) pointed out that the ecological role of the above-mentioned compounds is little understood, suggesting that they may be important for invertebrate parasitism and competition with other soil microorganisms. As some authors (Soares *et al.* 1979, Weiser & Pillai 1981) report, *T. cylindrosporum* can infect mosquito larvae. So its distribution in the soils studied might also be associated with the occurrence of a potential host (Chapter 10).

Fusarium sambucinum is a typical alpine fungus being quite common in low-altitude temporal and subtropic areas (Domsch *et al.* 1980). We isolated it frequently from the soils of all the communities studied except for FENs and RC. This fact can be explained by its limited tolerance to too high or too low pH values and/or to rather moist soils of fens.

A similar pattern of distribution is also observed for *Penicilium simplicissimum* and *P. lanosum* but they were also absent in soils that are too dry and poor in organic matter and plant biomass such as the SDS and ALH. These fungi are frequently reported in literature from another alpine, tundra, arctic and antarctic habitats (Gams 1959, Cooke & Lawrence 1959, Horak 1960, Latter & Heal 1971, Dal Vesco 1975, Janečková *et al* 1977, Bissett & Parkinson 1979a, Zabawski 1982, Kirtsideli *et al.* 1996, Kirtsideli & Tomilin

1997, Kirtsideli 1999a, 1999b, 2001a, 2001b) but still they are rather cosmopolitan (Domsch *et al.* 1980).

A group of species that were most frequent in fen communities comprises *V. psalliotae* and several more rare species not mentioned in Table 9.1. (*Acremonium tubaki, Oidiodendron truncatum*). *Acremonium berkeleyanum, Cylindrocarpon magnusianum, Verticillium lecanii* were also quite frequent in the FEN soils but beside that they were isolated from rather dry habitats such as SDS and ALH. This may be a consequence of their wide ecological amplitude for moisture and weak competitive ability.

Mortierella ramanniana var. *angulispora*, representative of the subgen. *Micromucor*, was abundantly isolated from the RC but was not found in any other site. *Mortierella ramanniana* is most commonly found in podzolic soils (Sewell 1959, Domsch *et al.* 1980, Mirchink 1988). Some signs of podzolisation have been found in the soils of the RC (Volkov 1999).

Some other species are confined to certain combinations of sites but these relationships are difficult to interpret.

9.1.4.2. Synecology

Ordination of the studied soil microfungal communities using the method of correspondence analysis showed that they fall into three distinctive groups (Fig. 9.1.). While most of the soils studied are close to each other in microfungal species composition fens and the RC are separate from each other as well as from the first group. The three first axes of ordination may be hypothetically interpreted as reflections of the following factors. The first axis may correspond to the organic matter content. At a larger scale this is evident since we can see the group of fens and the RC at the right end of the axis and the group of the rest of the sites at the left. However, the shift of the SDS and SBC to the positive end makes this assumption doubtful. Nevertheless, we can not deny it absolutely, as we have already mentioned above that some microfungal species seem to be suppressed by their competitors in their abiotic optimum. In this case a curve of the distribution of such a species along the abiotic factor should have two maxima. But since the mathematical procedure of correspondence analysis arranges the sites assuming that species distributions should approximate a Gaussian distribution the site sequence along this axis may be distorted. The second and third axes seem to reflect pH increase and duration of snow coverage respectively.

In SDS the subsite types did not seem to determine the species composition (Fig. 9.2A, B). However, in SMS the scores for "bare-soil" and "*Catabrosella variegata*" subsites appear separated in two clusters (for dilution plates – Fig. 9.2C) or were slightly overlapping (for cellulose baits– Fig. 9.2D).

Decrease of microfungal diversity with depth expressed in both number of fungal taxa and Shannon index has been shown by many authors for many habitats since the 1950s (Kjøller & Struwe 1982). However, in the peat soils of the fens these two parameters remained practically unchanged down to the mineral rock at 20-35 cm and even slightly increase in horizon T_3 of FEN_2 (Fig. 9.3.). This phenomenon may be explained by the fact that the organic matter and root content and, consequently, aeration also do not change significantly with depth within the limits of our investigation.

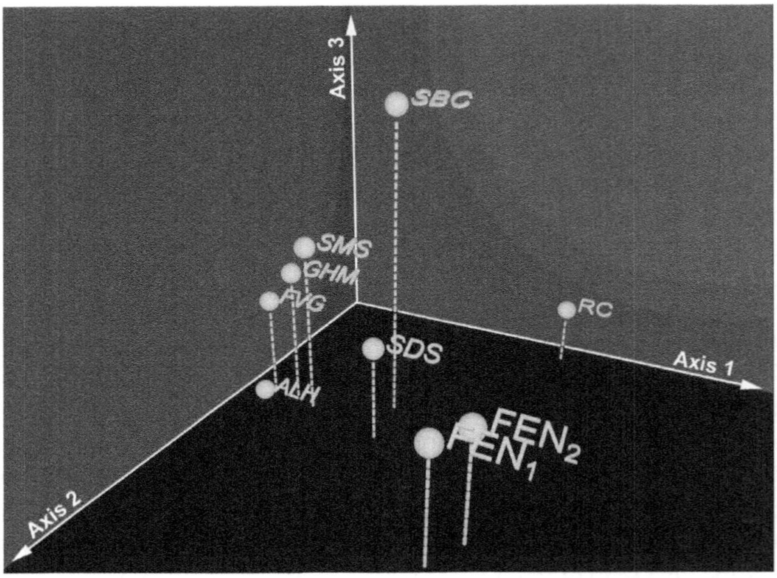

Figure 9.1. Correspondence analysis ordination diagram of microfungal communities of different habitats. Distribution along the three first axes. Plant communities (habitats): ALH – apine lichen heaths, FVG – *Festuca varia* grasslands, GHM – *Geranium gymnocaulon – Hedysarum caucasicum* meadows, SBC – snow bed communities, RC – *Rhododendron caucasicum* shrubs, FEN_1 and FEN_2 – fens, SDS – dry subnival scree, SMS – moist subnival scree.

9.1.4.3. Methodological experiments

The cellulose bait method allowed the isolation of certain soil microfungi, which are rare in the results of the dilution plate method (Table 9.2). These fungi comprise dematiaceous species known as decomposers of raw plant material and as weak plant parasites (*Alternaria alternata, Arthrinium arundinis, Cladosporium sphaerospermum,*

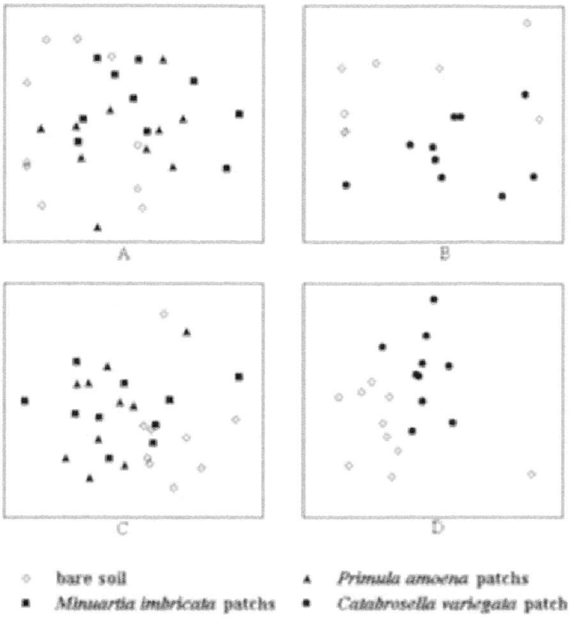

Figure 9.2. Results of multidimensional scaling analysis. Distances between scores reflect differences in the species composition of the microfungal community of different subsites within the subnival screes. Habitats: A, B – SDS, C, D – SMS. Methods of isolation: A, C – results of the plate dilution technique, B, D – results of cellulose bait technique

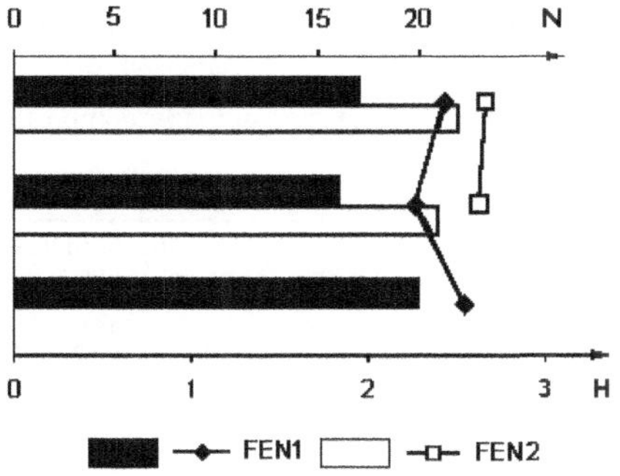

Figure 9.3. Number of taxa (N, plotted as histograms) and Shannon index (H, plotted as graphs) of soil fungi in different layers of fen soils

C. macrocarpum, Bipolaris spp., *Chaetomium* spp., *Curvularia* spp., *Epicoccum nigrum, Stachybotrys chartarum, Torula herbarum,* and *Ulocladium* spp.). *Cladosporium* spp., *Alternaria alternata* and *Epicoccum nigrum* are some of the commonest air-borne fungi (Gregory 1973). Warcup (1957) pointed out that conidia of *Cladosporium* are easily decomposed by other soil microorganisms or soon lose their viability in soil. Thus the fungi of this genus are considered not to be typical soil inhabitants. Nevertheless, Domsch and Gams, as summarized in Domsch *et al.* (1980) isolated *Cladosporium* spp. frequently from washed soil particles, indicating their active growth in soil. Possibly the dematiaceous fungi mentioned (except *Stachybotrys chartarum*) are easily suppressed on rich media by hyaline species, because of a higher competitive ability of the latter. This could result in the rare isolation of the former with the dilution plate method. In subnival screes such fungi dominate among the isolates whereas in the fens their proportion among the total microfungi is significantly lower. This seems to suggest that the cellulose bait method does not sufficiently recover fungi specific to a soil with a very low biological activity but it works quite well when applied to more biologically active soils.

We suppose that true soil mycobiota is at least partly rather specific for a certain soil, whereas fungi blown in from the air and from dead plant material and then surviving in the soil for a while in a dormant state are similar in different habitats. Therefore the capacity of a method to detect differences in microfungal community composition between plant communities might be taken as a measure of the suitability of the method. In both pairs (SDS - SMS, FEN_1 - FEN_2), one site differs from the other significantly in vegetation and soil properties. Therefore differences in soil microfungal communities had already been expected to occur at the onset of our research. Table 9.3. shows that for the screes a better differentiation was obtained with dilution plates but for the fens with cellulose baits.

Table 9.3. Similarities (CC) of microfungal community composition in two pairs of related sites as assessed by two methods

Compared habitats	Dilution plates	Cellulose baits
Dry scree – moist scree	0.53	0.76
FEN_1 – FEN_2	0.60	0.25

9.1.5. Conclusion and perspectives

The mycological investigations on high-altitude soils of the Teberda Reserve have covered all of the most important soil and vegetation types. A few taxonomically interesting isolates were occasionally obtained. However, these studies do not present an exhaustive inventory of the microfungal communities in each site and this may be the reason that relatively little specificity in the mycobiota of different soils and vegetation types was found. On the other hand, it is not impossible that the broad overlap in mycobiota of these habitats reflects the reality, that the ecological differences observed do not entail drastic changes in the microfungal communities. In further research a simultaneous and consistently repeated study on all sites will be necessary. This will reduce the factor of subjectivity of a researcher and allow assessing the influence of temporal dynamics. Moreover more suitable methods will have to be applied for the exploration of true soil mycobiota, such as soil washing or/and, if possible, molecular techniques.

We express our best gratitude to Prof Dr W. Gams (Centraalbureau voor Schimmelcultures, the Netherlands) for his help in identification of the isolates and useful discussion of the present chapter.

9.2. Phytopathogenic fungi

Parasitic microfungi play a relatively small role in alpine communities due to unfavourable abiotic conditions for their development and spore propagation (dry air, strong UV radiation, etc.) (Muller 1987, Kranz 1990, Burdon 1993). The number of phytopathogenic fungi species rapidly decreases with increasing altitude (Prikhod'ko 1991, Paulech et al. 1991, Hirsch & Braun 1992). Only a few species were found on alpine plants during our investigations (Table 9.4). Most of them were rare and appeared in autumn when leaves started to die.

Table 9.4. Phytopathogenic fungi on alpine plants in the Teberda Reserve

Fungi	Host plants	Location
Cintractia caricis (Pers) Magn.	*Carex umbrosa*	spikes
Entyloma veronicicola Lindz	*Veronica gentianoides*	leaves
Erysiphe polygoni DC.	*Rumex alpestris*	leaves
Exobasidium vaccinii Wor.	*Vaccinium vitis-idaea*	leaves
Melampsora salicine Lev.	*Salix kazbekensis*	leaves
Puccinia hieracii (Schum.) Mat.	*Hieracium laevigatum*	leaves
Puccinia montana Fchl.	*Centaurea cheiranthifolia*	leaves
Puccinia podospermii DC.	*Scorzonera cana*	leaves
Puccinia senecionis Lib.	*Senecio taraxacifolius*	leaves
Rhytisma salicinum (Pers.) Rehm st.con.	*Salix kazbekensis*	leaves
Schizonella melanogramma (DC.) Schroet.	*Carex umbrosa*	leaves
Urocystis pulsatillae (Bub) Moesz	*Pulsatilla aurea*	leaves
Uromyces geranii (DC.) Lev.	*Geranium gymnocaulon*	leaves
Uromyces hedysari-obscuri (DC.) Car. et Ricone	*Hedysarum caucasicum*	leaves

Cintractia caricis was the only species found which can damage inflorescences (spikes) of *Carex umbrosa* and decrease the seed production of the species. It is necessary to note, that from 30% to 50% of the generative shoots of the host were infected in *Festuca varia* grassland (FVG), where *Carex umbrosa* is quite rare. On the other side, the infected spikes were very rare on alpine lichen heath (ALH), where this sedge is one of the main dominant species. So, we can speculate that *Cintractia caricis* may cause a decrease in the performance of *Carex umbrosa* on relatively productive alpine meadows (FVG and GHM). As we showed by transplantation experiments (Chapter 7.1.*), Carex umbrosa* increased its abundance significantly when transplanted from ALH to FVG or GHM. The *Cintractia* infection may be a reason for the natural low abundance of this species in the alpine meadows.

9.3. Mycorrhiza

9.3.1. Introduction

Mycorrhiza are well represented in close alpine communities (Selivanov 1981, Read & Haselwandter 1981, Maksimova 1986, Shavkunova 1987, Blaschke 1991, Väre et al. 1997), although effective mycorrhizal symbiosis is lacking at the highest elevation in the subnival belt (Haselwandter & Read 1980, Selivanov 1981). Arbuscular mycorrhiza as well as ericoid mycorrhiza are common types for high mountain areas (Lesica & Antibus 1986, Stoyke & Currah 1991, Gardes & Dahlberg 1996). We studied mycorrhizal infection of 120 alpine species in the Teberda Reserve (Baikalova & Onipchenko 1988, Onipchenko 1996, Onipchenko & Zobel 2000). The aims of this study were: 1) to obtain quantitative estimations of mycorrhizal infection of alpine species; 2) to compare alpine communities according to mycorrhizal development of their species; 3) to investigate seasonal and ontogenetic dynamics of mycorrhizal infection for several alpine species; 4) to compare mycorrhizal infection for alpine species with different degrees of vegetative mobility; 5) to compare root infection rates and plant response to disturbance.

To study mycorrhizal infection we use Selivanov's methods (Selivanov 1981), which are not widespread in western countries, so we need to describe the methods in greater detail.

9.3.2. Methods of sampling and microscoping

Thin roots (less than 2 mm in diameter) from at least five plants in each community were collected and fixed in 4% formaldehyde solution for mycorrhizal analysis. The roots were collected during the second half of August 1984. Fixed roots were macerated in 20% KOH for 2-3 months. This period of maceration was long enough for soft colourless roots. The more rigid dark roots of several species (*Anemone speciosa, Sibbaldia procumbens*) were additionally heated in the same solution for one hour (Onipchenko & Zobel 2000). Then the roots were watered and stained with aniline blue (0.1 g aniline blue + 50 g lactic acid + 100 g water) for 1 hour. After removal of the stain solution the samples were put in pure lactic acid for half an hour. The root samples were then kept in glycerol. Thin roots (less than 0.5 mm in diameter) were studied microscopically where mycorrhizal infection was well represented. The total number of microscopic fields (x 120) was 130-150 for each sample. The amount of endophytic fungi was estimated on Selivanov's scale (Selivanov 1981) from absent (0) to very abundant (5 points):

Point	Description
1	Mycorrhizal fungi are present in few mesoderm cells
2	About ¼ of the mesoderm cells are occupied by fungi
3	About ½ of the mesoderm cells are occupied by fungi
4	About ¾ of the mesoderm cells are occupied by fungi
5	Practically all of the mesoderm cells are occupied by fungi

The rate of mycorrhizal infection (C) was calculated as:

$$\boldsymbol{C} = \boldsymbol{S} / (\boldsymbol{N} \times \boldsymbol{K})$$

where $\boldsymbol{S}$ = the sum of points for all microscopic fields, $\boldsymbol{N}$ = the number of microscopic fields, $\boldsymbol{K}$ = the highest point of the scale (5).

Four plant communities were the main objects of investigation: alpine lichen heath (ALH); *Festuca varia* grassland (FVG); *Geranium gymnocaulon - Hedysarum caucasicum* meadow (GHM); and alpine snowbed communities (SBC). Some species were collected in 4 other communities: Fens near a small alpine river (*Caro caucasici-Caricetum nigrae*, FENS), dwarf shrub heaths *(Polygono viviparum-Salicetum kazbekensis*, DSH), dry feldfields and screes (*Campanulo ciliatae-Chamaesciadietum acaulis,* DSC) and rock cracks (*Potentilletum divinae* and *Astragaletum levieri,* ROCK). Detailed description of the communities are represented in Chapters 1–4).

9.3.3. Types of mycorrhiza

Among 120 studied alpine species (90 genera, 37 families) in the present study area, 83 species had vesicular arbuscular mycorrhiza (VAM), four species had ericoid mycorrhiza, three species had orchid mycorrhiza, two species ectomycorrhiza and 28 species from 12 families were nonmycorrhizal (Table 9.5, Baikalova & Onipchenko 1988). Dark septate mycelium was found in *Carex umbrosa* and *C. sempervirens* roots, but we do not consider this endophyte as real mycorrhizal fungi. Thus, most of the species (69%) had VAM-type infection. In the four community types (ALH, FVG, GHM, SBC), almost equal percentages of species from the whole species list (74-77%) were VAM.

9.3.4. Comparison between communities

To compare rates of infection in different communities we used the following criteria (Selivanov 1981): 1) percentage of nonmycorrhyzal species, 2) unweighted mean mycorrhizal infection for all species (M) and for mycorrhizal species (Q) – these parameters were calculated separately for the dominants as well; 3) weighted rate of mycorrhizal infection (P):

$$P = \frac{\sum_{i=1}^{n}(B_i \times C_i)}{\sum_{i=1}^{n} B_i} \qquad (1)$$

where C_i = the mycorrhizal infection rate for species i, B_i = the abundance (weight) of species i, n = the number of species (including nonmycorrhizal). As abundance value we used the aboveground biomass of the species within the community (Chapter 5).

Mycorrhizal parameters of the communities are represented in Table 9.6. The mean rate of mycorrhizal infection did not differ between the communities. The percentage of nonmycorrhizal species (19-26%) was similar for all studied communities. Mycorrhizal infection was higher for the dominants than for other species. The better mycorrhizal development in the dominant plant species was observed in many native plant communities on the territory of the former Soviet Union (Selivanov 1981).

Mycorrhizal parameters for two other alpine areas were studied by the same method. The alpine communities of the Central Caucasus (Kazbegi) had higher mycorrhizal parameters than ours (Nozadze 1968), but alpine tundras and meadows in Kuznetskii Alatau (Altai, South Siberia) had similar or lower values (Maksimova 1985, 1986).

Table 9.5. Mycorrhizal infection rate (average ± standard error) of plants from alpine communities. Communities: ALH - alpine lichen heath; FVG - *Festuca varia*- grassland; GHM - *Geranium gymnocaulon - Hedysarum caucasicum* meadow; SBC - snowbed communities; FENS – Fens (*Caro caucasici-Caricetum nigrae*), DSH - dwarf shrub heaths *(Polygono viviparum-Salicetum kazbekensis*), DSC - dry feldfields and screes (*Campanulo ciliatae-Chamaesciadietum acaulis*); ROCK - rock cracks (*Potentilletum divinae* and *Astragaletum levieri*). Mycorrhizal types: VAM – vesicular arbuscular mycorrhiza, ECM – ectomycorrhiza, ER – ericoid mycorrhiza, OM – orchid mycorrhiza. Single "+" means presence of mycorrhiza without quantitative data (from Baikalova & Onipchenko 1988)

Species	Myc. type	Communities				
		ALH	FVG	GHM	SBC	others
Aconitum nasutum	VAM		53±4			
Aetheopappus caucasicus	VAM	39±2				
Ajuga orientalis	VAM		56±3			
Alchemilla caucasica	VAM	60±3				
Alopecurus ponticus	VAM			54±4		
Androsace albana	VAM	33±3				
Anemone speciosa	VAM	71±3				
Antennaria dioica	VAM	61±3	57±2			
Anthemis cretica	VAM	58±2	46±2	34±2		
Anthemis marschalliana	VAM	45±2				
Anthoxanthum odoratum	VAM		18±2	25±3	42±3	
Anthyllis vulneraria	VAM	38±3				
Arenaria lychnidea		0				
Asperula alpina	VAM					DSC:5±3
Aster alpinus	VAM	49±2				
Astragalus levieri	VAM					DSC:61±3
Botrychium lunaria	VAM					ROCK:+
Briza marcowiczii	VAM			61±3		
Bromus variegatus	VAM	54±3	56±4			
Bupleurum falcatum	VAM		77±2			
Campanula bellidifolia	VAM					DSC:51±3
Campanula ciliata	VAM					DSC:52±3
Campanula collina	VAM		40±3			
Campanula tridentata	VAM	80±3	60±3	36±3	74±2	
Carex atrata			0	0	0	
Carex oreophila					0	
Carex pyrenaica					0	
Carex sempervirens		0				
Carex umbrosa		0	0			
Carum caucasicum	VAM	62±3	72±3		76±2	
Carum meifolium	VAM			68±3	76±3	
Catabrosella variegata	VAM				10±3	
Centaurea cheiranthifolia	VAM		38±3			
Cerastium purpurascens			0			
Chamaesciadium acaule	VAM	78±3				
Cirsium simplex	VAM					FENS:64±2

Table 9.5. (Cont.)

Species	Myc. type	Communities				
		ALH	FVG	GHM	SBC	others
Coeloglossum viride	OM	29±3				
Corydalis conorhiza	VAM				28±3	
Daphne glomerata	VAM		46±3			
Deschampsia flexuosa	VAM			29±4		
Draba rigida						ROCK:0
Empetrum nigrum	ER	+				
Erigeron alpinus	VAM	59±3				
Erigeron caucasicus	VAM		64±3	59±3		
Eritrichium caucasicum		0				
Euphrasia ossica		0				
Festuca brunnescens	VAM		18±2		28±3	
Festuca ovina	VAM	21±2	43±3			
Festuca varia	VAM		51±3			
Fritillaria collina	VAM	15±2	43±2			
Gagea fistulosa	VAM				23±3	
Galium verum	VAM		8±3			
Gentiana aquatica	VAM	32±3				
Gentiana biebersteinii	VAM	54±3	53±4			
Gentiana pyrenaica	VAM	15±2	9±2			
Gentiana septemfida	VAM	35±2	33±3	22±3		
Gentiana verna	VAM	15±2				
Geranium gymnocaulon	VAM			57±3		
Gnaphalium supinum	VAM		59±3	38±3	70±3	
Gymnadenia conopsea	OM	48±3				
Hedysarum caucasicum	VAM			47±5		
Helictotrichon versicolor	VAM	45±3				
Hieracium umbellatum	VAM		33±3			
Huperzia selago						ROCK:0
Hyalopoa pontica	VAM				8±2	
Hypericum linarioides	VAM		49±3			
Juniperus communis	ECM	30±4				
Kobresia schoenoides						DSC:0
Leontodon hispidus	VAM		35±2	38±3		
Lloydia serotina	VAM	50±3				
Luzula multiflora				0		
Luzula spicata		0				
Matricaria caucasica	VAM		41±3	42±3		
Minuartia aizoides			0	0	0	
Minuartia circassica		0	0			
Minuartia recurva				0		
Myosotis alpestris			0			
Nardus stricta	VAM		50±3	45±3	54±3	
Oxytropis kubanensis	VAM	40±4				
Pedicularis comosa		0				
Pedicularis nordmanniana					0	
Phleum alpinum	VAM			40±4	25±2	
Plantago atrata	VAM	65±2	47±2			

Table 9.5. (Cont.)

Species	Myc. type	Communities				
		ALH	FVG	GHM	SBC	others
Polygonum bistorta		0				
Polygonum viviparum	ECM					DSH:+
Polypodium vulgare	VAM					ROCK:+
Potentilla crantzii	VAM			58±3	36±3	
Potentilla gelida	VAM	30±3				
Potentilla nivea	VAM	58±3				
Primula algida	VAM	30±2				
Primula amoena	VAM			39±2		
Primula ruprechtii		0				
Pulsatilla aurea	VAM			54±2		
Ranunculus oreophilus	VAM	50±2	53±2			
Rhododendron caucasicum	ER			42±3		
Rumex alpestris				0		
Salix kazbekensis	ECM					DSH:+
Saxifraga exarata						DSC:0
Saxifraga paniculata						DSC:0
Saxifraga sibirica						ROCK:0
Scabiosa caucasica	VAM	46±2				
Scorzonera cana	VAM	51±2	42±3	53±5		
Sedum tenellum				0		
Sempervivum caucasicum			0			
Senecio aurantiacus	VAM		53±3			
Senecio kolenatianus	VAM		30±3			
Senecio taraxacifolius	VAM			58±3		
Seseli alpinum	VAM		62±3			
Sibbaldia procumbens	VAM		59±3	58±3	55±3	
Silene saxatilis			0			
Solidago virgaurea	VAM			67±2		
Taraxacum stevenii	VAM	49±2	44±3		27±2	
Thymus nummularius	VAM		57±3			
Traunsteinera sphaerica	OM	23±3				
Trifolium polyphyllum	VAM	52±3				
Trisetum flavescens	VAM		27±3			
Vaccinium myrtillus	ER			+		
Vaccinium vitis-idaea	ER	30±2	35±2			
Veronica gentianoides	VAM	36±3	38±3	25±2		
Viola altaica	VAM		31±2			

9.3.5. Seasonal dynamics of mycorrhiza

Seasonal fluctuations in the mycorrhizal infection rate were studied for 10 species on three dates (July 5, July 30, August 25, 1983, Baikalova and Onipchenko 1988). The species can be combined into 4 groups

Table 9.6. Parameters of mycorrhizal infection for 4 alpine communities: M - mean unweighted infection rate for all species, including nonmycorrhizal, Q – mean unweighted infection rate for mycorrizal species only, P - weighted rate of mycorrhizal infection (see text). (from Baikalova and Onipchenko 1988)

Parameters	ALH	FVG	GHM	SBC
Number of studied species	49	48	29	19
Number of nonmycorrhizal species (%)	10(20%)	9(19%)	7(24%)	5(26%)
M –for all species	35,2%	36,8%	35,2%	32,3%
M – for dominants only	40,7%	50,5%	42,8%	41,2%
Q – for all mycorrhizal species	44,2%	45,3%	46,4%	43,8%
Q – for mycorrhizal dominants only	57,0%	50,5%	42,8%	51,5%
P – for all studied species	35,5%	45,8%	46,6%	39,8%

according to the observed seasonal changes of mycorrhizal infection (Fig. 9.4). 1) Species with a constant infection rate during summer –*Anemone speciosa, Trifolium polyphyllum, Hedysarum caucasicum*; 2) Species with a constant infection rate in July and an increased rate by the end of the growing season (*Festuca varia, Geranium gymnocaulon, Gnaphalium supinum, Sibbaldia procumbens*); 3) Species with a slight midsummer depression and an increased rate until the end of the growing season (*Antennaria dioica, Campanula tridentata*); 4) *Festuca ovina* was the only species that had a lower infection rate by the end of the season. The mean infection rate for 10 species was 43%, 43%, and 58% for 5 July, 30 July and 25 August respectively. Higher infection rates closer to the beginning of the autumn can be explained by several possible reasons: 1) better soil conditions (temperature and humidity) at this time; 2) end of seed maturation at the end of August can provide fungi with more assimilates than in the middle of the summer; 3) different growth rates for roots and mycorrhizal fungi (Alexander & Alexander 1984). Midsummer depression might be connected with intensive use of assimilates by flowering shoots at this time as well as nutrient shortage for mycorrhizal fungi.

9.3.6. Ontogenetic dynamics

Root samples of eight species were collected in different ontogenetic stages (seedlings, juvenile, vegetative and generative plants) (Table 9.7).

We can divide the eight species into two groups. Five species (*Anemone speciosa, Campanula tridentata, Carum caucasicum, Ranunculus oreophilus, Veronica gentianoides*), did not show significant changes in mycorrhizal infection rate between ontogenetic stages (Baikalova & Onipchenko 1988). Three other species (*Gagea fistulosa, Potentilla gelida, Taraxacum stevenii*) demonstrated a significant decrease in mycorrhizal infection from seedlings or juvenile plants to adult plants. So mycorrhizal infection develops during the earliest stage of ontogeny of the alpine plants. Mycorrhizal infection might decrease during the generative stage due to a change in the main direction of assimilate flow from roots to inflorescences.

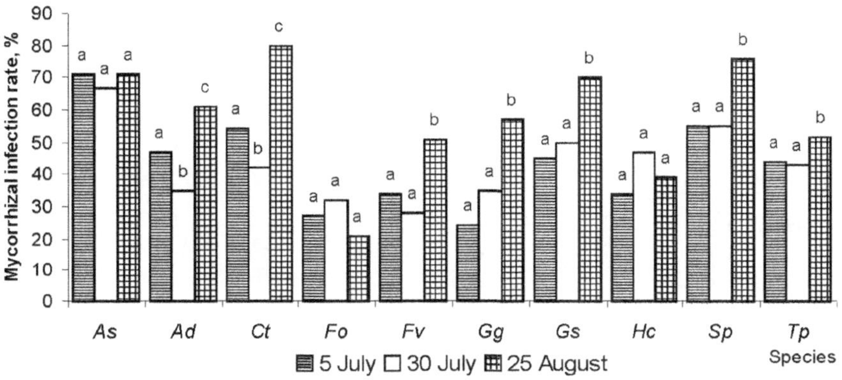

Figure 9.4. Seasonal dynamics of mycorrhizal infection. Different letters within species show significant difference between dates. Species: *As - Anemone speciosa, Ad - Antennaria dioca, Ct - Campanula tridentata, Fo - Festuca ovina, Fv - Festuca varia, Gg - Geranium gymnocaulon, Gs - Gnaphalium supinum, Hc - Hedysarum caucasicum, Sp - Sibbaldia procumbens, Tp - Trifolium polyphyllum*

9.3.7. Mycorrhizal infection and vegetative mobility

In natural communities, both mycorrhizal (Fitter & Nichols 1988, Read & Birch 1988, Smith et al. 1992) and clonal connections (Stuefer & Hutchings 1994, Stuefer et al. 1996) may enhance the establishment of young plants by supplying them with resources, which makes them more tolerant of environmental harshness and increases their competitive ability. However, since in both cases there is a considerable carbon cost either to feed mycorrhizal hyphae (up to 20% of assimilates – Finlay & Söderström 1992) or to establish rhizomes or stolons (up to 40% of its parent's dry weight – Callaghan 1987), one may assume that plants will invest more in just one of these possibilities. In other words, we may observe a trade-off between vegetative mobility and mycorrhizal infection (Onipchenko & Zobel 2000).

According to their morphological characteristics, species of the alpine communities were classified into 2 groups: 1) moderate or high vegetative mobility (annual rhizome/stolone increment more than 2 cm), 2) no or

Table 9.7. Mycorrhizal infection rate (%, mean and standard error) of alpine plants at different ontogenetic stages. Abbreviations: s – seedlings, j – juvenile plants, v – vegetative plants, g – generative plants (from Baikalova & Onipchenko 1988)

Species	Stage	Mean infection rate, %	Standard error
Anemone speciosa	j	58	2
	v	60	3
	g	67	3
Campanula tridentata	j	47	3
	g	41	2
Carum caucasicum	j	69	3
	g	62	3
Gagea fistulosa	s	41	3
	v	37	3
	g	23	3
Potentilla gelida	j	48	4
	g	30	3
Ranunculus oreophilus	j	51	2
	g	50	2
Taraxacum stevenii	s	50	4
	j	48	3
	v	17	2
Veronica gentianoides	j	34	3
	g	36	3

low vegetative mobility (increment less than 2 cm per year). *Minuartia aizoides* and *Anthoxanthum odoratum* are typical representatives of mobile species, while *Carex* spp., *Festuca* spp. or *Geranium gymnocaulon* represent the group of species with low mobility (see Chapter 5.1 for more detailed data).

Vegetative mobility showed a significant effect ($P<0.01$) on the rate of root mycorrhizal infection for all species, as well as for mycorrhizal species only (Onipchenko & Zobel 2000). The rate of infection was lower for mobile species (Fig. 9.5) and the mean root infection rate did not differ between the communities. Although there was no significant interaction between vegetative mobility and community type, it seems that the difference in the infection rate among species groups with different mobility was smallest in the GHM community (Fig. 9.5).

In order to check whether the community-level pattern of plant traits holds when only species with similar phylogeny are included, the differences in the rate of root infection between mobile and non-mobile/low mobility species were compared within two larger families, Poaceae and Asteraceae. In case of the Poaceae (4 mobile and 8 non-mobile/low mobility species) we found that the rate of root mycorrhizal infection was significantly lower (t-test, $P=0.020$) in the mobile species (20%) than in non-mobile/low mobility species (39%). The same tendency for non-mobile/low mobility species to have a higher infection rate than the mobile ones (39% and 52% respectively) was also observed in the Asteraceae (Onipchenko & Zobel 2000).

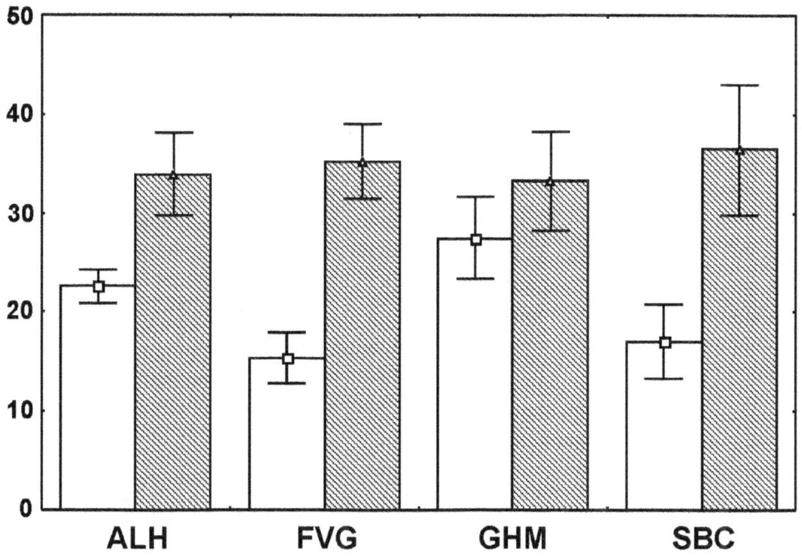

Figure 9.5. Mycorrhizal infection rate and vegetative mobility of alpine species in different communities (means and standard errors; all species, including non mycorrhizal ones). White columns - mobile species (more than 2 cm/year), dark columns - non/low mobility species. ALH - alpine lichen heath, FVG - *Festuca varia* grassland, GHM - *Geranium gymnocaulon* - *Hedysarum caucasicum* meadow, SBC - alpine snow bed community (from Onipchenko & Zobel 2000 with permission of *Folia Geobotanica*)

Thus, higher infection rates were observed in immobile species. Since these species depend only on seed reproduction and there are no functional connections between the mother plant and their offsprings, one may hypothesise that VAM may act like a clonal connector between them (Wilkinson 1998). The VAM network supports seedlings and young plants by providing organic components and nutrients, especially phosphorus, which are important in harsh (cold, nutrient poor) alpine environments. In natural communities, VAM infection may develop rather quickly (Fitter & Nichols 1988) and the early integration of plants into the mycelial network would provide them with an absorptive surface at a relatively low energy cost (Francis & Read 1994). We have seen above that seedling and juvenile alpine plants might have higher rates of mycorrhizal infection than the adult plants. The improvement of nutrient conditions through the rapidly developing mycelial network may serve as an alternative to the translocation of carbon which otherwise would be needed to build a root system (Onipchenko & Zobel 2000).

Also, VAM extramatrical hyphae may serve as alternatives for vegetative organs in foraging for soil resources over larger and often patchy areas. Under conditions when root growth is restricted, the fungi can contribute up to 80% of the phosphorus absorbed (Li et al. 1991). At the same time, absorption of phosphorus can take place a long distance (up to 7 cm) away from the roots (Rhodes & Gerdemann 1975). In this way, otherwise sessile plants may have the possibility to forage for nutrients beyond their own depletion zone, or just use the physical patchiness of the soil environment by foraging resources in more fertile microsites. Another possible strategy for plants to forage in patchy environments is clonal growth (Svensson & Callaghan 1988, Stuefer & Hutchings 1994, Stuefer et al. 1996). The pattern of the root AM infection described in alpine grassland communities in the Northwest Caucasus, supports the view that these two possibilities are alternatives (Onipchenko & Zobel 2000).

It is interesting to note that the difference in the infection rate for mobile and non-mobile/low mobility species was smallest for GHM communities. On the basis of our recent observations, this fact remains unexplained. It may be somehow related to the productivity of the plant community. These meadows are the most productive communities within the alpine area (Chapter 5.2). The microarthropod consumption of VAM fungi may lead to decreasing interplant connections (Finlay 1985). Density of all microarthropod groups (Oribatei, Trombidiformes, Gamasina, and Collembola) was the highest in GHM in comparison with other alpine communities (Netujilin et al. 1992). This type of meadow is subject to intense burrowing activity of small rodents, mainly vole *Pitymys majori* Thomas (Fomin et al. 1989), and the Caucasian mole (*Talpa caucasica* Satunin) (Zenyakin & Onipchenko 1997) as well as by large mammals like wild boars (*Sus scrofa* L.) and bears (*Ursus arctos* L.) (Onipchenko & Rabotnova 1994, see Chapters 5 and 10). Frequent disturbances may have a negative effect on the mycorrhizal network in the soil and may decrease the importance of mycorrhiza for plants. All of the above-listed factors may decrease the role of VAM connections between plants in GHM and lead to smaller differences in mycorrhizal infection rates between mobile and non/low mobile species (Fig. 9.5).

9.3.8. Disturbance and mycorrhizal infection

Mycorrhizal networks are sensitive to soil disturbance (Jasper et al. 1989, McLellan et al. 1995). On the other hand mycorrhiza seems to be less important for plants occupying disturbed areas where competition for nutrients and light is evidently reduced. For example, many species that can be classified as ruderals by Grime et al. (1988) belong to the Brassicaceae or Caryophyllaceae, and most of the species of these families are nonmycorrhizal (Harley & Harley 1987, Smith & Read 1996). One would expect that such a pattern is also valid on a fine scale, i.e. species which are more dependent on mycorrhiza are those which mostly grow in dense established stands, while those species which clearly prefer gaps are less dependent on mycorrhiza or even nonmycorrhizal (Onipchenko & Zobel 2000).

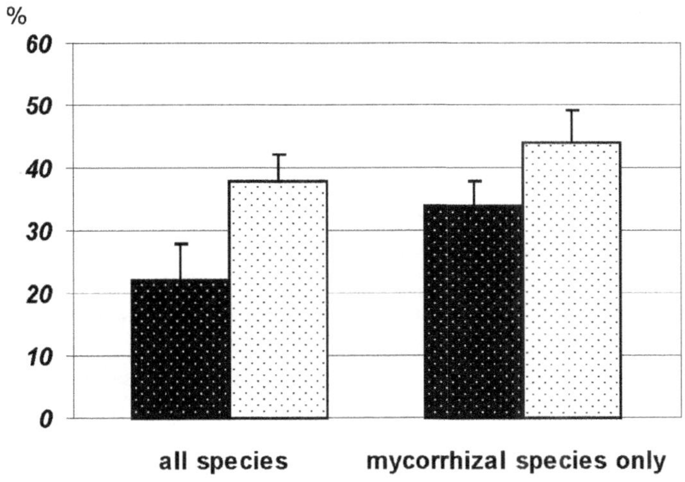

Figure 9.6. Mycorrhizal infection rate (%) and reaction to zoogenic disturbance, black bars - species whose frequency was significantly higher in gaps, white bars - species whose frequency was the same or lower in gaps (from Onipchenko & Zobel 2000 with permission of *Folia Geobotanica*)

Responses to natural zoogenic disturbance (gap formation) were investigated in GHM (Onipchenko & Rabotnova 1994, see Chapter 5). All species were divided into 2 groups: 1) species whose frequency was significantly higher in the meso- and/or microgaps in comparison with the control; 2) species whose frequency was the same or significantly lower in the meso-and/or microgaps. This information is used to check whether there was a difference in root VAM colonisation between the species of the two above mentioned groups (Onipchenko & Zobel 2000).

The mean rate of root mycorrhizal infection was lower for species preferring gaps (Fig. 9.6). According to the ANOVA, the difference was significant for all investigated species. The same tendency (but non-significant, $p= 0.0786$) was observed for mycorrhizal species only.

We conclude that at least in GHM, where disturbance processes are very important for the community structure (Chapter 5), there is a difference in mycorrhizal infection rate between species preferring gaps and closed patches of the community, respectively.

9.4 Soil fungal biomass and microbial activity

9.4.1 Introduction

Bacteria, actinomycetes, fungi and microfauna comprise the decomposition complex of an ecosystem. In many soils, the fungal component of soil microbial biomass exceeds the prokaryotic one and can reach 78-90% under grassland vegetation (Kjøller & Struwe 1982). Moreover fungi are the part of this microbial community which most effectively decomposes stable organic polymers, such as cellulose and lignin. However, the ecological function of soil microorganisms is not limited to decomposition. These organisms perform many other transformations of substances, e.g., nitrogen transformations that are especially important for plants, namely mineralisation, N_2-fixation and denitrification.

The studied plant alpine communities differ considerably in their production and nutrient status. We compared 4 communities (alpine lichen heath - ALH, *Festuca varia* grassland - FVG, *Geranium – Hedysarum* meadow - GHM and snow bed community - SBC, see description above) in terms of microbiological activity. Some other alpine communities (*Rhododendron caucasicum* shrubs – RC, FENs, dry and moist alpine screes – DSC, MSC) were studied for comparison as well. The following parameters were measured: fungal biomass, cellulose and wood decomposition rate (Leinsoo et al. 1991), potential and actual rate of soil respiration, N_2-fixation and denitrification (Stepanov & Onipchenko 1989).

9.4.2 Methods

All biomass and soil microbial activity measurements were performed for the upper (A_d) soil horizon in each community. Fungal biomass was estimated with a slightly modified membrane filter technique of Hanssen et al. (1974) (Leinsoo et al. 1991).

To estimate cellulose decomposition rate, pieces of filter paper (10×10 cm) were buried in the soil for one year. The following scale was used for the estimation: 1 – decomposition is invisible; 2 – scant fungal colonies are seen on specimen, which can be easily abstracted from the soil; 3 – abundant fungal colonies inhabit the specimen; looking more or less entire but thinned and breaks when being abstracted from the soil; 4 – small fragments are remained; 5 – complete or almost complete decomposition.

The relative wood decomposition rate was measured by using birch (*Betula litwinowii*) and pine (*Pinus sylvestris*) wood specimens (three per layer) of known weight (10-15g), which were buried. After two years of such exposition, the proportional weight decrease was measured.

The respiration rate was determined from CO_2 emission from the soil on the plots selected. Nitrogen fixation activity was established from the acetylene reduction rate (Hardy et al. 1968). Denitrifying activity was assessed from the N_2O release from the soil in the presence of acetylene (Yoshinari et al. 1977). The gas samples were collected at 1 h intervals from 200 ml diffusion chambers dug into the soil to the depth of 5 cm. The internal standard was propane. The gas samples were analysed with a gas chromatograph.

The potential activity of these processes was determined under laboratory conditions. For this purpose, 5g of air-dry soil was placed in 15 ml glass flasks, moistened to 60% of total field moisture capacity, and preincubated at 28° for three days in a desiccator at constant humidity. Respiration activity was determined from the CO_2 increment after addition of glucose (2.5 mg/g), while potential nitrogen-fixing activity was found by the method developed at the Soil Biology Department of Moscow State University (Zvyagintsev et al. 1980). Potential soil denitrifying activity was determined in the presence of acetylene (10% by volume) after addition of glucose (2.5 mg/g) and potassium nitrate (0.2 mg/g). Anaerobic conditions were created by replacing the air with argon. The gases were analyzed in gas chromatographs with thermal conductivity and flame-ionization detectors.

9.4.3 Results

9.4.3.1 Soil fungal biomass

Total fungal biomass was maximal in the soils of the FVG, RC and FEN and minimal in scree soils (Fig. 9.7). Hence, most favourable conditions for soil microfungi are observed in communities which are covered by a deep layer of snow in winter and become snowless not later than in the beginning of July.

Spores comprised only 1-2% of the total biomass in the FVG and GHM, but 24-37% in the communities with low fungi content, *e.g.* DSC, MSC and ALH. It is assumed (Gams 1998) that only mycelial forms may have considerable metabolic activity, while spores are dormant survival structures having low activity and limited importance in soil metabolism. Therefore, a lower spore percentage presumably indicates that a higher proportion of the total fungal biomass is physiologically active. There was no certain tendency in spore proportion changes depending on soil horizons. The relative content of dark hyphae increased downwards in the ALH, GHM and SBC, but the RC and FVG did not fit this pattern.

The relative content of dark and hyaline hyphae was also influenced by humidity. The proportion of hyaline mycelium increased in moister conditions.

Total fungal biomass was inversely related to depth (Leinsoo et al. 1991). However, steepness of the decrement curves differed considerably depending on the community type. Thus, the steepest decline was observed in ALH, SBC and RC soils. Given the comparatively slight change in the fungal content in FVG and GHM soils, it can be assumed that they have deeper profiles involved in biological processes.

Seasonal changes in total biomass content as well in the relative proportion of spores, dark and hyaline hyphae, did not show any significant tendency. Every community had its pattern of these changes and any attempts to make generalized conclusions would be premature (Leinsoo et al. 1991).

9.4.3.2 Soil respiration, N_2 – fixation and denitrification

Actual nitrogen fixation and denitrification activity and the respiration rate were measured in mid-August 1987, when the weather was relatively dry and hot and the soil had a low moisture content. The highest figures for actual nitrogen fixation, denitrification, and respiration at the time of the

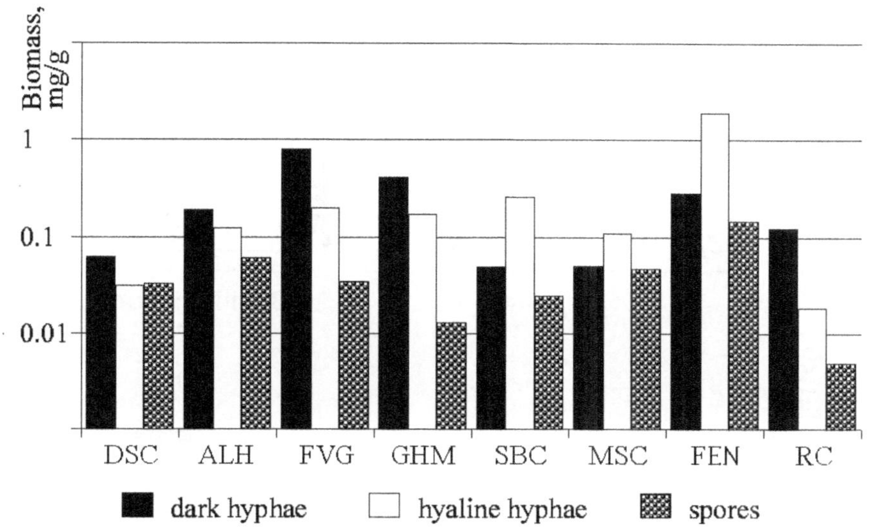

Figure 9.7. Soil fungal biomass in upper soil horizons. Means for 1986-1988 August measurements

observations were obtained in the SBC (Fig. 9.8), which had lost their snow only shortly before the measurements and had a higher soil moisture content (mean moisture content in the soil sod horizon during the observation period amounted to 42% for the SBC plots and 29% for the FVG and GHM plots).

Our estimates of actual nitrogen fixation activity somewhat exceeded those given for alpine associations in the Pyrenees (Tosca & Labrone 1981) and for the tundra and meadows of the Canadian high-latitude Arctic (Karagatzides et al. 1985). Nitrogen fixation activity values closer to our observations have been observed in alpine pastures in Australia (Henry & Svoboda 1986), while substantially higher levels have been reported for alpine clover meadows in New Zealand (Carr et al. 1980). The soil respiration rate in the alpine ecosystems of the Malaya Khatipara is high in comparison with that of other alpine communities in the Central Caucasus (Nakhutsrishvili et al. 1987).

9.4.3.3 Cellulose and wood decomposition

In spite of the fact that the study sites are situated far above the timberline (2300-2400 m a.s.l.), decomposition of conifer and deciduous wood was observed (Table 9.8). The pine wood decomposition rate was highest in

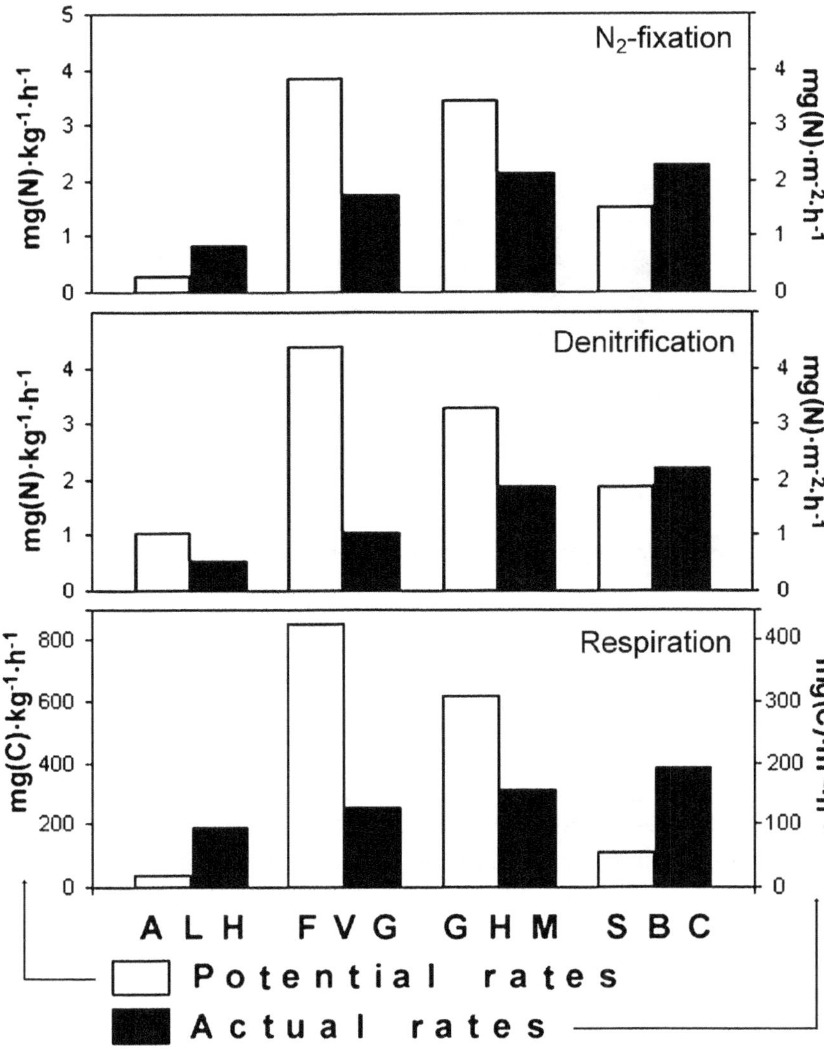

Figure 9.8. Actual and potential rate of respiration, N_2-fixation and denitrification in alpine communities

the soils of the GHM where it was almost four times as high as in the MSC. Birch wood decomposed approximately twice as fast as pine. Most intensive birch wood decomposition was observed in the FVG, SBC and GHM where it was almost five times as fast as in screes. Wood decomposition rates were similar in different soil horizons. So, the destruction processes differ slightly among closed alpine communities and within soil profiles. Even in the RC, where *Rhododendron* stem wood is abundant among natural decomposing materials, the rate of wood decomposition was low and similar to the same parameter in the alpine herbaceous communities.

The rate of cellulose decomposition was highest in the productive GHM and lowest in the DSC (Table 9.9). In case of cellulose, a decrease in the decomposition rate with soil depth was observed.

Table 9.8. Wood decomposition rate (weight loss for two years, %, mean and standard error) in different alpine communities and soils horizons

	Betula wood			*Pinus* wood		
Soil horizons	Ad	A1	B	Ad	A1	B
DSC	5.4 ± 1.1	-	-	3.0 ± 1.0	-	-
ALH	11 ± 3	8.6 ± 1.8	7.6 ± 1.6	5.6 ± 1.0	6.8 ± 2.3	5.1 ± 2.0
FVG	15 ± 3	9.2 ± 2.4	6.0 ± 1.8	5.6 ± 1.0	4.2 ± 1.5	6.0 ± 1.1
GHM	12 ± 4	18 ± 8	12 ± 3	9.2 ± 2.2	6.4 ± 1.8	4.3 ± 1.4
SBC	14 ± 5	6.5 ± 2.3	4.5 ± 2.2	6.5 ± 2.3	6.9 ± 2.2	4.0 ± 1.9
MSC	4.4 ± 0.6	-	-	2.3 ± 0.5	-	-
FEN	11 ± 3	-	-	6.0 ± 0.6	-	-
RC	11 ± 3	8.4 ± 2.0	-	6.4 ± 0.2	6.7 ± 1.4	-

Table 9.9. Estimation annual cellulose decomposition rate (points on a visual scale, see text)

Soil horizon	Communities							
	DSC	ALH	FVG	GHM	SBC	MSC	FEN	RC
Ad	1	2	4	5	3	3	4	3
A1	-	1	3	4	2	-	-	2

9.4.3.4 Conclusions

Analysis of the obtained data enabled us to conclude that:

1) The alpine meadows (FVG and GHM) were characterised by a higher potential activity for all of the processes studied compared to the ALH, SBC and screes. Thus, potential microbiological activity for the ALH soils was approximately an order of magnitude less than that for the alpine meadow soils.
2) Potential nitrogen fixation, denitrification, and respiratory activity were closely correlated with one another and the latter two showed a weak correlation with the fungal biomass. The correlation coefficient between nitrogen fixation and denitrification was +0.84 ($p>0.99$). Actual nitrogen fixation, denitrification, and respiratory

activity were closely correlated with one another, but neither with the potential parameters nor with the biomass. This difference results probably from the fact that the weather conditions during the measurement of actual parameters did not reflect the average annual conditions, which determine potential parameters (Fig. 9.9).

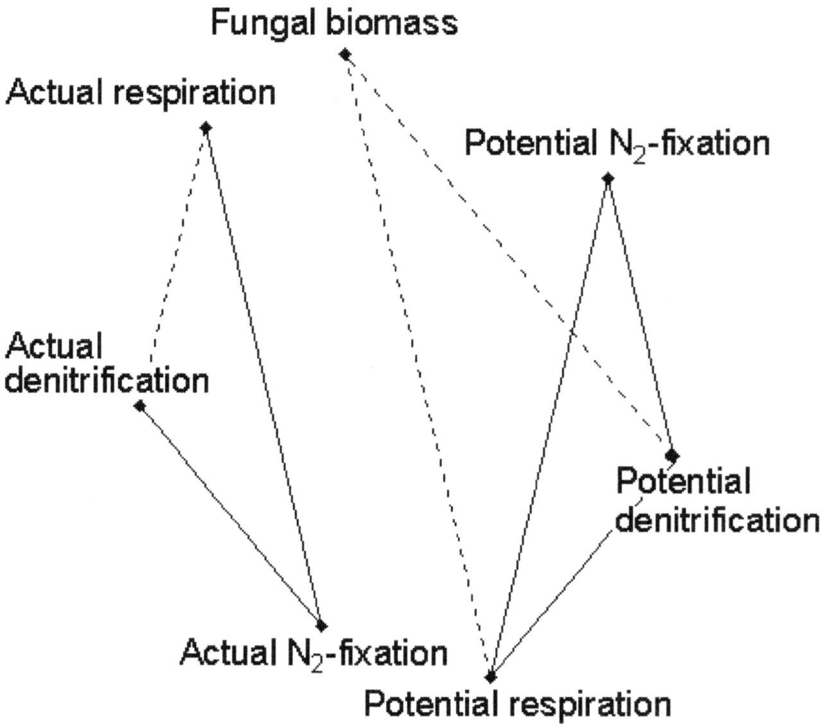

Figure 9.9. Correlation between studied microbiological parameters. Solid lines indicate a significant positive correlation coefficient ($p<0.05$), dotted lines - positive tendencies ($p<0.1$)

Fungal biomass is a questionable indicator of physiological activity. However, in our particular case these values were mostly correlated. Thus, the highest cellulose decomposition rate was observed for the soils of the GHM, FVG, and FEN. For the former two, a similar conclusion of a high microbiological activity had been demonstrated elsewhere. Thus, GHM communities have maximal values of aboveground and belowground plant biomass as well as relatively low mortmass values (Chapter 5) and hence a high microbiological activity. A high decomposition rate and a large fungal biomass in the mesotrophic FEN implies that even in a humid alpine area water availability may be a limiting factor for mineralisation.

Chapter 10

Main animal groups and their influence on plant communities

10.1. Wild ungulates

Seven native and one introduced (*Cervus nippon* Temmink, 1838) species of ungulates inhabit the Northwestern Caucasus (Sokolov & Tembotov 1993). Two of them (roe – *Capreolus capreolus* L. and elk – *Alces alces* L.) are extremely rare or absent in the vicinity of Teberda. Two other species (bison – *Bison bonasus* L. and red deer – *Cervus elaphus* L.) are typical woodland animals and they rarely occur in the alpine belt. So, only three species of ungulates, namely Caucasian goat - *Capra caucasica* Guldenst., chamois - *Rupicapra rupicapra* L. and wild boar - *Sus scrofa* L., spend significant parts of their life in the alpine area. These three species will be discussed in this chapter.

10.1.1. The Caucasian goat (ibex) (*Capra caucasica* Guldenst.)

This species is a symbolic ungulate for the alpine belt in the Great Caucasus (Fig. 10.1). These big animals reach 155 kg in weight. Their mean life span is 14-15 years, but individual animals live more than 20 years (Sokolov & Tembotov 1993). When it was protected from illegal hunting the density of the Caucasian ibex was about 63 animals per 10 km^2 in the Teberda Reserve (Tarasov 1973). Ibexes can be found from 800 to above 4000 m. They inhabit deciduous and dark-coniferous forests, and the subalpine, alpine and subnival belts, especially where cliffs and rocks provide sufficient refuge from carnivores. The total number of animals in the

V.G. Onipchenko (ed.), Alpine Ecosystems in the Northwest Caucasus, 303–353.

Teberda Reserve is about 500, but in the 1980s there were about 2000 ibexes in the Reserve.

Figure 10.1. Capra caucasica (Picture by V.M.Smirin from Sokolov & Tembotov 1993)

Availability of winter pastures is one of the most important limiting factors for the population density of that animal. In summer only a small part of the abundant plant biomass is consumed by ibexes within the Natural Reserve, but outside the protected area domestic stock is a serious competitor for summer feed.

The Caucasian ibex uses a various set of plants for food. His diet includes 195-256 plant species (Kotov 1968, Abdurakhmanov 1977). Grasses are the main component (80-90%) of the diet. A list of some alpine food species is represented in Table 10.1.

The most prefered food species are *Alopecurus* spp., *Bromus variegatus*, *Festuca* spp., *Hyalopoa pontica*, *Phleum* spp., *Senecio taraxacifolius* (Kotov 1968, Grishina et al. 1985, Sokolov & Tembotov 1993, Bobyr' et al. 1999).

Winter grazing by *Capra caucasica* was studied by K.Bobyr' et al. (1999). The authors observed a change in plant species composition from autumn to spring on a small (2.6 ha) windward snowfree slope and counted the number of animals during day hours from November to March (Fig. 10.2-10.3). Here ibexes consumed more than 70% of the aboveground phytomass (including dead parts of plants and litter) (Bobyr' et al. 1999). *Bromus variegatus, Festuca varia, Carex sempervirens, Anthyllis vulneraria, Campanula collina, Campanula tridentata, Hieracium umbellatum, Plantago atrata, Scabiosa caucasica,* and *Veronica gentianoides* decreased in aboveground biomass significantly during winter grazing. These species can be considered as the main food plants for ibexes in the observed plots. Ibexes were active during the whole day, but their activity decreased near noon (Fig.10.2). The density of grazing animals slightly increased from midwinter to early spring, probably due to increased snow depth in surrounding areas (Fig. 10.3). Using the average daily animal number and the length of the period of winter we could calculate that one Caucasian ibex consumed 3,5 kg of the hay (dry mass) at the studied pasture, which forms, approximately, half of his daily needs in winter. The remaining part of the fodder they got at the other plots, which are situated close by. We consider that the pasture was enough for maintaining 3-4 Caucasian ibexes during winter (Bobyr' et al. 1999).

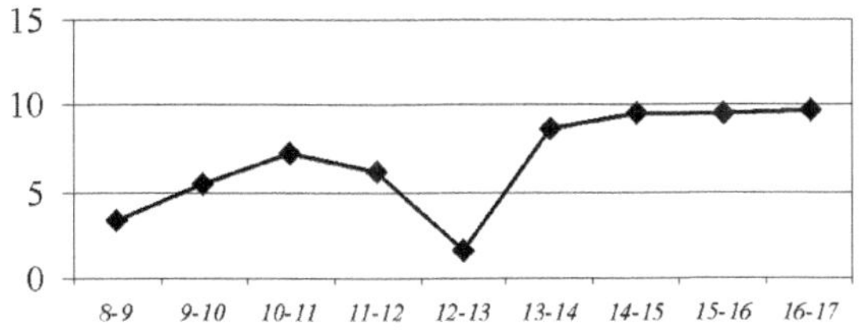

Figure 10.2. Mean number of ibexes (*Capra caucasica*) on a winter pasture permanent plot during a winter day. Ordinate – number of grazing animals, abscissa – time (hours) (from Bobyr' et al. 1999)

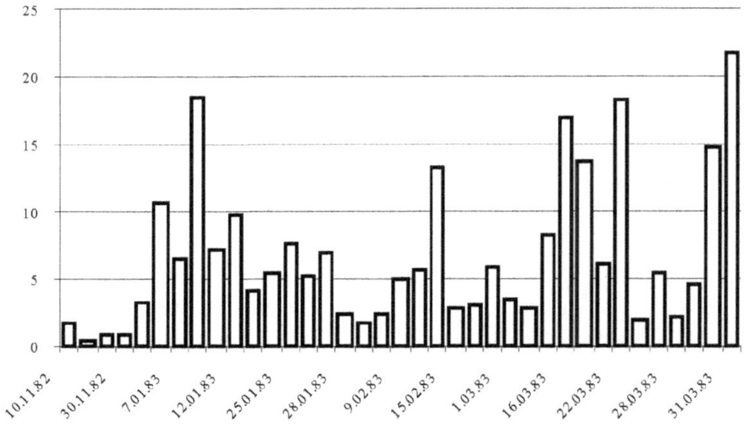

Figure 10.3. Mean number of grazing ibexes on a winter pasture during the period of observation (from Bobyr' et al. 1999)

Significant consumption of the aboveground standing crop (>70%) during winter grazing does not lead to degradation of the alpine meadows and heaths. Quite the reverse, removing old plant tissues and dead material increase the rate of decomposition and nutrient turnover. On snow protected areas of *Festuca varia* grassland litter might reach 900 g/m^2 and more (Chapter 5.2). So the role of ibexes in litter mineralisation is very important on snow free alpine areas. Similar conclusions were made for other grasslands (Abaturov 1984).

Alpine lichen heaths are widespread communities on windward crest and slopes (Chapters 1- 5). It was shown (Kotov 1968) that the main dominant of these communities, *Cetraria islandica,* is one of the important components in the ibex winter diet. These lichens might play the same role for high mountain ungulates as *Cladina* spp. and other arctic lichens for reindeers. Winter overgrazing of alpine lichen heaths can lead to degradation of the lichen cover.

Part of ibex population (especially young animals and females) migrate to the forest belt during winter. They use shrub and tree branches as a main winter food. Both parts of the population meet in subalpine meadows with fresh grass in spring (Sokolov & Tembotov 1993).

Table 10.1. Estimation of plant species preference by different herbivores in alpine communities. Abbreviations: Ds – domestic sheep, Ci – caucasian ibex, Ch – chamois, W – wild boar, Pm – *Pitymys majori*, Cn – *Chionomys nivalis*, Nc – *Nocaracris cyanipes*, Pt – *Podisma teberdina*, Sn – *Stenobotrus nigromaculatus*, Ov – *Omocestus viridulus*, Gs – *Gomphocerus sibiricus*. Scale points – see Chapter 10.2.3. – for voles, 10.5.1. – for grasshoppers. Based on Zharkov (1959), Kotov (1968), Larin et al. (1950-1956), Sokolov & Tembotov (1993), Magomedov et al. (2001), original observation by Fomin & Onipchenko (voles), Mikhailenko (grasshoppers).

	Ungulates				Voles		Grasshoppers				
	Ds	Ci	Ch	W	Pm	Cn	Nc	Pt	Sn	Ov	Gs
Aconitum nasutum					3		4	1	0	2	2
Aetheopappus caucasicus					3	4	5	3	2	2	3
Agrostis vinealis		3					4	4	5	5	5
Ajuga orientalis					2		5	1	0	1	1
Alchemilla caucasica	3	3	3	4	2	2	4	4	0	0	0
Alchemilla vulgaris aggr.							5	5	0	0	2
Anemone speciosa	1	2	3		3	4	3	2	0	0	0
Antennaria dioica	2				3	2	2	4	0	0	3
Anthemis cretica			3		4	5	3	3	0	1	0
Anthemis marschalliana		3	3		4		5	3	1	1	2
Anthoxanthum odoratum		4			4		1	3	3	4	3
Anthyllis vulneraria	3	5	2		5	5	3	2	2	1	1
Arenaria lychnidea		4			4	5	3	1	1	2	0
Aster alpinus	3	4			2	3	3	4	1	0	2
Betonica macrantha							4	0	0	0	2
Botrichium lunaria					5						
Briza marcowiczii							4	1	1	3	5
Bromus variegatus		5			4	4	2	2	5	5	4
Calamagrostis arundinacea		4					4	5	4	5	5
Campanula collina	3	5	4		5	5	4	1	1	1	4
Campanula tridentata	4	4	5	4	5	5	5	4	1	1	3
Carex atrata		3			4		3	4	5	5	4
Carex oreophila		3					1	0	4	4	4
Carex pyrenaica					4		3	3	3	4	5
Carex sempervirens	5	4			4	4	3	2	5	4	5
Carex umbrosa	3		3		4	4	2	0	3	3	3
Carum caucasicum	5	4	4	5	5	5	5	2	0	0	4
Carum meifolium		4		5	5		5	5	0	1	1
Catabrosella variegata					4		5	4	5	4	5
Centaurea cheiranthifolia			3		2		5	4	3	3	4
Cerastium purpurascens							4	5	0	4	2
Chaerophyllum roseum							3	2	0	0	3
Chamaenerion angustifolium				4			2	1	0	0	1
Chamaesciadium acaule		3					5	5	0	0	0
Coeloglossum viride					5		5	5	1	3	2
Corydalis conorhiza							5	1	0	0	1
Cruciata laevipes							3	5	0	0	0
Deschampsia flexuosa		5			3		2	3	4	4	4
Draba hispida							1	4	0	0	0

	Ungulates				Voles		Grasshoppers				
	Ds	Ci	Ch	W	Pm	Cn	Nc	Pt	Sn	Ov	Gs
Draba scabra							3	0	2	2	5
Empetrum nigrum				4			1	0	0	0	0
Erigeron alpinus	3	3	3		4	2	2	2	0	2	3
Erigeron caucasicus							5	3	0	0	0
Eritrichium caucasicum	2				3	2	1	2	1	1	
Euphrasia ossica					3	3	3	0	0	0	0
Festuca brunnescens		4			3		2	0	5	5	5
Festuca ovina	5	5			3		1	0	4	3	1
Festuca varia		3			3		4	3	5	5	5
Fritillaria collina	1	3			5	5	3				
Galium verum		3			3		3	5	0	0	0
Gentiana biebersteinii		2			2		5	0	0	1	0
Gentiana pyrenaica	1				3	3	3	2	0	0	1
Gentiana septemfida	1	3	5	4	2	2	5	3	3	0	0
Gentiana verna											
Geranium gymnocaulon					2	3	4	1			1
Gnaphalium supinum					3		5	3	0	0	0
Gymnadenia conopsea					4		1	1	0	0	2
Hedysarum caucasicum							5	4	1	0	1
Helictotrichon versicolor				4	4		4	4	1	1	3
Hieracium laevigatum	4	3			4	4	1	0	5	5	3
Hieracium umbellatum							5	5	0	2	4
Hyalopoa pontica		4			5		4	1	0	0	0
Hypericum linarioides		5			5		5	3	5	5	5
Juniperus communis							2	1	0	0	0
Kemulariella caucasica							0	2	0	0	0
Leontodon hispidus							4	3	0	1	1
Lloydia serotina					5		4	0	4	0	5
Luzula multiflora							4	1	0	0	1
Luzula spicata					4		5	3	5	3	4
Matricaria caucasica	2				4	4	2	2	2	4	1
Minuartia aizoides					4		3	3	0	1	1
Minuartia circassica		3			4		2	2	0	0	1
Minuartia recurva	2				4	4	3	1	1	1	1
Myosotis alpestris					4		3	5	0	1	3
Nardus stricta					3		5	3	0	2	0
Oxytropis kubanensis					3		0	2	4	3	3
Pedicularis caucasica		4	5		5	4	4	4	3	1	5
Pedicularis comosa	2	3			5	3					
Pedicularis condensata	2	3	3	5	4	5	3	1	1	0	2
Pedicularis nordmanniana					4		5	3	0	0	2
Phleum alpinum				4	5		1	2	0	0	1
Plantago atrata		5			4		5	5	5	5	5
Polygonum bistorta	5	4	5		4	4	5	4	1	1	2
Potentilla crantzii	2	3	3	4	3	2	2	3	1	1	4
Potentilla gelida					4	5	4	1	2	1	1
Primula algida	4	3			3	5	3	3	1	0	0
Primula ruprechtii	4	3			2	2	1	0	1	0	0

	Ungulates				Voles		Grasshoppers				
	Ds	Ci	Ch	W	Pm	Cn	Nc	Pt	Sn	Ov	Gs
Pulsatilla albana					3	3	1	2	0	1	1
Pulsatilla aurea							2	2	0	0	1
Ranunculus oreophilus			3		3		4	1	0	0	0
Rhododendron caucasicum	1	3	3		4	5	4	1	0	1	0
Rumex alpestris		3					1	1	0	0	0
Scabiosa caucasica			3		4		5	5	0	0	1
Scorzonera cana							5	5	0	0	1
Sedum tenellum	4				5	5	5	5	2	4	5
Sempervivum caucasicum					4	1	1	2	0	0	1
Senecio aurantiacus					3		4	2	0	2	1
Senecio kolenatianus					4						
Senecio taraxacifolius							5	5	0	2	4
Seseli alpinum					4		5	5	0	0	1
Sibbaldia procumbens							0	4	0	0	0
Silene saxatilis		3			1		4	4	0	0	2
Solidago virgaurea					5		4	4	2	3	4
Taraxacum confusum					4		2	5	1	1	3
Taraxacum stevenii							5	5	0	4	5
Thymus nummularius	4	3	3		5	5	5	4	1	1	0
Trifolium polyphyllum					3		1	3	0	0	0
Trisetum flavescens	4				5	5	3	2	3	2	2
Vaccinium myrtillus							5	4	5	5	5
Vaccinium vitis-idaea		3					2	4	0	1	0
Valeriana alpestris	1	4			1	1	1	3	1	1	0
Veronica gentianoides							5	0	1	1	2
Viola altaica	3	4	3		3	4	4	1	0	0	0
					4		2	0	0	0	0

10.1.2. Chamois (*Rupicapra rupicapra* L.)

The Caucasian chamois (Fig. 10.4) represents a separate subspecies from the European one (*Rupicapra rupicapra caucasica*). The species inhabits most parts of the Caucasus, but his optimal area is at the west Caucasus (Sokolov & Tembotov 1993). Recently the chamois is not a numerous species due to anthropogenic changes of the landscapes and illegal hunting. The total west Caucasian population was estimated at 5-6 000 animals, but at the middle of XIX century there were ten times more animals in the same territory (Duben' 1985).

The total number of chamois in the Teberda Reserve is about 250-300 animals. In contrast to the ibex, the chamois live preferentially in the subalpine belt, but they occur in alpine and subnival areas in summer and in the forest belts in winter. Their general altitudinal range is very wide – from 200 to 4000 m.a.s.l. Chamois prefer steep (20-40°) meadow slopes alternated with rocks and cliffs that can provide shelter for the animals against man and carnivores.

Figure 10.4. Caucasian chamois (Picture by V.M.Smirin from Sokolov & Tembotov 1993)

Caucasian chamois are relatively big animals. Mean adult male's weight is 36 kg, females are less (30 kg) (Sokolov & Tembotov 1993). They reach that weight at the age of 3 years. Life span is about 15-18 years, very rarely some individuals reach 25 years.

Lists of eaten plants include between 109 (Popkova 1967) to 203 (Zharkov 1959) species. It depends on the season. The chamois's summer diet includes mainly herbaceous plants from the subalpine and alpine plant communities. But in contrast to the ibex, forbs play a much more important role in the chamois diet (Table 10.1). Animals prefer to consume the generative shoots of forbs (*Rumex, Aquilegia, Pedicularis*, *Campanula, Cephalaria,* different *Asteraceae*). Sedges and grasses play a less important role. In spite of the relatively low percentage of the aboveground production that is consumed, the role of the chamois in decreasing the seed production of some forbs in the subalpine meadows may be important.

In winter and spring the chamois eat a lot of branches of woody plants. They consume trees (*Sorbus, Salix, Betula, Fagus, Acer, Quercus*), as well as shrubs and dwarf shrubs (*Lonicera, Rosa, Vaccinium myrtillus* and others). They can also eat the needles of pines, spruces and firs (Zharkov 1959, Popkova 1967, Nasimovich 1949). Fruticose lichens (epiphytic as well as epigeous) are important components of the chamois ration especially in winter time.

10.1.3. Wild boar (*Sus scrofa* L.)

The wild boar is a usual and widespread Caucasian mammal (Fig. 10.5.). It occurs in all areas in the Caucasus where woodlands are represented. The modern population density varies from 6.2 to 8.7 animals per 10 square kilometers. The total number of wild boars in the Caucasus exceeds 10 000. The relatively good conservation status of this species is related to its very high adaptive flexibility and to the Moslem tradition of the local people. The total number of wild boars in the Teberda Reserve fluctuates within 140-200 animals.

This big animal (male weight up to 240 kg, females – up to 148 kg, Sokolov & Tembotov 1993) inhabits different landscapes from sea level to the alpine belt (up to 2800 m a.s.l.). In mountain regions the wild boar has a regular season migration. In spring and early summer it migrates from the lower forest to the subalpine belt. In late summer and autumn some animals can be observed in the upper alpine belt. Later they go down to the broad-leaved forests where they spend the winter time. Fruits and nuts of wild trees and shrubs (*Fagus, Quercus, Malus, Pyrus, Crataegus*) are the main food of the wild boar in autumn and winter.

Plants are the main part of boar's diet, but small animals (rodents, frogs, insects, earthworms, etc.) build a steady component of his food (2-9% of the ration – Sokolov & Tembotov 1993). V.V.Durov (1987) listed 197 plant species consumed by wild boar in the west Caucasus. Among the alpine plants the boar consumes mainly species with underground storage in thick roots, rhizomes or bulbs (Table 10.1.), for example *Carum caucasicum, Carum meifolium, Chamaenerion angustifolium, Hedysarum caucasicum, Pedicularis* spp., etc.

But in spite of its low population density and occasional occurrence in the alpine belt, influence of rooting up activity on the alpine vegetation may be very important. In previous chapters (5.3, 5.6 and 6.3) we have discussed the recovery dynamics of alpine communities after zoogenic disturbances. Burrow activity of wild boars is one of most widespread causes of such disturbances. It leads to long-term recovery processes in the alpine lichen heath (Chapter 5.6). The transient area between the *Geranium gymnocaulon-Hedysarum caucasicum* meadow and the snow bed community is subjected to the most intensive rooting up activity in our study site. We suppose that some positive relationships exist between the wild boar and its main food

Figure 10.5. Wild boar (Picture by V.M.Smirin from Sokolov & Tembotov 1993)

plant of this area – *Carum meifolium*. This monocarpic *Carum* with significant seed production (about 1000 seeds per sq.m. – Logvinenko & Onipchenko 1999) benefits from the "ploughing up" activity destroying or reducing its competitors. This plant, in spite of its consumption by boar, increases its numbers several years after boar burrowing.

10.2. Burrowing and trophic activity of small mammals

10.2.1. Species diversity

Small mammals, especially rodents, are important components of the alpine ecosystems. They play a significant role in consuming primary production and strongly influence the structure of the plant cover. Their activity maintains species diversity in plant communities, intensifies nutrient turnover, affects the rate and direction of plant successions. Their burrowing activity has a very important influence on soil properties. It increases water permeability, improves aeration and air regime, leads to mixing of soil layers, accelerates weathering of parent rocks (Abaturov 1984). Zoogenic disturbance creates gaps in the plant canopy, which provide opportunity for establishment of seedlings and vegetative propagules of different species (see chapter 5.)

We found 8 species of small mammals in the vicinity of the alpine study area (Mt. Malaya Khatipara). Three of them belong to the order Insectivora, namely Satunin shrew (*Sorex satunini,* or *S. araneus* L. p.p.), Volnukhin shrew (*S. volnuchini* Ognev, or *S. minutus* L. p.p.) and Caucasian mole (*Talpa caucasica* Sat.). The other five species belong to the order Rodentia: birch mouse (*Sicista kluchorica* Sokol., Bask. et Kovalsk.), common field mouse (*Apodemus sylvaticus* L.), pine vole (*Pitymys majori* Thom.), snow vole (*Chionomys nivalis* Mart.) and Robert's vole (*Ch. roberti* Thom.) (Fig. 10.6).

The population density of all these species fluctuates significantly among years. One typical example of the small mammal community (except Caucasian mole) is represented in Table 10.2, where results over a period of three years are shown. These data were obtained in 1981-1983 by capturing small mammals in 5 hunting trenches (flute traps) (15- 25 meters) placed mainly in the ALH on the dry slopes. Robert's voles was not captured because the trenches were built far from their usual habitats that are banks of small rivers and river valleys with dense vegetation. In such places Robert's voles usually are rather common and even numerous. But this vole is a stricty stenotopic species and never occurs in any other habitats. Robert's vole is an endangered species that was included into International Red Data Book, so we tried not to disturb its habitats and not to catch the animals.

Table 10.2. Small mammal community structure in the alpine area of the Teberda Reserve (Results of flute trap capture, total for 3 year – 1981-1983, from Grishina et al. 1986)

Species	individuals	%
Sorex satunini	12	14
Sorex volnuchini	5	6
Sicista kluchorica	13	15
Apodemus sylvaticus	1	1
Chionomys nivalis	15	18
Pitymys majori	39	46
Total	85	100

The Caucasian mole is quite a rare species with a low population density. This animal is very cautious and it is nearly impossible to catch an adult specimen in a standard hunting trench. In spite of its small numbers, the mole is an industrious burrower, and it certainly affects and transforms soils and plant communities (see below). Earlier, the Caucasian mole was noted only below 2500 m a.s.l. (Sokolov & Tembotov 1989). We found its colonies and captured 2 specimens at the altitude of 2750 m (Zenyakin & Onipchenko 1997).

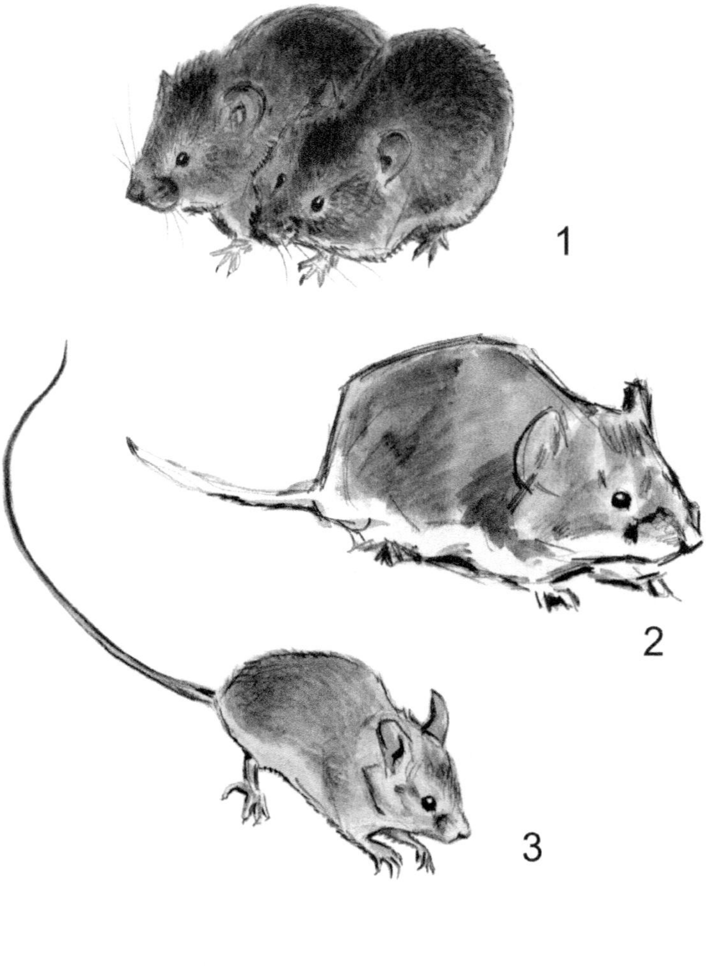

Figure 10.6. Alpine rodents: 1 - *Pitymys majori* , 2 - *Chionomys nivalis*, 3 - *Sicista kluchorica* (Picture by V.M.Smirin with permission of E.V.Zubchaninova)

Shrews and birch mice are very common and sometimes even abundant animals throughout the whole alpine area. Their total number may reach up to 1/3 of total small mammal number. But these animals are strictly insectivorous and they are not deeply involved in burrowing. So, we suppose that they do not have strong direct influence on the plant communities and soils.

The common field mouse does not permanent inhabit the alpine belt. Only in the years of very high population densities of this species some young individuals can migrate from the lower forests and reach the alpine area. During migrations they usually occur in specific habitats such as bottoms of valleys with rare willow shrubs. The animals try to avoid the open herbaceous communities. During the total period of our investigation we captured only a few field mice.

Based on population density the snow vole is the second species after the pine vole in the alpine area. Like Robert's vole, this species is strictly stenotopic and it inhabits only stable large stone screes. Snow voles are rather poor diggers and usually use hollows among rocks as their refuges. Snow voles may use all plant communities located near the screes for foraging. The less preferred communities are snowbeds (SBC). According to our observations the burrowing activity of snow voles is negligible. They might dig among rocks to enlarge their hollows, but we have never found any dug up heaps of soil on the surface in their settlements. Due to their very local distribution in the alpine belt, snow voles are significantly less numerous than pine voles.

The pine vole is the most common and abundant mammal in the alpine belt. It plays a most significant role as a key species in some alpine ecosystems. So we will consider the species in the next subchapters.

10.2.2. Pine vole population density and spatial distribution

To study the population density and spatial distribution of pine voles we used four hunting plots 50 x 100 m. Fifty Gero traps were distributed at an equal distance (10 m) from each other in each plot. We captured voles for 15 – 20 days up to the moment when voles stop to be caught in traps. The data of other authors (Tkachenko 1962) and our own observations show a rather low mobility of pine voles in the alpine area. So, the used method should give data on vole numbers that are very close to the absolute population density. The vegetation of each plot was mapped, so we had a possibility to estimate the density of voles in all studied plant communities. This study was done in 1981 when the population density of this species was low (Table 10.3). We continued to monitor animal numbers on one of the plots (N 3) during 1981-1983. These data are shown in the Table 10.4.

It should be noted that large variations in seasonal numbers are very typical for all vole species. According to I.M.Gromov and B.L.Vorobiov

(1955) seasonal variations in numbers of pine vole may be 5 – 10 times from early spring to late autumn. Our data were obtained in late summer and early autumn, therefore they describe the vole numbers, when the population density nearly reached its maximum annual values.

Table 10.3. Population density and biomass of pine voles in alpine communities (1981).

Plant community	Area, m^2	Captured animals for 4 days per 100 traps	Population density, animals/ha	Biomass (fresh mass, kg/ha)
Alpine lichen heaths	14000	1.3	15	0.23
Festuca varia grassland	2500	1.0	24	0.37
Nardus-Calamagrostis grasslands	400	0	0	0
Geranium-Hedysarum meadows	1500	22.5	146	2.30
Large stone screes	1200	12.5	42	0.65

Table 10.4. Dynamics of the pine vole population density and biomass

Plant community and studied area	Year	Captured animals for 4 days per 100 traps	Population density, animals/ha	Biomass (fresh mass, kg/ha)
Alpine lichen heaths	1981	3.6	31	0.48
($2900\ m^2$)	1982	4.6	43	0.67
	1983	12.5	129	2.61
Large stone screes with	1981	12.5	42	0.65
fragments of closed	1982	14.6	150	2.34
communities ($1200\ m^2$)	1983	16.5	208	3.24
Geranium-Hedysarum	1981	29.2	188	2.93
meadows ($900\ m^2$)	1982	25.0	444	6.93
	1983	60.0	943	14.74

Pine voles are most abundant in *Geranium gymnocaulon – Hedysarum caucasicum* meadows on northern slopes. These communities provide excellent forage, good wintering conditions and safe protection from aerial hunters with their high and dense vegetation. At their yearly peak density they may reach about 1000 animals per 1 hectare here.

In the large stone scree the numbers of pine voles are also high in spite of the presence of another species (snow vole) here and the probable competition between them.

Pine voles settle in the ALH only during spring, summer and early autumn. The density varies from 0 to 40 (up to 130 in peak years) animals per ha. Their biomass usually does not exceed 0.5 – 0.7 kg/ha, only in the peak years it may reach more than 2 kg/ha (average individual mass was 15.6 g). Animals, staying there in winter, most probably perish because of the strong freezing of the soil (Gromov & Vorobiov 1955). But many times we found winter foraging passages that lead from the FVG or GHM vole settlements to the ALH plots.

The SBC are the most unfavorable habitats for pine voles, mainly due to their very low summer protective quality by a short plant cover. The numbers of voles there were the lowest.

Summarizing, we believe that the most important factors that determine pine vole density in a certain plant community are the depth of the snow cover, the duration of the snowfree period and protection properties of the plant cover. The vole's preferences increase along the line: SBC < ALH < FVG < GHM. The number of entry holes in a peak year (1983) is conform this sequence (Fig. 10.7). It should be noted that in many areas within the alpine belt, sometimes even in the most preferable habitats, voles are totally lacking.

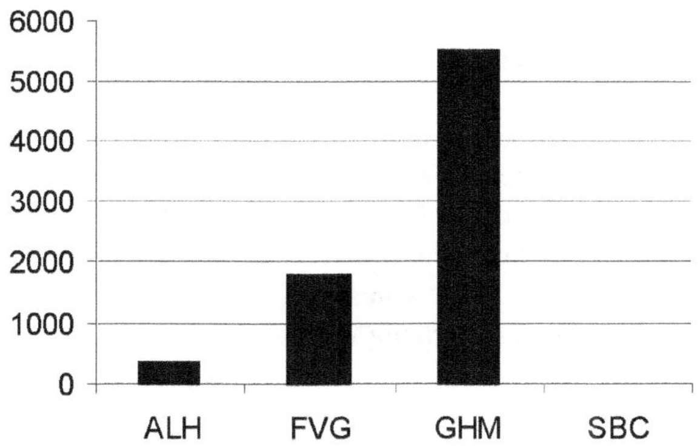

Figure 10.7. Number of pine vole entry holes in different alpine communities (holes/ha)

10.2.3. Snow and pine vole diets

Both studied species are specialized herbivores but not so strong as their nearest relatives – the common voles (*Microtus* spp.). Both species often live in the same places and it is interesting to compare their feeding preferences.

We offered a set of 3-4 plant species and oat grains (as emergency forage) to captured animals for 24 hours. Then we estimated feeding preference (FP) of all parts of the plant (leaves, stems, roots, flowers and fruits) on a 1 to 5 points scale: 1 –plants not eaten or less than 10% of the biomass was eaten, 2 – 11 – 25% of the biomass was consumed, 3 – 25 –

50%, 4 – 50 – 75%, 5 – more than 75% of the biomass was eaten. In this way we tested all alpine plants in 3 – 5 repetitions. In addition we studied many foraging traces and plant remnants in nature. Once, we found undamaged winter storage aboveground (but under snow!) in the SBC. It consisted completely of dry taproots of *Campanula tridentata*. In general we combined all our observations and published data in the FP list of alpine plants (Table 10.1.).

Only few dwarf shrub species (*Vaccinium vitis-idea, Sibbalia procumbens*) proved to be absolutely inedible. For the other plant species the degree of consumption was variable. The most preferable were species with high energy and/or latex contents (*Campanula tridentata, Carum caucasicum, Scorzonera cana, Taraxacum stevenii*), as well as high protein contents (Fabaceae: *Anthyllis vulneraria, Oxytropis kubanensis, Trifolium polyphyllum*). It is interesting to note, that pine voles eat some species that are hardly edible for cattle, sheep and some wild ungulates (*Anemone speciosa, Antennaria dioica, Anthemis marshalliana, Euphrasia ossica, Fritillaria collina, Ranunculus oreophilus*) (Grishina et al. 1986). Thus, the feeding activity of pine and snow voles may promote the improvement of the alpine pasture structure.

Our data showed that snow vole and pine vole diets are similar. It is necessary to note that seeds form a significantly larger part of the snow vole diet than the pine vole diet. *Arenaria lychnida* and *Potentilla crantzii* are much more preferred plants for snow voles than for pine voles. During the summer-autumn seasons snow voles feed much more on underground parts of plants (taproots, rhizomes) than pine voles.

10.2.4. Pine vole feeding and removal of primary production

We estimated the average feeding preference point for 9 plant families that include 4 or more studied species. They may be arranged in the next sequence with decreasing feeding preference of the pine vole:

Fabaceae > Caryophyllaceae > Cyperaceae > Scrophulariaceae > Asteraceae > Poaceae > Ranunculaceae > Rosaceae > Gentianaceae

The first five families have a preference point higher than the average (3.7) for all studied species.

To estimate the general forage preference of the whole alpine plant communities, we calculated the average preference points for plant species in each community (Fomin et al. 1989). Calculations were based on floristic composition only as well as species abundance (cover, biomass). The obtained values were quite close to one another and lie mainly in the interval 3.3 – 3.8. Thus, the different alpine communities are very similar in their foraging preference for pine voles. So, spatial distribution of pine voles among communities could not be explained by different food quality in the communities. These observations supported the idea that snow cover, production and protecting features of the plant cover are more responsible for pine vole distribution than food quality.

In all but the ALH communities the average preference values based on floristic composition were higher than those based on species abundance (weighted values). It is shown that dominant species of these communities are less preferable species than the more rare plant species. So dominants of the FVG, GHM and SBC might be favoured by the feeding activity of the pine vole which consumes the less abundant species. This phenomenon was most prominent in the SBC. Most of the SBC species are very edible for the voles, but the main dominant *(Sibbaldia procumbens)* was not eaten at all. But we could not show that the dominance of *Sibbaldia* in the SBC was caused by pine vole activity, because the voles are nearly absent there. Intensive former grazing by domestic cattle and sheep looks more like the main reason of *Sibbaldia* abundance in the studied SBC. In the ALH the dominant vascular plants, on the contrary, are consumed more than the other species. So, while the voles settle in the ALH, they can give rise to considerable changes in the ALH structure.

To estimate consumption of aboveground production by pine voles in different communities we used values of production (Chapter 5.2), animal's population density and the trophic index. The trophic index is the quotient of the weight of the eaten food per day to the body mass. This index was determined for pine voles by Larina & Tarasov (1980). We obtained approximate values of the annual biomass removal by pine voles of 5-7% of the annual aboveground production. Only in the most populated GHM in the peak years the total removal may reach 15% of the production. Similar values were recorded for this species in subalpine meadows by Larina & Tarasov (1980) - about 300 kg/ha.

Since our data are calculated on the basis of the trophic index that does not take into account traumatic litter (parts of plants that were removed but not eaten), the real removal values should be significantly higher. Usually traumatic litter varies from ½ to ⅔ of the removed plant biomass (Dobrinskii et al. 1983). So, real biomass removal should be twice as high, i.e. about 30% of the primary production. Such high rates of consumption must have significant effects on community structure and composition. It seems, that the low abundance of some widespread but highly palatable alpine species *(Carum caucasicum, Campanula tridentata)* in the GHM may be determined by pine vole consumption.

In the majority of terrestrial ecosystems natural herbivores consume from 0.2 to 15-20% of the primary production (Zlotin 1984), and voles usually consume about 1 - 4 % of the annual production (Dobrinskii et al. 1983). In general, our data for the GHM slightly exceed these values.

10.2.5. Burrowing activity of the pine vole

We studied pine vole burrowing activity in 1983 (at peak density) and in 1986 (low density). We counted burrow entrances (holes) and hills of erupted soil on the surface in plots of 3 x 3 and 5 x 5 m. Total numbers of plots were 10 to 20 for each community. The observations were done in late August – early September which allowed us to investigate nearly all soil eruptions arisen during the year. Due to little rain and a short vegetation season (2.5 – 4 months) it was easy to distinguish mounds from the current and previous year, because the latter started to be overgrown by plants. So, our data can be accepted as an estimation of the intensity of annual burrowing activity.

Two types of pine vole burrows can be distinguished in the alpine area, namely nest (reproductive) and shelter (protecting) burrows. We excavated 17 burrows to estimate their features (Table 10.5.).

Table 10.5. Comparative features of pine vole nest and shelter burrows (from Fomin et al. 1989)

Feature	Shelter burrows (n=12)		Nest burrows (n=5)	
	av. ± s.e.	range	av. ± s.e.	range
Number of entry holes	1.3±0.1	1-2	7.4±2.9	4-19
Total length of the burrow	20.5±3.1	10-40	181±72	95-465
Length per 1 entry	16±2	6-35	25±3	20-36

The shelter burrows were usually up to 40 cm long. They had no nesting chambers and only one or two entrances. The length of the nesting burrows was several meters, and a network of surface paths connected many entrances. Underground passages were mainly in the lower part of the turf (A) soil horizon at 7 – 8 cm depth. Average diameter of passages was 3.3 x 3.2 cm (Dobrolyubov 1986). Nesting chambers lay at 40 – 60 cm depth (Ognev 1950 and our own observations). Erupted soil from nesting chambers contained huge amounts of stones which is typical for the B or C soil horizons.

The data on the burrowing activity of the pine vole in the studied alpine communities are presented in Table 10.6. Maximal volume of eruptions in the peak year in the GHM was 0.7 m^3/ ha or about 1.5 tones/ha.

Table 10.6. Burrowing activity of pine voles in alpine communities, mean and standard error (from Fomin et al. 1989). Digging point means small hole (less than 10 cm) created as result of feeding activity.

	year	ALH	FVG	GHM	SBC
Number of entry holes	1983	390±180	1780±500	5500±1270	60±60
per 1 ha	1986	200±120	1200±320	1960±400	0
Number of digging points	1983	330±180	4780±730	14300±2850	780±210
per 1 ha	1986	40±40	1120±360	3760±760	0
Number of soil eruptions	1983	0	110±75	940±460	0
per 1 ha	1986	20±20	560±200	960±440	0
Volume of 1 eruption,	1983	-	370±30	750±90	-
cm^3	1986	-	105±15	70±12	-
Total volume of erupted	1983	-	41±29	705±350	0
soil, dm^3 per 1 ha	1986	8±8	59±23	67±32	0

Burrowing significantly influences many characteristics of the GHM soils (Chapter 2). Here, in comparison with FVG and SBC soils, water retention capacity is lower, but water permeability is higher. The radiocarbon age of fulvic acids in the lower soil horizons of the GHM is lower than in other communities due to soil mixing by pine voles (Grishina et al. 1987). Buried viable seeds in the GHM soils were found up to 10 cm depth, lower than in other communities. It is necessary to note that in the GHM voles are the only burrowing animals, since earthworms are nearly absent here unlike in ALH and FVG soils (see chapter 10.4).

The volume of pine vole burrowing is comparable with values for lemmings and long-tailed ground squirrels in the tundras of Alaska (McKendrick et al. 1980), and for subterraneus voles in the Carpathian Mts (Polushina 1975). But it is significantly lower than the burrowing volume reported for desert and steppe mammals (Grant et al. 1980, Dmitriev 1988).

10.2.6. Burrowing activity of the Caucasian mole

Talpa caucasica rarely occurs in the alpine belt. Its distribution there is restricted to the *Geranium – Hedysarum* meadows, sometimes it can be found in SBC. We found a small colony (area about 275 m^2) of Caucasian mole in the GHM in 1994. Parameters of borrowing activity were studied for this colony (Zenyakin & Onipchenko 1997). Due to low alpine soil depth the system of mole passages does not lay deep. Upper passages are in the lower part of the turf (A) horizon at the depth of 5-7 cm, its diameter is about 5 cm. Deeper passages lay in the B and C horizons at the depth of 25-30 cm, their diameter is about 6 cm. The high stone content of the soil (24 – 33 vol.% in GHM) does not prevent moles from burrowing. Fresh hills (2-3 weeks old) comprised only 14% of its total number. The volume of mole hills varied from 2.5 to 6 dm^3, its weight – from 2 to 4 kg. Total eruption volume according to our calculations was about 580 kg per study area, or 21 tones/ha, annual eruption volume was approximately 15 tones/ha (Zenyakin & Onipchenko 1997).

In comparison with the pine vole, the volume of mole eruption was one order of magnitude larger. But mole settlements in the alpine belt are very rare and they are always located far apart from one another, so that total volumes of mole burrowing activity are considerably lower than the ones of the pine vole. So, soil perturbation processes arising from Caucasian mole activity have a very local character in the alpine area, but locally they might be very heavy.

10.3. Predatory mammals

The fauna of predatory mammals of the North-Western Caucasus includes 14 species from 4 families. Mustellids (Mistellidae) (7 species) and Canids (Canidae) (4 species) are the most diverse families. Two other families are Cats (Fellidae) (2 species), and Bears (Ursidae) (one species). Predatory mammals of the region can be divided into four groups in accordance with the consistency of their occurrence in alpine belts (from the timberline to the nival belt):

1) species which constantly inhabit the alpine belt (*Mustella nivalis, Mustella erminea*);
2) species which visit the alpine area rather regularly (*Ursus arctos, Canis lupus, Lynx lynx*);
3) species visiting the alpine belt of mountains rather sporadically (*Martes martes, Martes foina, Vulpes vulpes*);
4) species of very rare occurrence in the alpine area (*Felis silvestris, Lutra lutra, Mustella lutreola, Canis aureus*).

Two species (badger, *Meles meles*, and introduced racoon dog, *Nyctereus procyonides*) were not noted in the alpine belt. Below we will describe predatory mammals from the first two groups in more detail, but other groups will be described briefly.

10.3.1. Weasel (*Mustella nivalis* L.)

10.3.1.1. Abundance and habitat distribution

The weasel (Fig. 10.8) is widely spread in all the territory of the North-west Caucasus. The animals inhabit plains situated in foothills as well as high-altitude places up to the subnival belt (3000-3500 m a.s.l.). The abundance of this species is not very high and has quite significant fluctuations from year to year. According to the data of perennial counts in the Teberda Reserve, quantity of traces along a route of 10 km varied in the forest belt from 0.2 to 9.8, whereas in the alpine belt (2200-3000 m above s.l.) there were in average 8.9 (from 0.5 to 14.6) traces. Weasel was recorded in all habitats in this region. Highest population densities were observed in areas with good feeding and protective conditions, namely tall herbaceous subalpine meadows (Table 10.7.). Generally, weasel was found in the belt of dark coniferous forests with significant abundance of small rodents and many windfallen trees.

Figure 10.8. Alpine Mustellids: 1 – ermine (in winter fur), 2 – weasel (in summer fur) (Picture by V.M.Smirin with permission of E.V.Zubchaninova)

Figure 10.9. represents more detailed data of weasel distribution in the alpine belt, obtained by catching the animals by live-traps in summer time. The highest number of entraps of weasel in 1985-1992 fell on the sites of meadows with former pastoral vegetation (tall ruderal), whereas the traps on screes were visited very seldom. The abundance of rodents in screes was quite high and stable but the competition with ermine does not allow weasel

Table 10.7. Habitat distribution of weasel and small rodents in the Teberda Reserve

Types of habitat	Altitude, m a.s.l.	Number of weasel traces per 10 km of route	Autumn abundance of small rodents (specimen per 100 trap-days)
Mixed deciduous forests in the valley of the Teberda river	1300-1600	2.8	-
Deciduous and mixed forests with *Corylus avellana* in glens of mountain streams	1300-2000	8.7	13.1
Dark-coniferous - broad-leaved forests	1400-1600	4.0	6.5
Dark-coniferous and dark-coniferous - birch forests	1500-1950	14.5	13.0
Dark-coniferous - deciduous forests with subalpine glades	1700-1900	24.7	-
Pine – deciduous forests with rocks and screes	1600-2000	13.2	-
Pine – deciduous valley forests with subalpine glades and screes	1900-2100	22.5	15.4
Berch elfin forest on the glacial moraines	1900	2.2	-
Subalpine and alpine meadows	2300-2900	8.9	2.0

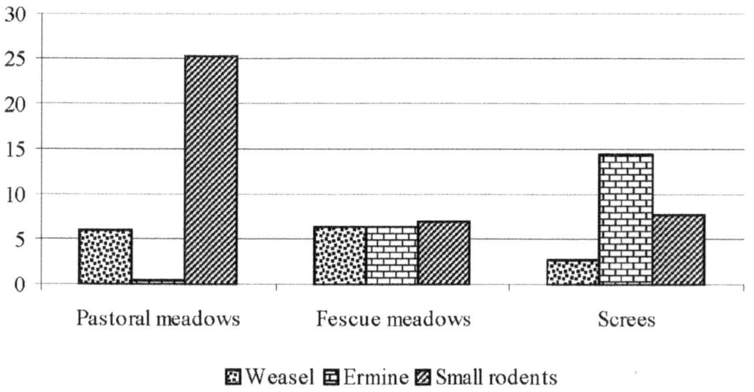

Figure 10.9. Abundance (captured animals per 100 trap-days) of weasel, ermine and small rodents in different alpine habitats

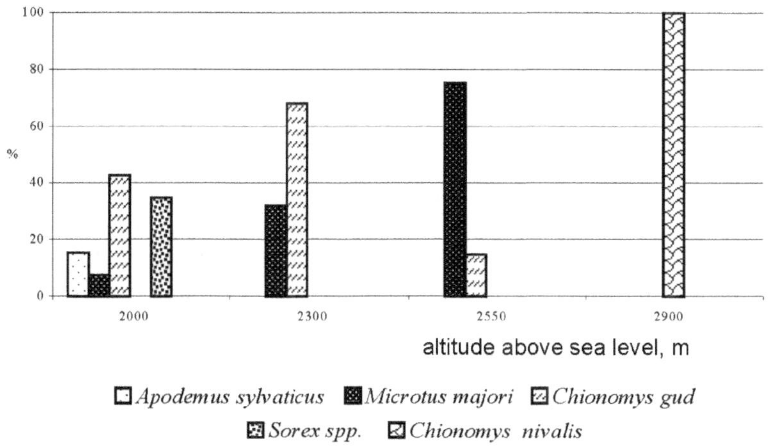

Figure 10.10. Species composition of small mammal communities at different elevation belts

using its food resources effectively. There places are inhabited by comparatively big voles (*Chionomys* spp., Fig. 10.10). It is difficult for weasel to catch them in comparison with pine voles (*Pitymys majori*).

The weasel as any predatory mammal has its own territorial system of use of its living space. The surface area and form of the habitat site of the animal are not constant and depend on many factors. Abundance of potential food is most important.

Our investigations have shown that the length of the daily route of weasels in the Teberda Reserve was variable and depended on the success of hunting. So in the winter of 1985/86 the length of the daily route of females varied from 15 to 500 m (227 m on average). The smallest length was observed when the animal, having caught the rodent, took it back to the refuge. Males, in contrast to females, are less adapted for hunting in burrows of rodents while their absolute demand for food is higher. This is a reason why the average length of their daily route was much larger. Observations in the winter periods of 1978/81 have shown that the length was 1.6 km on average (n=7). The area of the habitat site of females was not very large. According to our data two weasels in the winter of 1985/86 were hunting during 5 days on a site of 3.4-3.8 hectares. Males used a territory from 5.9 to 14.3 hectares (9.0 on average, n=4). Weasels can use their own habitat sites quite a long time. For example an adult male, marked on June 11, 1986, was being caught several times in the same area. The size of his hunting area was about 7.0 hectares.

10.3.1.2. Dwelling

In the mountains of the northwest Caucasus the weasel usually makes refuges in variable places protected from bad weather conditions and predatory animals. In the forest belt they make their refuges in niches under the roots of trees (37.5%), in hollows (25.0%) and conglomerations of stones (25.0%). In the alpine belt they do it mainly in the burrows of small rodents (90%).

The animal may use the same refuge several times. For example, on February 8, 1986 it was determined by the traces in the snow that a female weasel went out from the hollow in the trunk of a beech, caught a *Apodemus sylvaticus* (10 meters away) and came back. Only 24 hours after this incident the animal left her refuge.

In the alpine belt from ten refuges of weasel only four were visited several times.

10.3.1.3. Feeding

Small rodents (Table 10.8) are the main food for the weasel in the Teberda Reserve while the role of other food is insignificant (Dobrolyubov 1986).

In the forest belt predatory animals more often catch *Apodemus sylvaticus*, whereas in the alpine meadows they catch *Pitymys majory*. The latter species can be found in all the mountain habitats but dominates only in alpine meadow communities (Fig. 10.10). To resume, the relative abundance of different kinds of small rodents in the diet of the weasel is closely related with the extent of their abundance in the corresponding habitats.

From the insectivorous mammals only the mole was observed in the weasel's diet. Birds are represented in the diet by small species, some of which are caught by the predatory animal mainly in the nesting period. Attacks of weasels upon large birds were not registered but two unsuccessful cases of hunting on Caucasian heat-cock were described (Kotov & Ryabov 1963).

Observations in laboratory conditions show that the daily food demand of the weasel is rather high (Table 10.9.). It consists approximately of two body weights of pine voles (*Pitymys majori*).

10.3.1.4. Reproduction, sexual and age population structure

Weasel males achieve puberty at the age of one year. According to our observations, females have heat quite a long time: from April till July. In the same period the generative organs of males achieve their largest sizes and activity. Birth of young animals happens in different months. We have registered the birth of whelps at the end of May and at the beginning of August.

Table 10.8. Diet of the weasel in the Teberda Reserve. Percent from the number of investigated feces

Kind of feed	Forest belt		Alpine belt	
	Winter (snow period), n=10	Snowless period, n=7	Winter (snow period), n=14	Snowless period, n=110
Mammals including:	70	100	100	81.8
Small rodents	70	85.7	100	89.1
Apodemus sylvaticus	60	14.3	0	7.3
Pitymys majori	10	0	78.6	40.9
Chionomys spp.	0	0	7.1	21.8
Sicista kluchorica	0	0	0	1.8
Microtinae (undetermined)	20	0	0	9.1
Myomorpha (undetermined)	25	71.4	7.1	17.3
Moles (*Talpa* spp.)	0	14.3	7.1	0.9
Other mammals (undetermined)	0	0	7.1	1.8
Birds (Aves)	10	0	0	5.5
Reptiles (Reptilia)	0	0	0	0.9
Insects (Insecta)	0	0	0	1.8
Plant remnants	0	0	0	3.6

Table 10.9. Quantitative characteristics of the weasel' s food (based on observations under laboratory conditions)

Characteristics	Females (n=2)	Males (n=3)
Number of observation days	98	84
Damp weight of daily food (g)	30.4	37.2
Absolute dry weight of daily food(g)	9.13	10.6
Absolute dry weight of feces (g)	2.31	2.7
Absolute dry weight of assimilated food (g)	6.8	7.9

Young animals begin their individual life at the end of July or at the beginning of August. We know this from the results of weasel catching with the purpose of marking. The first young animals hit the traps at the end of July while in August their number sharply increased (Figure 10.11).

In different years males prevailed among the marked young animals in the studied area of Malaya Khatipara. They constituted up to 71.4%. The ratio between the sexes of adult animals was 68.2% males and 31.8% females (n=22). Prevalence of males was conformed statistically (χ^2 =12.5).

The catching of weasels as well as other martens is selectively. The bias towards a greater portion of males is due to the larger sizes of the hunting areas of males, their restricted prudence and higher activity. To obtain data on the sex ratio independent of the catching methods we registered the sex of the animal during observations it in wintertime by its traces in the snow. The sex ratio differed greatly from values obtained by catching and was close to 1 (Table 10.10).

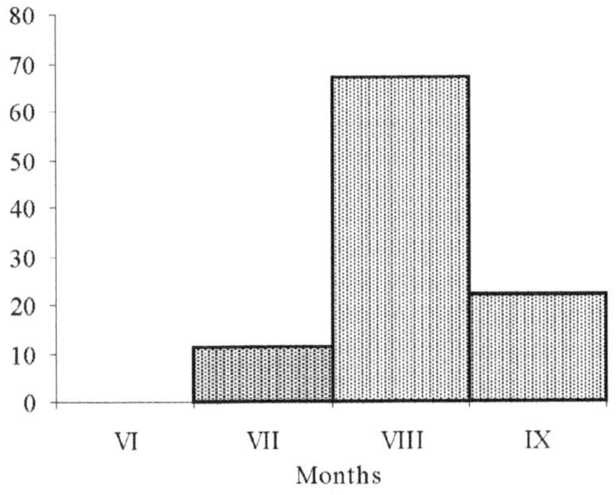

Figure 10.11. The catch dynamics of young (current year) weasels (%) in the alpine area of the Teberda Reserve

Table 10.10. Sex ratio of the weasel population in Teberda Reserve

Sampling method	n	Portion of sexes (%)		χ^2
		Males	Females	
Catching by live-traps	36	69.4	30.6	14.3
Catching by steel traps	40	72.5	27.5	19.4
Observations on traces on snow	45	57.8	42.2	2.1

We have studied a sample of 35 animals, caught in different years, to investigate the weasel age population structure. The age of the animals was determined by their annual teeth' s layers (Klevezal & Kleinberg 1967). Young (current year) animals consisted of 54.3%, while the animals of 1-1.5 years old represented 31.4%, 2–2,5 years old – 11.4%, 3 years old – 2.9%. Another measure of the proportion between adult and young animals was obtained by catching the animals by live-traps. Current year animals consisted of 38.9% from 36 weasels caught at the end of the summer and in autumn.

10.3.1.5. Parasites, diseases, enemies and competitors

Weasel helminthes were studied quite thoroughly. The list of helminthes includes 26 species from 22 genera and 18 families (Kontrimavichus 1969). Nematodes from genera *Filaroides* and *Skrjabingylus* are most widely spread and pathogenic. We revealed invasion by parasites belonging to *Filaroides* of 3.3% of the 35 weasel bodies caught from the forest and alpine belt of the Northwest Caucasus. The frontal sinuses' parasite *Skrjabingylus nasicola* was found in 71.4% of the studied animals. All adult animals were invaded by the nematode. It was revealed in 15.4% of the current year weasels. Such animals looked quite healthy and 81.8% of them had a good degree of fatness (3-4 marks from 5 as a maximum).

There are many potential feeding competitors of the weasel. They include all predatory mammals from ermine to wolf and carnivorous birds (tawny owl, Tengmalm's owl, buzzard). They also can be considered as potential enemies. The most serious competition is between weasel and ermine. Both are predatory and specialized on mice and voles. One or two weasels often live within the ermine's hunting area in the alpine belt and timberline zone.

It is well known that animals with keen competitive relations usually don't inhabit the same area. Therefore, a mechanism for a decreasing degree of food competition has evolved between weasel and ermine. Analysis of data from the alpine belt has shown the food specializations of these species. Usually weasel prefers to catch *Pitymys majori*, but the larger ermine prefers *Chionomys* spp.

10.3.2. Ermine (*Mustela erminea* L.)

10.3.2.1. Abundance and habitat distribution

The ermine inhabits valleys between the Bolshaya Laba and Terek rivers in the Northern Caucasus (Tembotov 1960, 1977, Baziev 1962, Tkachenko 1962, Kotov & Ryabov 1963, Zimina & Yasnyi 1977, Naniev 1986, Dobrolyubov 1988, 1994). The species was noted for the southern slopes of the Great Caucasian Range (Transcaucasus) as well (Spassov & Barichnikov 1986).

The ermine's main habitat areas are located in the alpine belt from the timberline to the subnival belt (2000-3500 m a.s.l.). It is comparatively rare in the forest belt, but sometimes goes down to the altitude of 1400 m.

Ermine is a common and numerous species in the Teberda Reserve and adjacent valleys. We found that the ermine's winter abundance in typical habitats (pine-deciduous forests, alpine meadows, screes, and avalanche fans) varies from 3.3 to 44 traces per 10 km of route

The ermine's favorite habitat sites are screes in pine-deciduous mountain valley forests and in birch elfin forests (Tkachenko 1962, Baziev 1962, 1977, Dobrolyubov 1985, 1986). The animals were registered in the Teberda Reserve at the altitude of 1600 m up to 3500 m a.s.l.

The spatial distribution of ermine was determined firstly by the general availability of its food objects (small rodents). Pray population density played a secondary role (Figure 10.9). High population densities of *Pitymys majori* in tall herbaceous subalpine meadows did not lead, in contrast to the weasel, to ermine abundance. The reason is that for the ermine it is difficult to penetrate into pine vole's burrows. The ermine seems too large to visit the burrow, but it can catch voles in holes among stones. Therefore ermines did not hunt in tall meadows, but they were hunting near and in screes. We found that ermines can obtain food if the hole's diameter is equal to 3.3 cm as a minimum for males, but for a weasel the size of the hole can be less: 2.9x2.9 cm for a male; 2.6 x 2.1 cm for a female. According to our observations, an average size of a entry hole of *Pitymys majori* was 3.3 x 3.2 cm.

Ermines use holes among stones in screes and cavities in rocks as refuges.

10.3.2.2. Feeding

Mainly small rodents serve as the ermine's food. *Chionomys gud* and *Chionomys nivalis* inhabit screes, therefor they are more often caught by ermines. Pine vole (*Pitymys majori*) takes a second place in the animal's diet.

Table 10.11. Ermine diet in the Teberda Reserve

Kind of food	Occurrence of the food in the diet (% from number of investigated feces)	
	Forest belt (2000-2400 m) n=189	Alpine belt (2400-2900 m) n=137
Small rodents (Myomorpha), including	95.8	96.4
Apodemus sylvaticus	6.9	3.6
Pitymys majori	19.6	24.8
Chionomys spp.	39.7	27.0
Microtinae (undetermined)	13.8	23.8
Sicista kluchorica	1.6	1.5
other Muridae (undetermined)	14.3	16.1
Glis glis	0.5	0
Sorex sp.	2.1	1.7
Insectivora (undetermined)	0.5	0
other small mammals (undetermined)	1.6	2.2
Birds (Aves)	5.8	5.8
Lizards (Lacertidae)	0	1.5
Insects	2.1	2.9
Plant remnants	1.0	0

Ermines catch small birds occasionally; they consists of around 6% on average of the ration (9.2% - in summer and 5.4% - in spring). Availability of nesting birds in the alpine belt in summer may explain the seasonal differences in proportion of birds in the ermine diet. Insects (*Orthoptera, Carabidae*) and reptilians are sometimes eaten by ermines (Table 10.11.).

The daily food demand of the ermine was found to be equal on average to one adult *Chionomys gud* (observations under laboratory conditions) (Table 10.12).

Table 10.12. Quantitative description of the ermine's food

Parameter	Females (n=2)	Males (n=3)
Number of observation days	18	70
Damp weight of daily food (g)	39.3	51.7
Absolute dry weight of daily food (g)	12.4	16.0
Absolute dry weight of feces (g)	2.98	3.36

10.3.2.3. Habitat sites and use of territory

The ermine is a settled mammal. Our results of ermine marking at the study area (M. Khatipara) conform that. For example, a young male was caught in August 1985, after which it regularly got into the same traps till October 1986. It was not found in spring 1987. Probably it died in the winter, but we could not exclude migration to another place.

Settled behavior during a year and longer is typical for females as well. One of them was observed at the same area for about 3 years from 1983 till 1986. This animal was caught in live traps in the autumn of 1983 at an altitude of 2500 m a.s.l. In the next years its habitat was located at 700 m from the first, at an altitude of 2700 m.

10.3.2.4. Reproduction, sexual and age population structure

The ermine's heat begins in May in the Northwestern Caucasus. Sexual activity decreases to the middle of August. Birth of young animals usually happens from the second part of April to the beginning of May. Lactating females are being caught till the third decade of June. From the females of one year old and older 55.6% of the animals had their own broods. There were on average 3.4 current year animals per every parturient female at the beginning of August when dispersion of the young animals started.

In August 1985 we have caught and marked a young female at the study area (M.Khatipara). In spring 1986 when we caught it again it was suckling. So reproduction of the female ermine is possible even during the first year of its life.

The young (current year) animals composed 32% from 53 animals caught in different years by steel and live traps. This data are consistent with ones for North-West Russia (41.1%) published by P.I.Danilov and I.L.Tumanova (1976). There were equal proportions of males and females (50%) among the adult ermines, whereas females prevailed among the current year animals (64.7%).

10.3.2.5. Parasites, diseases, enemies and competitors

Nematodes from the genera *Filaroides* and *Skrjabingylus* are often parasites on ermines. Investigated animals had low frequencies of invasion by *Filaroides* (4.3%) and much higher invasion frequencies of frontal sinuses parasites (*Skrjabingylus*) – 12.1%. Invaded animals comprised 33% of the current year animals and 100 % of the adults. 46% of the invaded animals had cranium perforations of the frontal sinus area.

10.3.3. Bear (*Ursus arctos* L.)

Predatory animals as bear, wolf and lynx belong to the *second* group. These species are able to cover long distances and they use all mountain belts. They are mainly attracted by the forest belt, though often they use alpine areas to procure food, to have rest or to bring up their broods.

10.3.3.1. Spreading, spatial distribution and abundance

The Caucasian bear is spread over a large area of the Northwest Caucasus. It was observed in all mountain belts. The animals show seasonal migration; they move from broad-leaved forests to alpine meadows during a year. They occur in the upper forest belt among rock outcrops and avalanche fans, in tall herbaceous communities (*Mulgedio-Aconitetea*), or in forest gaps with a dense cover of raspberry-bush (*Rubus idaeus*). Bear migrations depend on seasonal food availability. From the beginning of the herbaceous plant growth season till to middle of July the animals dwell in the subalpine belt. They prefer forest gaps, elfin birch woodland and valleys of mountain rivers (Bobyr' 1977, Bobyr' & Kudaktin 1991). Bears are often found in areas with active avalanche processes (Bobyr' 1979). The animals visit the alpine belt rather seldom, they go up to snow beds and glacial circuses (corries) with alpine lakes through valley bottoms and avalanche canals.

In the Teberda Reserve bear abundance varied in different years from 80 to 120 animals, with a mean population density of about one animal per 1000 hectares (Bobyr' 1986).

10.3.3.2. Dwellings

Most bears go in hibernation at their lairs (dens) in winter. They use natural refuges for lairs. In the Teberda Reserve lairs were mostly situated in rock niches and under big stones, sometimes among tree roots (Bobyr' 1985). The bear dug out the entrance in order to easily enter into the lair. Dry leaves, moss, lichens and grasses cover the floor of the lairs.

In summer bears rest in dense forests near the timberline under crowns of trees, on rock ledges, and in *Rhododendron* scrub.

10.3.3.3. Feeding

The diet of bears is diverse and shows a seasonal pattern. After leaving the lairs (in the middle of March) bears look for dead ibexes (*Capra caucasica* L.) and chamoises (*Rupicapra rupicapra* L.) in avalanche snow depositions. At the same time they use fruits of the dog rose (*Rosa* spp.) and nuts of the beech from last year's crop. Then, in the end of April and at the beginning of May, bears start feeding on young grass and forb tillers in the forests and subalpine meadows. Avalanche places are especially important for animals at this time, because the areas support new growth of grasses and forbs along the retreated snowmelt border till the middle of summer. The main food plants for bears are grasses (especially *Millium effusum*), and tall Apiaceae (*Anthriscus sylvestris, Heracleum* spp.). Sometimes bears procure food in the subalpine and alpine meadows by digging *Hedysarum caucasicum* roots (Bobyr' 1987).

In the second part of the summer bears start eating fruits (*Rubus idaeus, Vaccinium myrtyllus, Prunus divaricata, Pyrus communis*). Before hibernation the animals use food rich in calories to put on weight. At this time nuts of the beech (*Fagus orientalis*), acorns (*Quercus robur*), berries (*Vaccinium vitis-idaea, Sorbus aucuparia*) and dog rose fruits are preferentially used by bears. While procuring the food, bears often break the twigs of trees and brush. About 30% of the oaks were damaged by bears in a mast year.

10.3.4. Wolf (*Canis lupus* L.)

10.3.4.1. Spatial distribution, abundance and dwellings

The wolf is widely spread in the Northwestern Caucasus, but at foothills it has been rare in the last decades. There were no special censuses of wolfs, but according to expert estimations there are about 400-460 animals here (Kudaktin 1977).

The abundance of the species in the Teberda Reserve is low: 2 or 3 families with 10-15 animals in total. Wolves use all mountain belts to procure food, to have rest and to bring up the brood but they occur more often in the forest belt.

The dens of the wolf on the Teberda Reserve were not registered. But the presence of young animals in the Bolshaya Khatipara valley (personal communication of G.Bobyr') suggests successful breading of the species.

The young animals are not very careful and draw attention to themselves by howling.

10.3.4.2. Feeding

The wolf's main prey in the mountains of Teberda are ungulates: firstly, ibex (*Capra caucasica*), then wild boar (*Sus scrofa*), sometimes chamois (*Rupicapra rupicapra*) and red deer (*Cervus elaphus*).

Small animals are not rare in the diet of the wolf. The frequency of occurrence of small animals consisted of up to 12.6% in the remains from 305 investigated feces (Bobyr 1997). The role of small mammals in wolf feeding shows a seasonal variation. In winter and in spring the wolf procures ungulates more often (69.7 % of occurrence) than rodents (7.2 % of occurrence). We suppose this decrease of small rodents is due to its decreased availability under snow. At that time ibexes concentrate in comparatively small areas with a shallow snow cover. In such places they are more vulnerable to wolves.

In spring after the avalanches have gone down wolves become competitors to bears. They look for ungulates, which have been killed by the avalanches and covered by snow. When wolves have found a dead ibex they do not leave the body till they have completely consumed it. So, in May of 1985 in the Ullu-Murudzhu valley we registered the place where wolves, found a dead ibex, and were staying next to it for several days. We found a lot of wolf couches in pine-forest at a distance of 50-100 m from the prey. In another case, we have registered the happenings when a group of predatory animals found an ibex in the Khadzhibei valley deep under the snow. They reached it not from the snow surface but from the bottom through the snow tunnel created by a stream.

Wolves feed on other kinds of animals, namely squirrels, hares, birds, sometimes reptiles, as well as domestic animals. Sheep are found in wolf diets more often then cows, horses and dogs.

10.3.5. Lynx (*Lynx lynx* (L.))

10.3.5.1. Spatial distribution and abundance

The lynx is wide spread in the Northwestern Caucasus and inhabits mainly the middle and upper parts of the forest belt. Sometimes it occurs in the alpine belt when it travels from one valley to another or hunts on hares and Gallinaceae birds. The lynx itself and its traces were often found in dense forests with rocks and cliffs.

The lynx is not a numerous animal in the Teberda Reserve, where about 20 animals live. During the winter censuses of 1985-1992 in the forest belt we have never registered traces of more than one individual per day.

10.3.5.2. Feeding

Young animals of ibex and chamois, as wel as small rodents prevail in the lynx diet (Table 10.13).

Table 10.13. Lynx diet in the Northwestern Caucasus

Kind of food	Occurrence in the diet (% from number of investigated feces)	
	Caucasian Reserve (from Kotov & Ryabov 1963, n=136)	Teberda Reserve (n=6)
Small rodents (Myomorpha)	34.5	37.5
Insectivores (Insectivora)	0	12.5
Ungulates (Artiodaktila)	51.3	25.0
Birds (Aves)	7.8	0
Insects (Insecta)	6.3	25.0

Like most of the predatory mammals feeding on small rodents, the lynx procures Microtinae more often because they are not as quick and more abundant in the forest and the alpine belts.

10.3.6. Stone and Common martens (*Martes foina* Erxl., *M. martes* L.)

Stone marten, pine marten and fox can be consider among the third group of mammals. Some individuals of these species occur at high areas quite regularly, appearing at the timberline and sometimes in subalpine communities.

10.3.6.1. Spatial distribution, abundance and dwellings

Martens are wide spread in the Northwestern Caucasus. They are especially typical for the foothills and deciduous and dark-coniferous forests (Bakeev 1973). In the Caucasian Reserve they can be found in high densities up to the altitude of 1000-1200 m a.s.l. and in higher altitudes they are rare. According to our observations and data of Kurapova (1949) in the Teberda Reserve the stone marten is more numerous than the common marten even in the dark-coniferous forests. The animals can be found up to 2000 m a.s.l. and sometimes higher.

Winter precipitation is one of possible reasons for the differences in altitudal distribution of the stone marten in the Northwestern Caucasus. Precipitation decreases along the Great (Glavnyi) Caucasian Ridge from west to east. The Caucasus Reserve is situated to the west from Teberda.

The stone marten in comparison with the common marten is less adapted to snowy winters. That's why hillsides and rocks without snow are favorable for this species. The common marten more often inhabits the forest belt, higher than 2000 m a.s.l., where the snow cover is thicker and lasts for longer (Table 10.14). The altitudinal distribution of martens is continual. We have repeatedly observed that stone and common martens occur in each others neighborhood, and that their habitats overlap.

Table 10.14. Altitudinal distribution of stone and common martens in the Teberda Reserve

Altitudinal range (m a.s.l.)	Occurrence (% of total number of registration)	
	Stone marten	Common marten
1300-1500	57.1	0
1500-1700	14.3	0
1700-1800	28.6	0
1800-2000	0	33.3
2000-2200	0	66.7

Due to the abundance of food and good protective conditions, martens often inhabit dark-coniferous (*Abies nordmanniana, Picea orientalis*) and dark-coniferous - deciduous forests, whereas pine forests are not so attractive for the species. But the most preferable marten habitats are elfin birch woodlands with hazel (*Corylus avellana*) and rowan (*Sorbus aucuparia*) underbrush avalanche channels and fans. The habitats are rich in rowan fruits and they usually support the highest population density of small rodents. We have not registered seasonal differences in habitat distributions of martens. Relative abundance of martens in the Teberda Reserve was estimated at 10.5 traces per 10 km of winter route (average for 11 years of observation).

In the Caucasus like in any other part of its area martens use different natural hollows as refuges: hollows in the tree trunks, cracks in rocks, niches in screes. Many times we found stone martens, resting in the daytime in hollows among screes.

Common marten prefers hollows in firs, maples and other trees.

10.3.6.2. Feeding

Small mammals, insects (mainly *Carabidae*) and fruits are the main components of the marten diet in the Teberda Reserve (Table 10.15). In some years insects predominate in the spring and summer diet of martens. Fruits of rowan are of importance too. In rowan mast years martens consume only its berries from December to April. Fruits of many other shrubs and dwarf shrubs were found in marten feces (*Amelanchier rotundifolia, Crataegus monogyna, Juniperus* spp., *Padus avium, Vaccinium myrtillus, Viburnum* spp.).

Table 10.15. Martens diet in the Teberda Reserve (occurrence of the food in investigated feces, %, n= 626)

Kind of food	Occurrence
Mammals (Mammalia) including:	67.1
Rodents (Rodentia)	58.6
Apodemus sylvaticus	6.8
Pitymys majory	13.4
Chionomys gud	3.3
Glis glis	5.3
Squirrel (*Sciurus vulgaris*)	18.1
Insectivores (Insectivora)	6.6
Birds (Aves)	19.2
Egg shell	2.1
Insects (Insecta) including:	69.6
Carabidae	55
Cerambicidae	0.8
May beetle (*Melolontha* sp.)	1.0
Snout beetles (Curculionidae)	1.8
Other beetles (Coleoptera)	6.5
Hymenoptera including:	6.4
Wasps (Vespidae)	4.2
Bumble-bees (*Bombus*)	0.6
Bees (Apidae)	0.6
Orthoptera	1.1
Lepidoptera (caterpillars)	2.9
Mollusks	1.1
Plants, including fruits of:	35.1
Sorbus aucuparia	17.4
Vaccinium vitis-idaea	3.0
Rosa spp.	2.9
Rubus idaeus	1.9
Pyrus communis	2.1
Spruce and fir needles	1.6

Due to the wide spectrum of possible food, stone martens suffer very seldom from lack of food, changing successfully from one kind of food to another. For example, there was a depression in the small rodent population density in 1981. The rodents took a very small part in the diet of martens that year (Fig. 10.12), but the occurrence of insectivorous mammals, birds and *Carabidae* remains in feces increased. It seems, that the increase or decrease of marten consumption of these groups of animals depends not on their abundance, but on the dynamics in the availability of the main food component (small rodents).

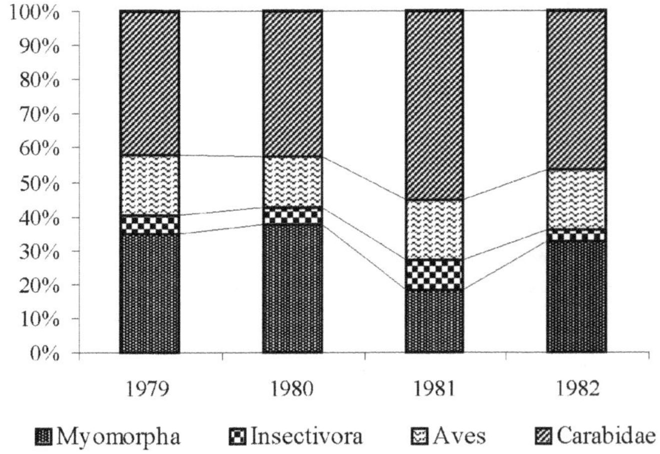

Figure 10.12. Dynamics of animal food consumption by stone marten in the Teberda Reserve during 1979-1982

In winter and spring there were some observations of marten feeding on meat of ibexes, which had died in avalanches, as well as wild boars and chamois, which were killed by wolves.

The daily demand for food was determined only for the stone marten. According to our data under laboratory conditions the animal consumed about 200 g of animal food daily. This value is close to data obtained for the species in the Ukraine (Abelentsev 1973).

10.3.7. Red fox (*Vulpes vulpes* L.)

10.3.7.1. Spatial distribution and abundance

The red fox is widespread in the Teberda Reserve and in the region as whole. It occurs in all mountain belts excluding the subnival and nival belts. But the deciduous and mixed forests situated in the wide valley of the Teberda River are preferable habitats in the Reserve (Fig. 10.13). Looking for food the fox goes up to the subalpine and alpine meadows. For example, we have observed large males on alpine meadows in the circus of the M. Khatipara.

The fox population density in the Teberda valley is rather high. We counted on average about 20 traces per 10 km in the forest belt. The highest concentration of animals in winter was observed near people's settlements.

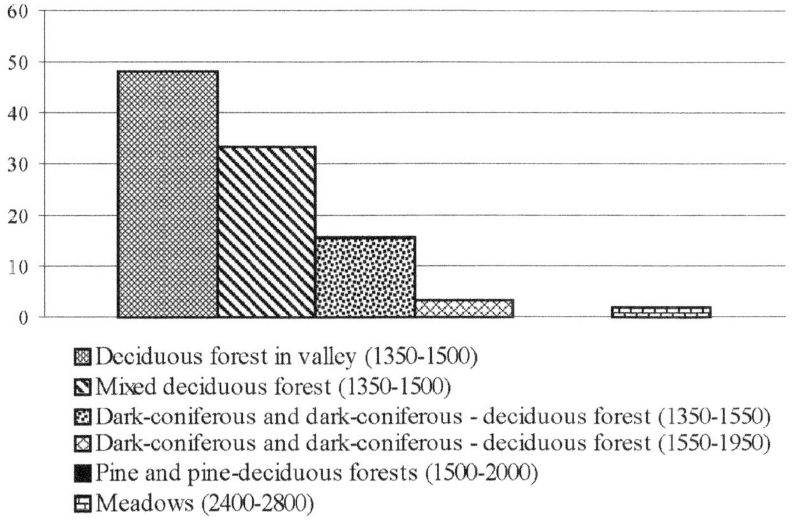

Figure 10.13. Habitat and altitudinal distribution of the fox in the Teberda Reserve (number of traces per 10 km route)

10.3.7.2. Dwelling

All observed fox burrows were situated in the Teberda valley and on mountain slopes not higher than 1700 m a.s.l. Animals partly dig the refuges themselves, especially in places with sandy soils. Sometimes they use badger's burrows. There is a seasonal pattern of burrow usage. In summer the family lives in the burrow until the whelps grow and have learnt to procure food themselves. As a rule, the fox doesn't use burrows in winter. Depending on weather conditions the animals rest in the daytime under stones, crowns of trees and in rock caves.

10.3.7.3. Feeding

The basis of the fox diet consists of small mammals, firstly small rodents. Foxes eat shrews (*Sorex* spp.) unwillingly, and only if they are very hungry. In spring and in summer carabid-beetles constitute an important part of the fox diet (Table 10.16.). In mountain forests these beetles are numerous. They are spread over all the habitats, and that is why they can be easily procured even by young foxes.

Table 10.16. Fox diet in the Teberda Reserve in spring and summer time (occurrence in investigated feces, %, n=23)

Kind of food	Occurrence
Mammals:	69.6
Rodents (Myomorpha)	30.4
Pitymys majori	8.7
Chionomys gud	4.3
other small rodents (Myomorpha)	8.7
Insectivores (Insectivora)	4.3
Hare (*Lepus europeus*)	4.3
undetermined mammals (Mammalia)	13.0
Birds (Aves)	34.8
Insects (Insecta) including:	87.0
Carabid beetles (Carabidae)	56.5
Orthoptera	4.3
Plants (fruits of *Pyrus communis*)	13.0

10.3.8. Other predatory mammals

The European wild cat, Eurasian otter and European mink are regarded as species, rarely or occasionally visiting the alpine belt of the Teberda Reserve and adjacent areas.

Eurasian otter (*Lutra lutra* L.)

We have observed that otters from the Dzhemadat river valley went upstream to the alpine area and traveled through the pass to the neighboring valley of the Daut River.

European mink (*Mustela lutreola* L.)

Individuals of this species are regularly registered in the lower part of the Teberda River, but they seldom penetrated in higher areas. O.A.Vitovitch and P.P.Podsvirov (personal communication) have found mink traces and have seen the animals in the Nazlykol valley at the altitude of 2000 m a.s.l.

Wild cat (*Felis silvestris* Schreb.)

This species is very rare, so it is difficult to observe its penetration into the upper part of the forest belt. At the end of April 1984 we found a dead female in M. Khatipara on the border between pine forest and subalpine meadow. The animal, probably, was stifled by lynx. In the cat's stomach we found the remains of 14 bodies of *Pitymus majori* in different degrees of digestion. So, we suggest that the wild cat regular hunts in the subalpine meadows where population density of pine voles is high.

Jackal (*Canis aureus* L.)

Not long ago this species was absent in the studied area. It was first recorded in 1970th near Teberda town. Now it may be regarded as a common species. According to observations of local people, the jackal was noted in the alpine area of the Gonachkhir valley near a temporal herder hut.

10.4. Large soil invertebrate communities (mesofauna)

10.4.1. Introduction

Soil invertebrates play a significant role in different ecosystem processes, especially in decomposition and mineralisation of plant dead matter (Gilyarov & Chernov 1975, Striganova 1980). Their role may be very important in arctic and alpine communities because those ecosystems have a slow biological turnover and a small production. Our knowledge about alpine soil invertebrates in the Caucasus is poorer than in the Alps (Franz 1979, 1981, Cuendet 1984, Dethier 1985).

The purpose of our study was to compare communities of large soil invertebrates ("mesofauna" according to Gilyarov 1975, or "macrofauna" according to Petersen & Luxton 1982) of four alpine ecosystems of the Northwestern Caucasus (alpine lichen heath - ALH, *Festuca varia* grassland - FVG, *Geranium - Hedysarum* meadow- GHM, and snow bed community-SBC) in terms of density and biomass (Onipchenko & Zhakova 1997).

We obtained invertebrates by means of handsorting the soil samples (Gilyarov 1975). The size of samples were 25 x 25 cm to a depth of 20-30 cm. Twenty samples in each plant community were investigated (Onipchenko & Zhakova 1997).

10.4.2. Total density and biomass

The general biomass of large soil invertebrates decreased in the order ALH-FVG-GHM-SBC from 14,2 to 0,4 g/m^2 ("fresh" mass) because of the decreasing amount of Lumbricidae which dominates in the ALH and FVG ecosystems (Table 10.17.). The total density of the animals have the same tendency (Table 10.18). The maximal invertebrate biomass without Lumbricidae was found in the most productive ecosystem (GHM).

In all but the FVG ecosystems more than 95% of the individuals making more than 90% of the total biomass were found in upper 0-10 cm soil layer. In the FVG these values were 92% and 77%, respectively. In the SBC soil large invertebrates were not found deeper than 10 cm. Apparently, more frequent periods of soil desiccation in the FVG lead to the distribution of invertebrates in deeper soil layers.

As Lumbricidae-dominated communities the ALH and FVG are similar to temperate grasslands (Petersen & Luxton 1982) and alpine tundras of the North Urals (Olshwang & Fileva 1982, Fileva 1983). The invertebrate biomass in soils of alpine *Nardus-Poa* grassland of Karabach is similar to that of the ALH, but Diplopoda are dominant in the former and absent in all our investigated communities (Striganova & Loginova 1984). Biomass of soil invertebrates in the ALH and FVG exceeds that in high mountain steppes and deserts of West Tian-Shan, but is less than that in meadows of that region (Zlotin 1975).

Table 10.17. Biomass of large soil invertebrates in alpine ecosystems (g/m^2 "fresh" mass, average and standard error, italic bold values represent dry mass) (from Onipchenko & Zhakova 1997 with permission of *Oecologia Montana*)

	ALH	FVG	GHM	SBC
Lumbricidae	13.6±4.3	2.9±1.0	0.3±0.3	0
	3.3	***0.7***	***0.07***	
Enchytraeidae	0.18±0.03	0.006±0.003	0.004±0.002	0
	0.03	***0.001***	***0.001***	
Aranei	0.017±0.007	0.34±0.10	0.29±0.08	0.009±0.003
	0.005	***0.1***	***0.08***	***0.003***
Chilopoda	0.13±0.10	0.46±0.08	0.19±0.04	0.004±0.004
	0.02	***0.08***	***0.03***	***0.001***
Coleoptera	0.12±0.06	0.57±0.12	0.98±0.21	0.28±0.09
including:	***0.03***	***0.15***	***0.32***	***0.06***
Carabidae	0.06±0.05	0.02±0.02	0.06±0.03	0.015±0.008
	0.02	***0.006***	***0.02***	***0.005***
Staphylinidae	0.008±0.003	0.016±0.008	0.07±0.06	0.014±0.007
	0.003	***0.006***	***0.03***	***0.006***
Elateridae	0	0.21±0.10	0.48±0.18	0
		0.08	***0.19***	
Tenebrionidae	0	0	0.04±0.04	0.10±0.07
			0.01	***0.02***
Curculionidae	0.05±0.03	0.32±0.07	0.33±0.08	0.15±0.06
	0.01	***0.06***	***0.07***	***0.03***
Diptera	0.31±0.15	0.15±0.11	0.40±0.31	0.08±0.06
	0.06	***0.03***	***0.08***	***0.016***
Lepidoptera	0	0.05±0.04	0.02±0.02	0.02±0.02
		0.005	***0.002***	***0.002***
Hymenoptera	0	0.008±0.005	0.03±0.02	0
		0.002	***0.01***	
Total biomass	14.3±4.3	4.4±1.0	2.1±0.5	0.37±0.11
	3.45	***1.07***	***0.59***	***0.08***
The sum without Lumbricidae	***0.15***	***0.37***	***0.52***	***0.08***

Meyer (1981) investigated the structure and biomass of soil invertebrate communities in two alpine ecosystems in the Austrian Alps: *Curvuletum* (ecologically analogous to ALH) and *Salix herbacea* snowbed (ecologically analogous to our SBC). The total biomass was estimated as 925 and 480 mg/m^2 (dry mass), respectively, so the difference between snow-free and snow-bed communities is less than in our region. The density and biomass of Coleoptera were higher than in the Caucasus ecosystems, while the abundance of Lumbricidae was lower. Meyer (1981) did not find Chilopoda in alpine communities, although this group is very common in our ecosystems.

Table 10.18. Density of large soil invertebrates in alpine ecosystems (individuals/m^2, average and standard error, n=20) (from Onipchenko & Zhakova 1997 with permission of *Oecologia Montana*)

	ALH	FVG	GHM	SBC
Lumbricidae	39±10	25±6	2±2	0
Enchytraeidae	338±52	7±4	4±1	0
Aranei	5±2	72±21	36±7	10±3
Chilopoda	23±19	100±17	46±9	2±2
Coleoptera	15±5	53±8	46±7	20±4
Carabidae	3±2	4±2	4±2	3±1
Staphylinidae	5±2	10±3	3±1	3±1
Elateridae	0	4±2	8±3	0
Tenebrionidae	0	0	2±2	2±1
Curculionidae	7±4	35±7	29±5	12±4
Diptera	190±136	12±4	81±64	5±2
Lepidoptera	0	4±2	1±1	1±1
Hymenoptera	0	2±1	2±1	0
sum	610±148	275±30	218±67	38±6

10.4.3. Abundance and composition of taxonomic groups

Only one widely distributed Caucasian species of Lumbricidae was found in our communities - *Dendrobaena schmidti* Michaelsen, 1907. The biomass of this species differs significantly between the studied communities, being the highest in the ALH (Onipchenko & Zhakova 1997). In the GHM and SBC, on the other hand, earthworms are practically absent, probably due to too short snow free seasons in these ecosystems. To complete the life cycle the earthworms need more time than the snow free season lasts in the area of relatively cold alpine soils.

The abundance of *D.schmidti* in subalpine meadows of the Central Caucasus (Kazbegi) was not great and did not exceed the value for the ALH (Kvavadze 1985). The high biomass of Lumbricidae distinguishes the ALH and FVG communities from typical lowland arctic tundras, where the role of this group is not significant (Petersen & Luxton 1982). Five species of Lumbricidae were found in calcareous alpine areas of Switzerland where their density and biomass were much more than in our ALH (Cuendet 1984). Lumbricidae were found in Swiss snow bed alpine communities (*Salicetum herbaceae*) but with low abundance.

Chilopoda was the most abundant group in the FVG. In other communities the density and biomass of Chilopoda are similar to tundra soils (Petersen & Luxton 1982). The high mountain soils of Tian-Shan contain more Chilopoda (Zlotin 1975) in comparison with the Caucasian communities. Among Lithobiomorpha only one species (*Monotarsobius sseliwanoffi* Gard.) was found, being usual in all studied ecosystems. Among Geophilomorpha the most abundant species was *Strigamia acuminata* (Leach, 1814) (found in ALH, FVG, GHM). *Fagetophilus elegans* Folk, 1956 also was usual (ALH, FVG), while *Geophilus* sp., found only in the FVG, was more rare.

We have not found any Diplopoda in our samples, although this group dominated some soil invertebrate communities of the alpine meadows of the Small Caucasus (Striganova & Loginova 1984) and it is common in subalpine meadows of Kazbegi (Kohia 1987). We suggest that the absence of Diplopoda was connected with the low level of Ca-content in alpine soils and in its parent materials (granites, biotite schists), because species of Diplopoda need significant amounts of Ca for making their shells (Pokarzhevskii 1985). Maybe the absence of Mollusca in our communities is for the same reason.

Aranei are very abundant and diverse in alpine communities. Using handsorting, however, we could not estimate correctly their density and biomass, because most large Aranei (especially Lycosidae) are very mobile. Apparently, the most underestimated density of Aranei was in the ALH and SBC where plant height and protective properties of the herb layer were small. The list of Aranei species is presented in Onipchenko & Zhakova (1997). This group was most diverse in dense FVG and occupied relatively warm sites. Species of Linyphiidae were more frequent, especially *Trichoncus hispidosus* Tanasevic, 1990 (Tanasevich 1990). Only Erigonidae species were founded in SBC.

Various Carabidae were usual in all communities. Just as for Aranei we underestimated their biomass and density due to their high mobility. *Carabus koenigi* Gglb. and *Pterostichus swaneticum* Rtt. are common species in our communities. Usual in soil samples also are: *Amara* sp. (ALH), *Amara (Lejocnemis) sabulosa* Dej. (GHM), and *Bembidion (Testedium) bipunctatum rugiceps* Chb. (SBC, more rare in GHM). More rare were *Bradycellus* sp. (FVG), *Dyschirius lederi* Rtt. (FVG), and *Calanthus (Neocalanthus) melanocephalus* L. (FVG).

Species of Staphylinidae were found in all communities. The following taxa were identified: *Geostiba (Ditroposipalia)* sp. (*betubereulata* group) (FVG), *Micralymma caucasicum* (Melichar) (FVG), *Philonthus frigidus svanetiensis* Coiffait. (FVG, GHM, SBC), *Geodromicus latiusculus* Eppelsheim (SBC), *Othius stenocephalus* Eppelsheim (FVG), *Tachyporus* sp. (ALH, FVG), *Mycetoporus ruficornis* Kraatz (FVG), and *Staphylinus* sp. (GHM).

Larvae of Elateridae were found in low densities only in grassland ecosystems (FVG, GHM). Among soil herbivores (rhizophagous) the Curculionidae species were the most common. They occurred in all studied ecosystems, being the most abundant in alpine meadows (FVG, GHM). All of them were identified as members of the genus *Otiorrhynchus*: *O.carbonarius* Reitt.(ALH, more rare in GHM), *O.circassicus* Reitt. (FVG, more rare in GHM), *O.bidentatus* Tourn. (FVG, GHM), and *O.cinereus* Stierl. (rare in GHM, SBC). Their density and biomass are represented in Table 10.17 and 10.18 as totals for all life stages (larvae, pupae, imago).

The total density and biomass of Coleoptera in alpine grasslands (FVG, GHM) exceeds values for the *Curvuletum* in the Tirolian Alps (Schatz 1981).

Diptera larvae occurred in soils of all studied communities (Table 10.19). Sciaridae species had the highest density. As these species have extremely aggregate distributions but the precise data are difficult to obtain. We have found two aggregates in the ALH and one in the GHM. Empididae larvae were the most frequent in the investigated soils, while Tipulidae were relatively rare.

Table 10.19. Diptera larvae in alpine soils (total number for 20 samples) (from Onipchenko & Zhakova 1997 with permission of *Oecologia Montana*)

	ALH	FVG	GHM	SBC
Sciaridae	200	0	80	0
Tipulidae	1	1	1	0
Empididae	15	4	17	2
Brachycera-Cyclorrhapha	7	2	0	2
Therevidae	1	0	0	0
Dolychopodidae	2	0	0	0
Cecidomyidae	0	2	0	0
Tabanidae	0	0	0	1

10.4.4. Trophic groups

Among the trophic groups saprotrophs dominate in terms of biomass because of the high abundance of Lumbricidae (Fig. 10.14). Dominance of saprotrophs was noted for high mountain communities in the Tirolian Alps (Schatz 1981, Meyer 1981). The most significant role of phytophagous groups was recorded for the GHM. In this ecosystem the ratio of plant necromass to plant biomass was the least, so the litter decomposition rate is the highest here. The proportion of phytophages is great in snowbeds (SBC), but the total invertebrate biomass was extremely low there. A similar peculiarity was noted for analogous communities in the Swiss Alps (Dethier 1985). The proportion of carnivores is relative great in more productive grasslands (FVG, GHM). This data support the "exploitation ecosystem" hypothesis (Oksanen et al. 1981).

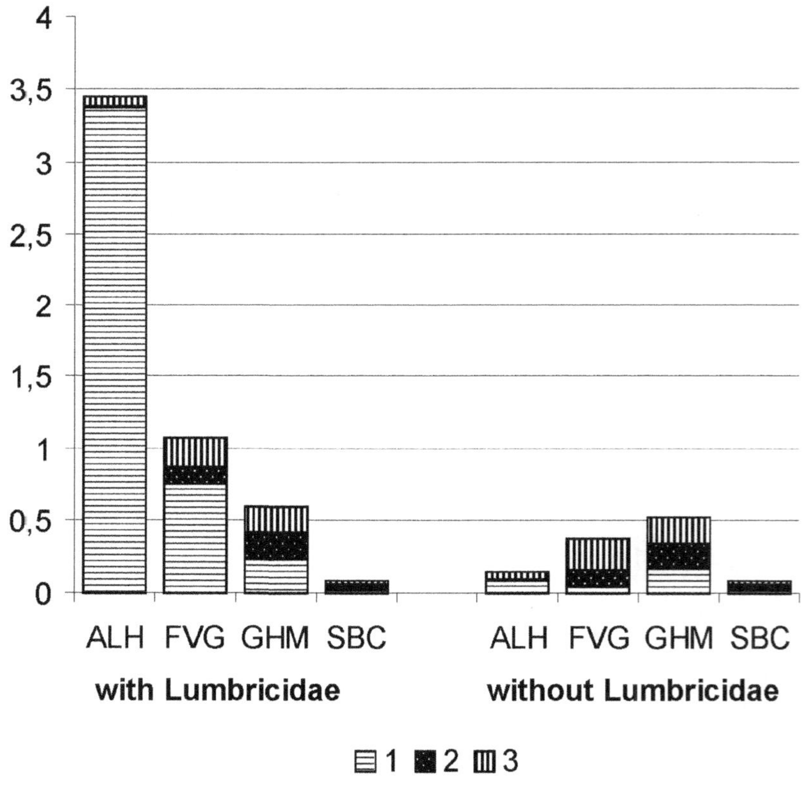

Figure 10.14. Trophic composition of the total large soil invertebrate biomass (g/m^2 dry mass). Trophic groups: 1 - saprotrophs, 2 - phytophages, 3 – carnivores (from Onipchenko & Zhakova 1997 with permission of *Oecologia Montana*)

10.4.5. Conclusions

The total dry biomass of large soil invertebrates varied from 0.08 to 3.45 g/m^2 decreasing with increasing snow depth. Each ecosystem had a specific composition of the soil invertebrate community. Studied alpine communities differed from that of arctic tundras by a smaller abundance of Enchytraeidae and Tipulidae and by a larger role of Chilopoda and Curculionidae.

The composition of the soil invertebrate community of the ALH was similar to that of plain meadows and Ural alpine tundras. The community of alpine grasslands (FVG, GHM) is closer to alpine communities of the Tian-Shan and the Alps. Apparently, the type of parent soil material (silicate or carbonates) has a strong influence on the composition of soil invertebrate communities, especially for such Ca-demanding groups as Diplopoda or Mollusca.

10.5. Grasshoppers

10.5.1. Introduction and methods

Grasshoppers are important hexapod herbivores in many grasslands including alpine ecosystems. But their role in the regulation of plant species composition is not clear. To understand their role it is necessary to study the abundance of different grasshopper species and food specialization between them. Little information is known about food specialization in Orthoptera (Stolyarov 1975; Sergeev et al. 1995).

The grasshoppers were studied in four alpine communities: alpine lichen heath (ALH), *Festuca varia* grassland (FVG), *Geranium-Hedysarum* meadow (GHM), snow bed community (SBC). Samples of 120 plant species were collected for epidermal cell microscoping. Grasshoppers were collected for investigation of their gut contents and excrements as well as for plant preference studies in cages. Food preference was determined for 110 plant species in 2-3 replications. Estimation of plant species preferences was made on a 6 - point scale: 0 – absolutely non eaten, 1 – very poor eaten (<2% plant material was consumed), 2 – weakly eaten (2-5%), 3 - satisfactorily eaten (5-20%), 4 - well eaten (20-50%), preferable (>50%).

The biomass and population density of grasshoppers in the ALH were monitored on permanent plots with a total area of 62 m^2 during 19 years (1981-1999). The observations were made at the end of August or beginning of September. To compare grasshopper communities in different ecosystems, the population density was determined in 100-120 m^2 plots in each community in August 1999.

10.5.2. Species composition

We found seven species of grasshoppers (Acridoidea) and two species of bush-crickets (Tettigonoidea) in studied plant communities during a season 1999:

Fam. Tettigoniidae
  subfam. Phaneropterinae
    Polysarcus zacharovi Stshelk.
  subfam. Tettigoniinae
    Decticus verrucivorus L.
Fam. Pamphagidae
  subfam. Pamphaginae
    Nocaracris cyanipes F.-W.
Fam. Acrididae
  subfam. Catantopinae
    Podisma teberdina Rmme.
  subfam. Acridinae
    Stenobotrus nigromaculatus H.-Sch.
    Omocestus viridulus L.
    Gomphocerus sibiricus caucasicus Motsh.
    Chorthippus apricarius major Pyln.
    Chorthippus porphyroptera Voronzovsky

We divided all species according to their population density into dominants (more than 1000 individuals/ha), common species (200 - 1000 individuals/ha), and rare species (less than 200 individuals/ha). Six species were found in the ALH (Fig. 10.15). Three species (*Nocaracris cyanipes*, *Stenobotrus nigromaculatus* and *Gomphocerus sibiricus*) were considered as dominants, two *(Podisma teberdina* and *Omocestus viridulus*) were common, one (*Chorthippus porphyroptera*) was rare.

Chorthippus apricarius was absent in the ALH. The population density and biomass of grasshoppers fluctuated greatly between years (Fig. 10.16). The abundance of grasshoppers increased in dry years and decreased in wet years. *Podisma teberdina* was a dominant species in most years (Fig. 10.16).

Five species were found in the FVG (Fig. 10.15). *Nocaracris cyanipes*, *Omocestus viridulus* and *Gomphocerus sibiricus* were dominants there. *Stenobotrus nigromaculatus* was a common species and *Chorthippus apricarius* was rare. Two other species were not found there.

The total population density of grasshoppers was highest in the GHM. It has about 20 000 animals per m^2. (several times more than in other communities). Two species (*Gomphocerus sibiricus* and *Omocestus viridulus*) were dominant there (Fig. 10.15). Other species were much less frequent. So this high productive alpine plant community was very attractive for some grasshoppers. It may be connected not only with plant production, but with the abundance of bare soil patches (gaps) which are convenient for egg development.

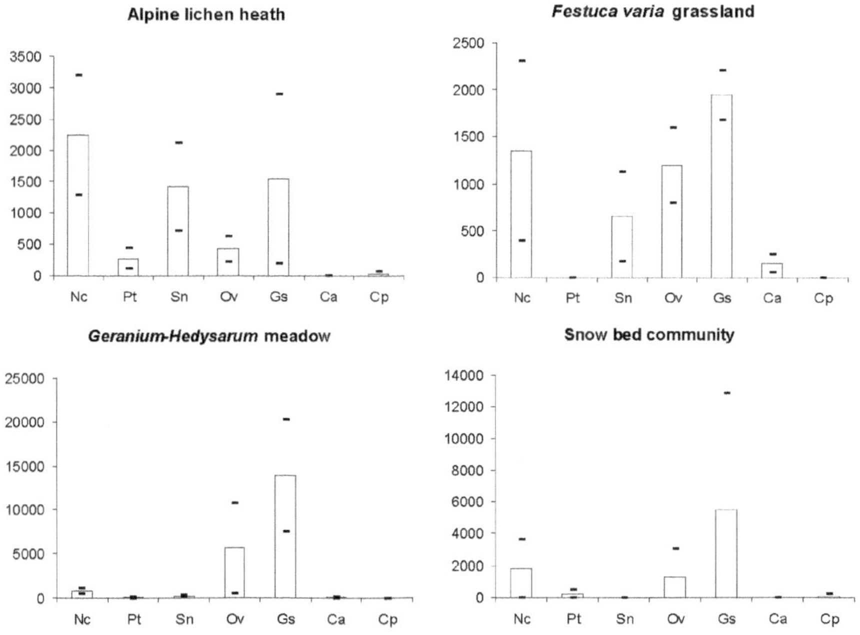

Figure 10.15. Density of grasshoppers in alpine plant communities (individuals per ha, average ± st. er.). Abbreviations: Nc - *Nocaracris cyanipes*, Pt - *Podisma teberdina*, Sn - *Stenobotrus nigromaculatus*, Ov - *Omocestus viridulus*, Gs - *Gomphocerus sibiricus caucacicus*, Ca – *Chorthippus apricarius major*, Cp - *Chorthippus porphyroptera*

Three species (*Nocaracris cyanipes, Omocestus viridulus* and *Gomphocerus sibiricus*) were noted as dominants in the SBC (Fig. 10.15). Two other species were rare.

Only few individuals of bush-crickets were found. *Polysarcus zacharovi* was noted only in the GHM, and *Decticus verrucivorus* only in the FVG.

So, we conclude that there are four main species of grasshoppers (dominants) only (*Nocaracris cyanipes, Stenobotrus nigromaculatus, Omocestus viridulus* and *Gomphocerus sibiricus*).

Podisma teberdina was a common species. It is necessary to note, that at least in the ALH the population density of *Podisma teberdina* was much lower, but the population density of *Gomphocerus sibiricus* was much higher

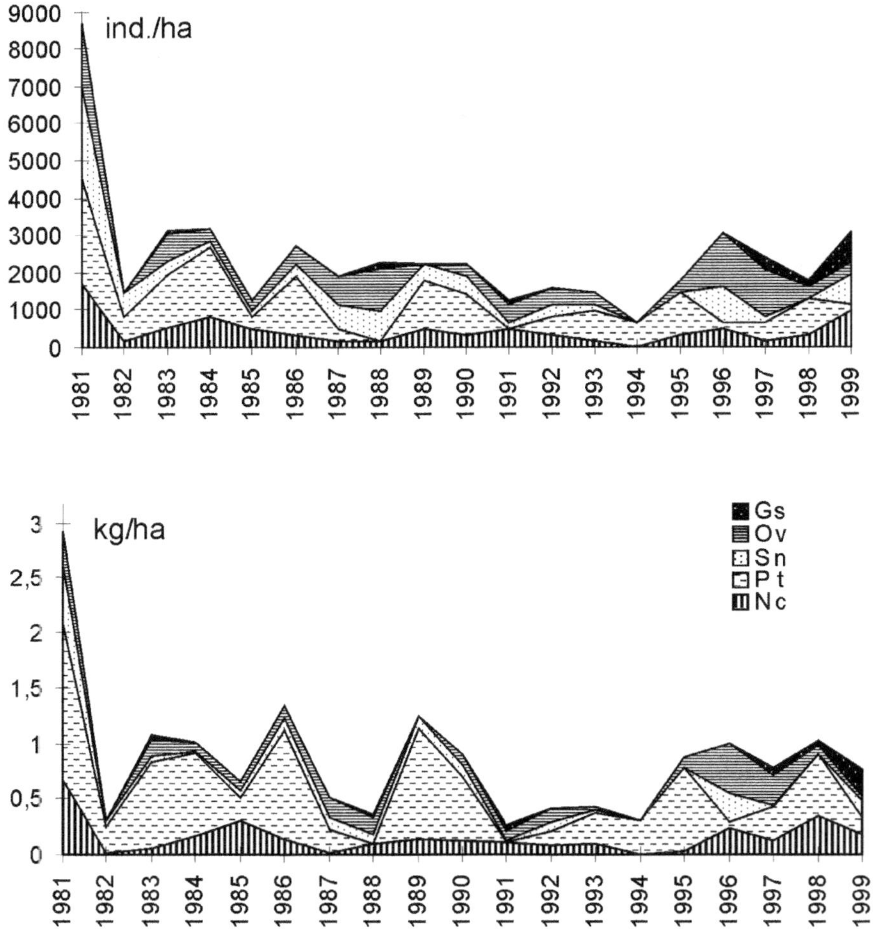

Figure 10.16. The perennial dynamics of abundant grasshopper species on permanent plots of ALH, ind./ha and kg/ha (fresh mass). Abbreviations: Nc - *Nocaracris cyanipes*, Pt - *Podisma teberdina*, Sn - *Stenobotrus nigromaculatus*, Ov - *Omocestus viridulus*, Gs - *Gomphocerus sibiricus caucacicus*

in 1999 than in other years of observation (Fig. 10.16). Three *species (Nocaracris cyanipes, Omocestus viridulus* and *Gomphocerus sibiricus*) are abundant enough in all habitats in spite of a great difference between plant communities. So, the species showed high ecological plasticities. Other species have shown the certain degree of habitat preference. The spatial distribution in the habitats of *Stenobotrus nigromaculatus* has demonstrated a preference to less snowy communities (ALH and FVG). *Podisma teberdina* and *Chorthippus porphyroptera* prefer habitats with a lower plant cover, *Chorthippus apricarius* was found in tall and closed plant communities.

10.5.3. Plant species preference by grasshoppers (Acridoidea)

Analysis of species consumption by grasshoppers showed that all studied alpine species can be eaten by some of the grasshoppers but at different rates (Tables 10.1, 10.20). In Table 10.1. we represent an estimation of plant species preference by grasshopper species in comparison with other alpine herbivores. Mean indexes of plant group preference are represented in Table 10.20. We divided the species of 3 families into two groups: Poaceae into hard-leaved (*Agrostis vinealis, Festuca* spp., *Helictotrichon versicolor, Nardus stricta*) and soft-leaved grasses (*Anthoxanthum odoratum, Briza marcowiczii, Bromus variegatus, Calamagrostis arundinaceae, Catabrosella variegata, Deschampsia flexuosa, Hyalopoa pontica, Phleum alpinum, Trisetum flavescens*), Caryophyllaceae into broad-leaved (*Cerastium purpurascens, Silene saxatilis*) and narrow-leaved species (*Arenaria lychnidea, Minuartia* spp.), Scrophulariaceae into hemiparasitic (*Euphrasia ossica, Pedicularis* spp.) and others (*Eritrichium caucasicum, Myosotis alpestris, Veronica gentianoides*).

Table 10.20. Estimation of plant group preference by grasshoppers (mean values for families with more than 3 species). Abbreviations: n – number of studied plant species, Nc – *Nocaracris cyanipes*, Pt – *Podisma teberdina*, Sn – *Stenobotrus nigromaculatus*, Ov – *Omocestus viridulus*, Gs – *Gomphocerus sibiricus*. For species data see Table 10.1, for scale points – Chapter 10.5.1.

Plant groups	n	Nc	Pt	Sn	Ov	Gs
Cyperaceae (*Carex* spp.)	5	2.2	1.8	4.0	4.0	4.2
Poaceae: hard-leaved	6	2.0	1.5	4.7	4.3	3.7
Poaceae: soft-leaved	9	3.7	2.3	4.1	4.4	4.6
Apiaceae	5	3.6	3.6	0	0.2	1.6
Asteraceae	20	3.9	3.6	0.7	1.2	2.4
Caryophyllaceae: broad-leaved	2	4.0	4.5	1.0	3.5	3.0
Fabaceae	4	3.5	3.0	2.3	1.3	2.8
Gentianaceae	4	4.3	1.5	1.0	0.3	0.5
Lamiaceae	3	3.3	1.3	0	0.3	1.0
Ranunculaceae	5	3.4	1.4	0	0.6	0.6
Rosaceae	5	4.0	3.3	0.6	0.2	1.0
Scrophulariaceae: non parasitic	3	3.3	1.6	0.2	0.3	0.8
Broad-leaved forbs	71	3.6	2.7	0.5	0.7	1.6
Scrophulariaceae: hemiparasitic	4	3.0	1.5	0.3	0	1.3
Caryophyllaceae: narrow-leaved	4	2.8	2.3	0.5	1.0	1.3
Ericaceae	3	1.3	2.7	0.3	0.7	0

Differentiation between grasshopper species was obvious. It appears, that all studied species, including *Nocaracris cyanipes* and *Podisma teberdina*, eat grasses and others monocotyledons. However, *Stenobotrus nigromaculatus, Omocestus viridulus* and *Gomphocerus sibiricus* prefer

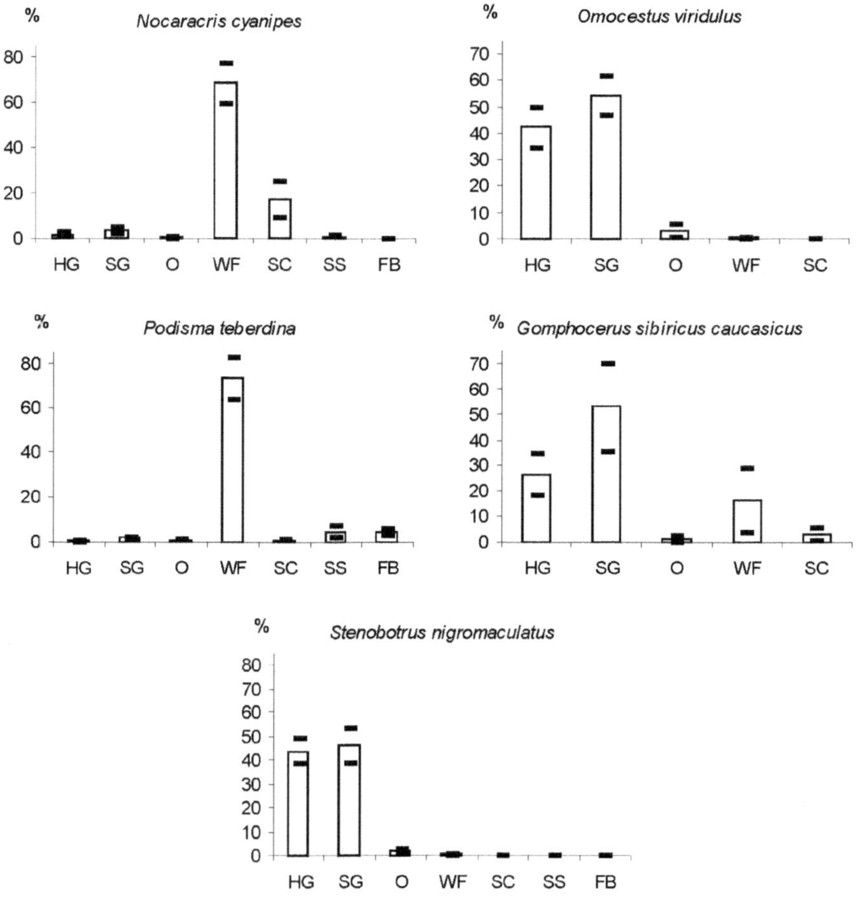

Figure 10.17. The spectra of plants really eaten by abundant grasshopper species in nature (average ± st. er.) based on an analysis of excrements and gut contents. Plant groups: HG - hard-leaved grasses, SG - soft-leaved grasses, O - other monocotyledons (*Carex* spp.), WF - broad-leaved forbs, SC - narrow-leaved Caryophyllaceae (*Arenaria, Minuartia* spp.), SS - hemiparasitic Scrophulariaceae (*Euphrasia, Pedicularis* spp.), FB – *Vaccinium vitis-idaea.*

graminoids. The latter species were considered as graminoid-eating species in previous investigations as well (Chernyakhovskii 1968, 1970). *Nocaracris cyanipes* prefered soft-leaved grasses and broad-leaved forbs, and *Podisma teberdina* willingly ate dwarf shrubs (*Vaccinium* spp.) and forbs under cage conditions.

The cage method has some disadvantages due to the unnatural conditions in cages and the limited number of plants to choose from (Stebaev & Pshenitcina 1984). The real set of consumed plants may be determinated from the analysis of excrements and gut contents (Sergeev et al. 1995). This analysis was made for grasshoppers from the ALH and GHM. Seven groups of plants were identified: hard-leaved grasses, soft-leaved grasses, other monocots, broad-leaved forbs, narrow-leaved Caryophyllaceae, hemiparasitic Scrophulariaceae, evergreen dwarf shrub (*Vaccinium vitis-idaea*) (Fig. 10.17).

The grasses and sedges are well represented in excrements and gut contents in all studied species. But their role was different for various grasshopper species. More than 90% of all plant rests were grasses for two specialised species – *Stenobotrus nigromaculatus* and *Omocestus viridulus*. On the other hand, broad-leaved forbs were the absolutely dominant food plants for the large non-flying grasshoppers *Nocaracris cyanipes* and *Podisma teberdina* (Fig. 10.17). The species ate leaves of hemiparasitic Scrophulariaceae and evergreen *Vaccinium vitis-idaea* as well. *Gomphocerus sibiricus* was a less specialized feeder with some grass preference.

So, as a whole the results of two methods for determination of plant preferences show rather similar results. They allowed us to obtain a real picture of species consumption by grasshoppers in alpine communities.

Chapter 11

Human activity and nature conservation problems

11.1. Human activity: history of land use

The early history of human activity in the Northern Caucasus is known from the peat, dung and fossil analysis of a few sites in North Ossetia and Karachaevo-Cherkessiya (Savinetskii 1992). The earliest period with proven livestock grazing comprises 4000-2000 yrs BP, when pig raising in the broadleaf forests became common at lower elevations. The second period was characterized by the greater prominence of sheep herding beginning ca. 2000 yrs BP. Particularly severe pressure on the region's forests become noticeable especially since ca. 700 yrs BP, when the Alans and related tribes had to intensify grazing in response to the Tatar-Mongol invasion which pushed local tribes into the mountains and disrupted trade with the lowlands.

In 13-14th centuries as a result of the Tatar-Mongol invasion as well as Tamerlan's conquests, the ancestors of the modern Karachai moved from the plains and foothills of the Northern Caucasus into the upper reaches of the Kuban and Teberda Rivers to poorly accessible mountain valleys. While the exact ethnic origin of Karachai is unclear and is likely a mix of the Polovtsy, Alans, Khazars, and native Caucasian tribes, their lifestyle and economy to a great extent were determined by their geographical environment. Rich alpine pastures of the Greater Caucasus provide an excellent base for cattle breeding, an activity that became most important for the inhabitants of the region. Other types of economical activity were regarded as less valuable. The long experience of the Karachai with cattle breeding can be seen also in their folklore, where the main heroes deal with the livestock, and is also evidenced by the well developed cattle-related vocabulary in the Karachai-Balkarian language.

V.G. Onipchenko (ed.), Alpine Ecosystems in the Northwest Caucasus, 355–362.

For centuries the cattle breeding by the Karachai was of the extensive type, i.e. livestock was fed year round only on green forage. Growth in human population demanded increasing numbers of livestock, and despite the lack of forage, epizooties, high rent fees, military conflicts and some other negative factors, the number of domestic ungulates in the Karachai highlands grew more and more. Unfortunately no historical data regarding livestock numbers prior to the middle of the 19th century are available, but the comparison of numbers obtained in 1880 (Petrov 1880) and 1908 (Abramovskaya Land Commission 1908) serves as a strong evidence of increasing quantities of livestock (Table 11.1).

Table 11.1. Livestock number in Karachai in 1880 and 1908. (from Petrov 1880, Abramovskaya Land Commission 1908)

Type of livestock	Number	
years	1880	1908
Horses	18,500	33,758
Cattle	32,200	125,027
Sheep and goats	231,800	497,471
Donkeys	2,500	1,460

Extensive livestock required permanent change of pastures, and prior to 1917 the semi-nomadic way of life was typical for the Karachai (Fig. 11.1). The majority of them followed their herds in seasonal migrations from highlands to foothills and back. In the middle of May the herdsmen moved their livestock to the upper-slope summer pastures, while in the valleys and lower parts of adjoining slopes new green grass appeared at that time. Horses and cows with newborn calves were moved to the fresh pastures first, while other cattle and sheep were driven there later.

Seasonal migrations covered long distances: sometimes summer pastures were located 60-65 km away from the villages. Herding was not an easy job: in the spring shepherds had to be careful of late avalanches and mudflows, with narrow streams turning into rushing rivers during the snowmelt. Spring migrations significantly aggravated the health of calves, and the weakest lambs and goat kids were transported in baskets bound to donkeys and horses.

In the highlands, the herds were not kept at the same places for long periods to avoid overgrazing and to take an advantage at different plants blooming at different times on various altitudes: gradually herdsmen moved higher and further into the mountains until in July they found themselves at alpine meadows very close to the snowline.

The herds were driven back to the highlanders' villages situated in the river valleys, after the middle of August until mid-September, when the first frosts came to the highlands. By that time the highlanders usually were done with haymaking and green after-grass appeared in the hayfields. After spending some weeks at home, the Karachai started moving their flocks and herds downstream — to the foothills and further into the plains, where they usually leased winter pasturs from the Kabardins.

Figure 11.1. Old road in Kuban' valley (from Raev, *ca* 1910)

Thus, the environment of the region could be considered as favourable for livestock raising, but not to cultivation of crops. That is why the latter always played a secondary role in the Karachai economy. Despite poor quality of the soils, narrow and cold valleys and unfavourable climate conditions, the highlanders from ancient times have also cultivated plants, and step by step transformed stony mountain terrain into small patches of tillage. Pasturing cattle in the valley forests resulted in a deforestation process that enlarged the area of land fit for the plough. Sometimes the Karachai tried to establish terrace agriculture, but mud floats and downpours destroyed most of terraces. Very often irrigation was used; some of the irrigation systems constructed 3–4 centuries ago still remain in use today.

Most of the population were poor carpenters, and therefore the timber usage was not rational: in the process of logging a lot of timber waste was usually left to rot in the woods. Timber did not provide any considerable income for the Karachai. Water resources were sometime used to drive water mills (Fig. 11.2).

Figure 11.2. Deforested bottom of the Teberda valley (from Raev, *ca* 1910)

Figure 11.3. Dombai Glade before the tourist camps were built at the beginning of the XXth century (from Raev, *ca* 1910)

Gradually the population increased, and the utilisation of the alpine and subalpine meadows as well as forests became more and more intensive. Some attempts to establish a rational and sustainable use of forest resources were neutralized by persistent clear-cutting. Processes of deforestation combined with overgrazing resulted in the appearance of vast badlands, which cannot be used even as goat pastures today. Lack of skill in using fertilizers sometimes led to chemical contamination of soils. A good deal of the valleys was lost after construction of the regional hydroelectric power station.

At the same time, numerous mineral water springs, picturesque mountains and favourable conditions (Fig. 11.3) for mountain skiing stimulated the development of resorts and recreational industry in the republic. Many hotels and tourist camps were built, especially in Dombai and Teberda. Construction was followed by the creation of a well-developed, at least by the Soviet standards, tourist-oriented infrastructure (hotels, ski lifts, resorts, kiosks, bus tours and guides, etc.).

In 1943, during World War II, the Karachai faced allegations on numerous crimes linked to the alleged collaboration with the Nazis and therefore were subjected to deportation to Central Asia and Kazakstan. This horrible example of Stalin's national policy was a great tragedy for the people. In 1957 fairness triumphed, and the people were allowed to come back. Ironically, during the same period, the anthropogenic impact upon local landscapes decreased significantly, and many disturbed ecosystems started to regenerate.

11.2. Nature conservation

11.2.1. Conservation priorities and approach

The Caucasus as a whole is a globally threatened ecoregion and a hotspot with high biodiversity and endemism in a variety of ecosystems (Myers et al. 2000, Krever et al. 2001). Much of the region has been impacted by human activities for thousands of years. Pristine areas of nature remain only in remote areas. Conservation strategies for the Caucasus take into account the high human density and long history of resource use, while not veering away from the three fundamental objectives required to conserve biodiversity: to ensure representation of native ecosystem types and their full range of variation; to conserve viable populations of animals and plants with abundant space and resources; and to sustain ecological processes, such as natural disturbance regimes, nutrient cycling, and predation. Cultural aspects of local Karachai communities are to be also considered in carrying out conservation strategies.

There are six main conservation priorities in the Karachai Region:
Institutional strengthening and capacity building;
Improving protected areas network;
Conserving endangered species and species of special concern;
Conserving priority biomes: forests and alpine ecosystems;
Promoting sustainable resource use;
Promoting public awareness and environmental education.

11.2.2. Institutional strengthening and capacity building

Since the collapse of the Soviet Union 10 years ago, many of the institutions and laws governing the environment have lacked the necessary tools and mechanisms to guarantee conservation and sustainable use of biodiversity.

Priorities in institutional development and capacity building include assessment and reform of environmental legislation, training of natural resource managers, creating mechanisms to protect endangered species and ecosystems (TRAFFIC, CITES), implementing the obligations of key conventions, and developing an environmental information network for the region.

11.2.3. Short history of the Teberda Strict Nature Reserve

The northern slopes of the Main Caucasus mountain range in the south of the Karachai-Cherkessia Republic were designated the Teberda Strict Nature Reserve (*Teberdinskii Zapovednik*) in 1935 to cover 851 sq. km. The Reserve consists of two sites, 30 km from each other. The main Teberda site (ca. 650 sq. km) is in the basin of the river Teberda, the other site, Arkhyz (ca. 200 sq. km), is in the river Kizgych basin. The river Teberda is formed within the Reserve at the confluence of the rivers Amanauz and Gonachkhir, which, in their turn, run down from the mountain glaciers. Part of the Reserve has long ago had a status of a national ski resort — Dombai, where facilities are available for skiers and hikers.

The Teberda Reserve is unique from many points of view. First, it is the only *zapovednik* in the Russian Federation used as a resort, with a lot of visitors coming here every year. Actually Dombai and neighbouring territories are used *de facto* as a national park, rather than as a scientific Reserve. Secondly, the Teberda Reserve is the only in the Caucasus where livestock grazing has been strictly forbidden since 1943.

Gradually the research unit of the Reserve developed, and now it is a well-known research center, where numerous investigations, surveys and studies have been carried out. The present book is a result of some of these studies.

Sixty-five years of strict conservation policy have not fully protected the ecosystems from massive anthropogenic impact, such as fires, pasturing, felling trees. It used to be a very popular tourist destination: in the late 1980s up to 700, 000 tourists came to visit Teberda and Dombai per year. It produced a really negative effect on the ecosystems. Now there are fewer visitors, by an order of magnitude or more.

From the ecosystems' standpoint, however, the decline in tourism was not all that beneficial. Since tourism used to provide most earnings for the local communities, now they are much more dependent on the natural resources in the area. Environmental impact produced by tourists has been replaced by one caused by the local population. Development of ecological tourism can become one of the possible scenarios for improvement as it will both generate local income and diminish the disturbance of unique ecosystems.

11.2.4. Creating a network of protected areas

Conservation of biodiversity can be best guaranteed by setting aside unique and representative ecosystems in a variety of protected areas with different management objectives. Protected areas can be national parks, strict nature reserves or multiple use areas that connect protected areas, where traditional or sustainable use of resources is permitted in order to provide benefits for the local people. A network of protected areas is going to be established in the Caucasus Ecoregion. It will set aside representative and endangered ecosystems, and create links between existing protected areas through corridors and stepping stones with varying types of management regimes. Corridors are important for facilitating migration and recolonization of large animals in fragmented landscapes. Using GIS tools and a gap analysis approach, the discontinuities in the protected areas system can be identified and steps to conserve key habitats for linking the system can be taken. Conservation practices in existing protected areas need to be examined to ensure that they are appropriate for ensuring biodiversity conservation.

There are two main protected areas in the Karachai-Cherkessia Republic: the Teberda Reserve characterised above and a piece of the Caucasian Strict Nature Reserve in the southwest. The proposed Arkhyz National Park will be the third key element of the network at the republican level. Together with numerous corridors and buffers in between they will create the crucial knot in the future Caucasus-wide network ensuring necessary protection for both biodiversity and landscapes.

REFERENCES

Aarssen, L.W. & Epp, G.A. 1990. Neighbour manipulations in natural vegetation: a review. Journal of Vegetation Science 1: 13-30.

Abaturov, B.D. 1984. Mammals as a component of ecosystems [Mlekopitayushchie kak komponent ekosistem]. Nauka, Moscow (in Russian).

Abdaladze, O.G. & Voznesensky, V.L. 1984. On photosynthesis in high-altitude plants of the Central Caucasus. Izvestiya AS GSSR, Biology series 10(5): 313-318. (in Russian).

Abdul-Fatih, H.A. & Bazzaz, F.A. 1979. The biology of *Ambrosia trifida* L. I. Influence of species removal on the organization of the plant community. New Phytologist, 83: 813-816.

Abdurakhmanov, M.G. 1977. Caucasian ibex. pp. 186-200. In: Dormidontov, R.V. et al., Kopytnye zveri. Lesnaya promyshlennost', Moskow (in Russian).

Abelentsev, B.I. 1973. Stone marten. pp. 202-213. In: Nasimovich, A.A. (ed), Sobol', kunitsy, kharza. Nauka, Moscow (in Russian).

Abramovskaya Land Commission 1908 [Trudy Abramovskoi zemleustroitel'noi komissii]. Vladikavkaz (in Russian).

Abrosov, N.S., Kovrov, B.G. & Cherepanov, O.A. 1982. Ecological mechanisms of coexistence and species regulation [Ekologicheskie mekhanizmy sosushchestvovaniya i vidovoi regulyatsii]. Nauka, Novosibirsk (in Russian).

Acosta, A., Diaz, S. & Cabido, M. 1991. Patch structure in natural grasslands of Cordoba mountains (Argentina) in relation to different rock substrates. Coenoses 6: 21-27.

Agakhanyants, O.E. 1986. Botanical geography of the USSR. Vysheishaya shkola, Minsk (in Russian).

Albov, N.M. 1896. Essay of Colchic vegetation. Zemlevedenie 1: 1-78 (in Russian).

Alekhin, V.V. 1936. Steppes. pp. 505-559. In: Walter, H. & Alekhin , V.V. Principles of botanical geography [Osnovy botanicheskoi geogragfii]. Biomedgiz, Moscow, Leningrad (in Russian).

Alexander, C. & Alexander, I.J. 1984. Seasonal changes in population of the orchid *Goodyera repens* Br. and its mycorrhizal development. Trans. Bot. Soc. Edinburgh 44: 219-227.

Allen, E.B. & Forman, R.T.T. 1976. Plant species removals and old-field community structure and stability. Ecology 57: 1233-1243.

Aksenova, A.A., Onipchenko, V.G. & Blinnikov, M.S. 1998. Plant interactions in alpine tundra: 13 years of experimental removal of dominant species. Ecoscience 5: 258-270.

Andreyashkina, N.I. 1981. Store and regrowth of aboveground phytomass in tundra plant communities, the Transural Mountains. pp. 3-11. In: Gorchakovskii, P.L. & Danilov, N.N. (ed.) Structure and function of ecosystems in Priob' North [Struktura i funktsionirovanie biogeotsenozov Priobskogo Severa]. Sverdlovsk.

Angers, D.A. & Caron, J. 1998. Plant-induced changes in soil structure: processes and feedbacks. Biogeochemistry 42: 55-72.

Anthipov-Karataev, I.N., Anthipova-Karataeva, T.F. & Simakova, L.G. 1936. Mountain meadow and mountain forest soils of the Teberda Reserve at the Northern Caucasus. In: Proceed. of the Dokuchaev` Soil Science Institute. [Trudy pochvennogo instituta im.V.V.Dokuchaeva] Izdatelstvo AN SSSR, Moscow.-Leningrad. 13: 367-398. (in Russian).

Archibold, O.W. 1984. A comparison of seed reserves in arctic, subarctic and alpine soils. Canadian Field Naturalist 98: 337-344.

Arft, M., Walker, M.D., Gurevitch, J., Alatalo J.M., Bret-Harte, M.S., Dale, M., Diemer, M., Gugerli, F., Henry, G.H.R., Jones, M.H., Hollister, R.D., Jonsdottir, I.S., Laine, K., Levesque, E., Marion, G.M., Molau, U., Molgaard, P., Nordenhall, U., Raszhivin, V., Robinson, C.H., Starr, G., Stenstrom, A., Stenstrom, M., Totland, O., Turner, P.L., Walker, L.J., Webber, P.J., Welker, J.M. & Wookey, P.A. 1999. Responses of tundra plants to experimental warming: meta-analysis of the international tundra experiment. Ecological Monographs 69: 491-511.

Arrhenius, O. 1921. Species and area. Journal of Ecology 9: 95-99.

Atkin, O.K. & Collier, D.E. 1992. Relationship between soil nitrogen and floristic variation in late snow areas of the Kosciusko Alpine region. Australian Journal of Botany 40: 139-149.

Bab'eva, E.N. & Sizova, T.P. 1983. Micromycetes in Soils of Arctotundra Ecosystem. Pochvovedenie (10): 98-101.

Baikalova, A.S. & Onipchenko, V.G. 1988. Mycosymbiotrophism of alpine species in the Teberda Reserve. pp. 93-107. In Onipchenko, V.G. & Petelin, D.A. (eds.) Opyt issledovaniya rastitel'nikh soobshestv v zapovednikakh. CNIL Glavokhoty RSFSR, Moscow (in Russian).

Bakeev, N.N. 1973. Stone Marten. Caucasus. pp. 213-219. In: Nasimovich, A.A. (ed), Sobol', kunitsy, kharza. Nauka, Moscow (in Russian).

Baloyan, S.A. 1987. Seasonal dynamics of alpine plant communities in the Mt Aragats. VINITI 3639-B87, Moscow (in Russian).

Barsegyan, A.M., Ziroyan, A.N. & Afrikyan, K.G. 1987. Bioproductivity of main plant communities in Sevan basin. Sbornik nauchnykh trudov Armyanskogo otdel. Vsesoyuz. botan. obshchestva 10: 143-146 (in Russian).

Baziev, J.H. 1962. Ermine from the Great Caucasus. Zoologicheskii Zhurnal 41: 121-124 (in Russian).

Bazilevich, N.I., Davydova, M.V. & Yashina, A.V. 1987. Plant community production of subnival, alpine, subalpine and after-forest meadows in the Caucasus. pp. 50-69. In: Kotlyakov, V.M. & Yashina, A.V. (eds), Transformation of mountain ecosystem of the Great Caucasus under antropogenic activity [Transformatsiya gornykh ekosystem Biol'shogo Kavkaza pod vliyaniem khozyaistvennoi deyatel'nosti]. Inst. Geography AN SSSR, Moscow (in Russian).

Bell, K. 1974. Autumn, winter and spring phenology of some Colorado alpine plants. Amer. Midland Natur. 91: 460-464.

Bell, K.L. & Bliss, L.C. 1979. Autecology of *Kobresia bellardii*: why winter snow accumulation limits local distribution. Ecological Monographs 49: 377-402.

Berendse, F. 1998. Effects of dominant plant species on soils during succession in nutrient-poor ecosystems. Biogeochemistry 42: 73-88.
Berezina, N.A. & Tyuremnov, S.N. 1969. Use of spore-pollen analysis of soils in palaeoecological research. Vestnik MGU seriya 16 Biologia 3: 65-70 (in Russian).
Bieberstein, M.F.A. 1808-1819. Flora taurico-caucasica, exhibens stirpes phanerogamas in Chersoneso taurica et regionibus Caucasicis sponte crescentes. T.1-3. Charkoviae.
Bigwood, D.W. & Inouye, D.W. 1988. Spatial pattern analysis of seed banks: an improved method and optimized sampling. Ecology 69: 497-507.
Billings, W.D. 1973. Arctic and alpine vegetation: similarities, differences, and susceptibility to disturbance. BioScience 23: 697-704.
Billings, W.D. 1974. Adaptations and origins of alpine plants. Arctic and Alpine Research 6: 129-142.
Billings, W.D. & Bliss, L.C. 1959. An alpine snowbank environment and its effects on vegetation, plant development and productivity. Ecology 40: 388-397.
Billings, W.D. & Mooney, H.A. 1968. The ecology of arctic and alpine plants. Biol. Rev. 43: 481-529.
Bissett, J. 1983. Notes on *Tolypocladium* and related genera. Canadian Journal of Botany 61: 1311-1329.
Bissett, J. & Parkinson, D. 1979a. The distribution of fungi in some alpine soils. Canadian Journal of Botany 57: 1609-1629.
Bissett, J. & Parkinson, D. 1979b. Fungal community structure in some alpine soils. Canadian Journal of Botany 57: 1630-1641.
Bissett, J. & Parkinson, D. 1979c. Functional relationships between soil fungi and environment in alpine tundra. Canadian Journal of Botany 57: 1642-1659.
Blaschke, H. 1991. Distribution, mycorrhizal infection, and structure of roots of calcicole floral elements at treeline, Bavarian Alps, Germany. Arctic and Alpine Research 23: 444-450.
Blinkova, O.V. & Onipchenko, V.G. 2001. Analysis of interannual dynamics in an alpine lichen heath, Teberda Reserve. Byull. Mosk. o-va ispyt. prir. otd. biol. 106(5): 66-73 (in Russian).
Blinnikov, M.S. 1994. Phytolith analysis and Holocene dynamics of alpine vegetation. Veröff. des Geobot. Inst. der ETH, Stiftung Rübel. Zürich 115: 23-40.
Blinnikov, M., Busacca, A. & Whitlock, C. 2002. Reconstruction of the late Pleistocene Columbia Basin grassland, Washington, USA, based on phytolith records in loess. Palaeogeog., Palaeoclim., Palaeoecol. 177: 77-101.
Bliss, L.C. 1971. Arctic and alpine plant life cycles. Annual Review of Ecology and Systematics 2: 405-438.
Bliss, L.C. 1985. Alpine. pp. 41-65. In: Chabot, B.F. & Mooney, H.A.(eds.). Physiological ecology of North American plant communities. Chapman & Hall, New York & London.
Bobyr', G.Y. 1977. Habitat distribution and population density of brown bear in the Teberda Reserve. pp. 27-28. In: Pokrovskii, A.V. (ed), Ekologiya, methody izucheniya i organizatsiya okhrany mlekopitayushchikh gornykh oblastei. Sverdlovsk (in Russian).
Bobyr', G.Y. 1979. Meaning of snow avalanches for brown bear in the Northwestern Caucasus. pp. 211-212. In: Sokolov, V.E. Ekologicheskie osnovy okhrany i ratsional'nogo ispol'zovaniya khishchnykh mlekopitayushchikh. Nauka, Moscow (in Russian).
Bobyr', G.Y. 1985. Winter ecology of the brown bear in the Northwestern Caucasus. pp. 43-56. In: N.N.Polivanova (ed), Ecological investigations in the Northwestern Caucasus [Ekologicheskie issledovaniya na severo-zapadnom Kavkaze]. CNIL Glavokhoty RSFSR, Moscow (in Russian).

Bobyr', G.Y. 1986. On abundance of brown bear and tendency of its change in the Northwest Caucasus. 4th Congress of VTO (all-Union mammalogists society), Moscow, 2: 226-227 (in Russian).
Bobyr', G.Y. 1987. On the problem of brown bear influence on the vegetation in the Teberda Reserve. pp. 26-28. In: Bolshakov, V.N. (ed), Ekologiya i okhrana gornykh vidov mlekopitayushchikh (Ecology and protection of mountain species of mammals). M. UMS (in Russian).
Bobyr', G.Y. 1997. On the problem of investigation of wolf influence on the Western-Caucasian goat population in the Teberda Reserve. pp. 28-37. In: Khokhlov, A.N. (ed), Nauchnoe nasledie N.Ya.Dinnika i ego rol' v razvitii sovremennogo estestvoznaniya. Stavropol' (in Russian).
Bobyr', G.Y. & Kudaktin, A.N. 1991. Brown bear in the Northern Caucasus. pp. 34-40. In: Pazhetnov, V.S. Medvedi SSSR – sostoyanie populyatsii. Rzhev (in Russian).
Bobyr', K.G., Onipchenko, V.G. & Bobyr', G.Y. 1999. Ibex (*Capra caucasica*) winter grazing on an alpine grassland in the Teberda Reserve, the Northwest Caucasus, Russia. pp. 161-165. In: Pavlov, V.N. (ed.), High-mountain ecosystems of the Teberda Reserve. Transactions of Teberda State Reserve [Trudy Teberdinskogo Gos. Zapovednika.]. Moscow, 15 (in Russian).
Borisova, I.V. 1972. Seasonal dynamics of a plant community. Polevaya Geobotanika 5: 5-94 (in Russian).
Bormann, F.H. & Likens, G.E. 1979. Pattern and process in a forest ecosystem. Springer, New York.
Bouma, J. & van der Plas, L. 1971. Genesis and morphology of some Alpine Pseudogley Profiles. Journal of Soil Science 22: 81-93.
Bouma, J., Hoeks, J., van der Plas, L. & van Scherrenburg, B. 1969: Genesis and morphology of some Alpine Podzol Profiles. Journal of Soil Science 20: 384-398.
Bowman, W.D. 1992. Inputs and storage of nitrogen in winter snowpack in an alpine ecosystem. Arctic and Alpine Research 24: 211-215.
Bowman, W.D., Schardt, J.C. & Schmidt, S.K. 1996. Symbiotic N_2-fixation in alpine tundra: ecosystem input and variation in fixation rates among communities. Oecologia 108: 345-350.
Bozarth, S.R. 1992. Classification of opal phytoliths formed in selected dicotyledons native to the Great Plains. pp. 193-214. In: Rapp, G, Jr. & Mulholland, S. C. (eds), Phytolith Systematics: Emerging Issues. Advances in Archaeological and Museum Science
Braun-Blanquet, J. 1975. Die Schneebodengesellschaften (Klasse der Salicetea herbaceae). Fragmenta Phytosociologica Raetica 1. 195/196: 42-71.
Brown, D.A. 1984. Prospects and limits of a phytolith key for grasses in the central United States. Journal of Archaeological Sciences 11: 345-368.
Brown, R.W. & Johnston, R.S. 1979. Revegetation of disturbed alpine rangelands. pp. 76-94. In: Johnson, D.A. (ed.). Special management needs of alpine ecosystems. Denver.
Brown, R.W., Johnston, R.S. & Johnson, D.A. 1978. Rehabilitation of alpine tundra disturbances. Journal of Soil and Water Conservation 33: 154-160.
Brzoska, W. 1973. Dry matter production and energy utilization of high mountain plants in the Austrian Alps. Oecologia Plantarum 8: 63-70.
Bulatnikova, I.V. & Makarov, M.I. 2000. Primitive soils on rocks of Teberda Reserve. Moscow University Soil Sci. Bull. [Vestnik Movskovskogo Universiteta. Pochvovedenie] (4): 10-15 (in Russian).
Burdon, J.J. 1993. The structure of pathogen populations in natural plant communities. Annual Review of Phytopathology 31: 305-323.

Busch, E.A. 1909. Data for the flora of the Karachai and Teberda valley. Trudy Tiflis. botan. sada 11(1): 1-140 (in Russian).
Busch, E.A. & Busch, N.A. 1909. Plantae caucasicae novae vel minus cognitae. Trudy Botan. sada Yur'ev Univ. 10(1): 1-7.
Busch, N.A. 1897. Preliminary report of the trip to the Northwest Caucasus in 1896 with aim to investigate glaciers and flora. Izvestiya Russkogo geogr. o-va 33: 1-33 (in Russian).
Busch, N.A. 1898. Preliminary report of the second trip to the Northwest Caucasus in 1897. with aim to investigate glaciers and flora. Izvestiya Russkogo geogr. o-va 34: 519-589 (in Russian).
Busch, N.A. 1935. Botaniko-geograficheskii ocherk Evropeiskoi chasti SSSR i Kavkaza. Izd-va AN SSSR, Moscow, Leningrad (in Russian).
Buslik, V.A. 1990. Ecological and floristic classification of subalpine meadows in the Baksan valley. Trans. High Mountain Geophys. Inst.[Trudy vysokogornogo geophizicheskogo instituta] 79: 116-142 (in Russian).
Callaghan, T.V. 1987. Plant population processes in arctic and boreal regions. pp. 58-68. In: Sonneson M. (ed), Research in Arctic Life and Earth Sciences: Present Knowledge and Future Perspectives. Ecological Bulletins, 38.
Callaghan, T.V. & Emanuelsson, U. 1985. Population structure and processes of tundra plants and vegetation. pp. 399-439. In: White, J. (ed), The Population Structure of Vegetation. Junk, Dordtrecht e.a.
Callaway, R.M. 1995. Positive interactions among plants. Botanical Review 61: 306-349.
Carlsson, B.A. & Callaghan, T.V. 1991. Positive plant interactions in tundra vegetation and the importance of shelter. Journal of Ecology 79: 973-983.
Carr, D.J., Carr, S.G.M. & Papst, W.R. 1980. Field studies of nitrogen fixation of Australian alpine plants and soils. Australian Journal of Ecology 5: 211-220.
Carreiro, M.M. & Koske, R.E. 1992. Effect of temperature on decomposition and development of microfungal communities in leaf litter microcosms. Canadian Journal of Botany 70: 2177-2183.
Caswell, H. 1982a. Optimal life histories and the maximisation of reproductive value: a general theorem for complex life cycles. Ecology 63: 1218-1222.
Caswell, H. 1982b. Stable population structure and reproductive value for populations with complex life cycles. Ecology 63: 1223-1231.
Caswell, H. 1986. Life cycle models for plants. Lectures on mathematics in the life sciences 18: 171-233.
Chambers, J.C. 1993. Seed and vegetation dynamics in an alpine herb field: effects of disturbance type. Canadian Journal of Botany 71: 471-485.
Chapin, F.S.III. 1980. The mineral nutrition of wild plants. Annual Review of Ecology and Systematics 11: 233-260.
Chapin, F.S.III & Korner, C. 1994. Arctic and alpine biodiversity: patterns, causes and ecosystem consequences. Trends in Ecology and Evolution 9: 45-47.
Chapin, F.S. & Shaver, G.R. 1985. Individualistic growth responses of tundra plant species to environmental manipulations in the field. Ecology 66: 564-576.
Chapin, F.S. & Shaver, G.R. 1996. Physiological and growth responses of arctic plants to a field experiment simulating climate change. Ecology 77: 822-841.
Chapin, F.S.III, Shaver, G.R., Giblin, A.E., Nadeloffer, K.J. & Laundre, J.A. 1995. Responses of arctic tundra to experimental and observed changes in climate. Ecology 76: 694-711.
Cherednichenko, O.V. & Yurtseva, O.V. 1999. On potential ability of taproot-rhizomatous plants to vegetative propagation. pp. 106-117. In: Pavlov, V.N. (ed.), High-mountain ecosystems of the Teberda Reserve. Transactions of Teberda State Reserve [Trudy Teberdinskogo Gos. Zapovednika.]. Moscow, 15 (in Russian).

Chernyakhovskii, M.E. 1968. Types of feeding and mandibula structure of different life forms in Acridoidea. Zoologicheskii zhurnal 47: 238-248 (in Russian).

Chernyakhovskii, M.E. 1970. Morphological and functional pecularities of grasshopper life forms. Uchenye zapiski MGPI 394: 47-63 (in Russian).

Chiang, C.L. 1984. The life table and its applications. Robert E. Krieger Publishing Co., Malabar, Florida.

Christensen, M. 1981. Species diversity and dominance in fungal communities. pp. 201-232. In: Wicklow, D.T. & Carroll, G.C. (eds), The fungal community: its organization and role in the ecosystem. Marcel Dekker, Inc., New York, Basel.

Christensen, M. 1989. A view on fungal ecology. Mycologia 81: 1-19.

Chuyo, H. 1985. Patternd grounds and vegetation in the alpine area of Mt. Ontake, central Japan. J. Sci. Hirosima Univ., Ser. 2B, 19(2): 171-236.

Clements, F. E. & Weaver, J. E. 1924. Experimental vegetation. Carnegie Institute, Washington, DC.

Connell, J.H. 1990. Apparent versus "real" competition in plants. pp. 9-26. In: Grace, J.B. & Tilman, D. (eds), Perspectives on plant competition. Academic Press, San Diego e.a.

Cooke, W.B. & Lawrence, D.B. 1959. Soil mould fungi isolated from recently glaciated soils in South-eastern Alaska. Journal of Ecology 47: 520-549.

Costin, A.B. 1955. Alpine soils in Australia with reference to conditions in Europe and New Zealand. Journal of Soil Science 6: 35-50.

Costin, A.B., Hallsworth, E.G. & Wood, M. 1952. Studies in pedogenesis in New South Wales. III. The Alpine Humus Soils. Journal of Soil Science 3: 190-218.

Cox, D. R. & Oakes, D. 1984. Analysis of survival data. Chapman and Hall, London.

Cuendet, G. 1984. Les peuplements lombriciens des pelouses alpines du Mont La Schera (Parc national suisse). Rev. suisse zool. 91: 217-228.

Cui, M. & Caldwell, M. M. 1997. Shading reduces exploitation of soil nitrate and phosphate by *Agropyron desertorum* and *Artemisia tridentata* from soils with patchy and uniform nutrient distributions. Oecologia 109: 177-183.

Czerepanov, S.K. 1995. Vascular plants of Russia and adjacent states (the former USSR). Cambridge Univ. Press, Cambridge.

Dal Vesco, G. 1974. Fungi del suolo di un pianoro acquitrinoso in valle de Cogne (Aosta). Allionia 20: 81-92.

Danilov, P.I. & Tumanov, I.L. 1976. The Mustellids of Northwest of the USSR [Kun'i severo-zapada SSSR]. Nauka, Leningrad (in Russian).

Davis, P.H. (ed.) 1965-1980. Flora of Turkey and the East Aegean Islands. Edinburgh Univ.Press, Edinburgh, V. 1-10.

de Graaf, M.C.C., Bobbink, R., Roelofs, J.G.M. & Verbeek, P.J.M. 1998. Differential effects of ammonium and nitrate on three heathland species. Plant Ecology 135: 185-196.

de Ruiter, G.A., van Bruggen-van der Lugt, A.W., Rombouts, F.M. & Gams, W. 1993. Approaches to the classification of the *Mortierella isabellina* group: antigenic extracellular polysaccharides. Mycological Research 97: 690-696.

del Moral, R. 1983a. Competition as a control mechanism in subalpine meadows. American Journal of Botany 70: 232-245.

del Moral, R. 1983b. Vegetation ordination of subalpine meadows using adaptive strategies. Canadian Journal of Botany 61: 3117-3127.

del Moral, R. 1985. Competitive effects of the structure of subalpine meadow communities. - Canadian Journal of Botany 63: 1444-1452.

del Moral, R., Clampitt, Ch. A. & Wood, D.M. 1985. Does interference cause niche differentiation? Evidence from subalpine plant communities. American Journal of Botany 72: 1891-1901.

Demin, A.P. 1970. Underground phytomass of meadow vegetation in flood-lands of Oka and effects of fertilization. Byull. Mosk. o-va ispyt. prir. otd. biol. 75(6): 79-85 (in Russian).
Dethier, M. 1985. Distribution des larves d'Insectes Pterygotes dans une pelouse alpine. Bull. Soc. Linneenne de Lyon 54(3): 64-76.
Diemer, M. & Korner, C. 1998. Transient enhancement of carbon uptake in an alpine grassland ecosystem under elevated CO2. Arctic and Alpine Research 30: 381-387.
Diemer, M., Korner, C. & Prock, S. 1992. Leaf life spans in wild perennial herbaceous plants: a survey and attempts at a functional interpretation. Oecologia 89: 10-16.
Diemer, M. & Prock, S. 1993. Estimates of alpine seed bank size in two Central European and one Scandinavian subarctic plant communities. Arctic and Alpine Research 25: 194-200.
Dierssen, K. 1984. Vergleichende vegetationskundliche Untersuchungen an Schneeboden. Ber. Deutsch Bot. Ges. 97: 359-382.
Dietl, W. 1986. Pflanzenbestand und Ertrag von subalpinen Wiesen in Hinterrheingebiet, Graubunden Veröff. des Geobot. Inst. der ETH, Stiftung Rübel. Zürich 87: 260-274.
Dinesman, L.G. (ed.) 1992. Historical ecology of wild and domestic ungulates: History of pasture ecosystems. [Istoricheskaya ekologiay dikikh i domashnikh kopytnykh]. Nauka, Moscow (in Russian).
Dmitriev, P.P. 1988. Change of soil profile under activity of borrowing mammals. Pochvovedenie (11): 75-81 (in Russian).
Dobrinskii, L.N., Davydov, V.A., Kryazhimskii, F.V. & Malafeev, Y.M. 1983. Functional relationships between small mammals and vegetation in meadow communities [Funktsional'nye svyazi melkikh mlekopitayushchikh s rastitel'nost'yu v lugovykh biogeotsenozakh]. Nauka, Moscow (in Russian).
Dobrolyubov, A.N. 1985. Problems of ecology of the Mustellids in the Northwest Caucasus. pp. 57-73. In: N.N. Polivanova (ed), Ecological investigations in the Northwestern Caucasus [Ekologicheskie issledovaniya na severo-zapadnom Kavkaze]. CNIL Glavokhoty RSFSR, Moscow (in Russian).
Dobrolyubov, A.N. 1986. Marten family in mountain-meadow communities (population density, biotopic preference and feed activity). pp. 67-73. In: Grishina, L.A. et al., Sostav i struktura biogeocenozov alpijskih pustoshei. Izd-vo Moskovskogo Universiteta, Moscow (in Russian).
Dobrolyubov, A.N. 1988. Ecology of ermine in North-Western Caucasus. pp. 68-70. In: Khokhlov, A.N. (ed), Resursy zhivotnogo mira Severnogo Kavkaza. VOOP, Stavropol. (in Russian).
Dobrolyubov, A.N. 1994. Ermine in the Caucasus. Okhota i okhotnich'e khozyaistvo (12): 16-17 (in Russian).
Dolukhanov, A.G. 1966. Plant cover of the Caucasus. pp. 223-255. In: Gerasimov, I.P. et al. (eds), Kavkaz. Nauka, Moscow (in Russian).
Domsch, K.H., Gams, W. & Anderson, T.-H. 1980. Compendium of soil fungi. Vol. 1-2. Academic Press, London.
Dowding, P. & Widden, P. 1974. Some relationships between fungi and their environment in tundra regions. pp. 123-150. In: Holdings, A.J., Heal, O.W., MacLean, S.F. & Flanagan, P.W. (eds), Soil organisms and decomposition in tundra. Tundra Biome Steeting Comitee. Stockholm.
Dreyfuss, M.M. & Chapela, I.H. 1994. Potential of fungi in the discovery of novel low-molecular weight pharmaceuticals. pp. 49-80. In: Gullo, V.P. (ed), The discovery of natural products with therapeutic potential. Butterworth-Heinemann, Boston.
Dreyfuss, M., Härri, E., Hofmann, H., Kobel, H., Pache, W. & Tscherter, H. 1976. Cyclosporin A and C, new metabolites from *Trichoderma polysporum* (Link ex Pers.) Rifai. European Journal of Applied Micriobiology 3: 125-133.

Duben', V.V. 1985. Structure and density of chamois populations in relation with some ecological factors. pp. 31-49. In: Kurazhkovskii, Y.N. (ed), Ekologicheskie issledovaniya v Kavkazskom biosfernom zapovednike. Izd-vo Rostov. Universiteta, Rostov na Donu.

Durov, V.V. 1987. Wild boar in the West Caucasus [Kaban zapadnogo Kavkaza]. Dissert. kand.biol. nauk. Moscow (in Russian).

Dzybov, D.S. 1979. On bioproductivity of some microassociations of high mountain meadows in Karacaevo-Cherkessia. Problemy botaniki 14(1): 147-151 (in Russian).

Ebersole, J.J. 1989. Role of seed bank in providing colonizers on a tundra disturbance in Alaska. Canadian Journal of Botany 67: 466-471.

Eek, L. & Zobel, K. 1997. Effect of additional illumination and fertilization on seasonal changes in fine-scale grassland community structure. Journal of Vegetation Science 8: 225-234.

Egerton, J.J.G. & Wilson, S.D. 1995. Plant competition over winter in alpine shrubland and grassland, Snowy Mountains, Australia. Arctic and Alpine Research 25: 124-129.

Egorova, L. N. 1986. Soil fungi of Far East: Hyphomycetes [Pochvennye griby Dal'nego Vostoka: gifomitsety]. Nauka, Leningrad (in Russian).

Elberse, W.T. & Berendse, F. 1993. A comparative study of the growth and morphology of eight grass species from habitats with different nutrient availabilities. Functional Ecology 7: 223-229.

Elenevskii, R.A. 1940. High mountain landscape profile of the north slope in the West Caucasus. Byull. Mosk. o-va ispyt. prir. otd. biol. 49 (2): 79-89. (in Russian).

Ellenberg, H. 1986. Vegetation Mitteleuropas mit Alpen in ökologischer Sicht. 4 Aufl. Ulmer, Stuttgart.

Ellmauer, Th. 1993. *Calluno-Ulicetea*. S. 402-419. In: Mucina, L. et al. (eds.) Die Pflanzengesellschaften Osterreichs. Gustav Fischer, Jena e.a., Teil 1.

Englisch, T. 1993. *Salicetea herbaceae*. S. 382-401. In: Grabherr, G. & Mucina, L. (eds.) Die Pflanzengesellschaften Osterreichs. Gustav Fischer, Jena e.a., Teil 2.

Epp, G.A. & Aarsen, L.W. 1989. Predicting vegetation patterns from attributes of plant growth in grassland species. Canadian Journal of Botany 67: 2953-2959.

Erschbamer, B. 1996. Growth dynamics and nutrient contents of the alpine sedge, *Carex curvula* subsp. *rosea*, on different soils. Flora (Jena) 191: 121-129.

Erschbamer, B., Buratti, U. & Winkler, J. 1998. Long-term population dynamics of two *Carex curvula* species in the Central Alps on native and alien soils. Oecologia 115: 114-119.

Erschbamer, B. & Winkler, J. 1995. Shoot and leaf demography of *Carex curvula* ssp. *curvula* and *C. curvula* ssp. *rosae* in the Central Alps. Jornal of Vegetation Science 6: 593-598.

Evans, G.R. 1980. Phytomass, litter, and nutrients in montane and alpine grasslands, Craigieburn Range, New Zealand. Techn. Pap. Forest Res. Inst. N.Z. Forest Serv. 70: 95-109.

Facelli, J.M. & Pickett, S.T.A. 1991. Plant litter: its dynamics and effects on plant community structure. Botanical Revew 57: 1-32.

Faegri, K., Kaland, P.E. & Kzywinski, K. 1989. Textbook of pollen analysis. Wiley. New York.

Feher, D. 1933. Untersuchungen über die Mikrobiologie des Waldbodens. Springer, Berlin.

Fenner, M. (ed) 1985. Seed Ecology. Chapman & Hall, London.

Fileva, O.N. 1983. Investigation of the mountain tundra invertebrates communities in the North Urals. pp. 55-56. In: Bogacheva, I.A. (ed) Fauna i ekologiya nasekomykh Urala. Sverdlovsk (in Russian).

Finlay, R.D. Interactions between soil micro-arthropods and endomycorrhizal associations of higher plants. pp. 319-331. In: Fitter, A.H et al. (eds), Ecological interactions in soil. Blackwell, Oxford e.a.

Finlay, R. & Söderström, B. 1992. Mycorrhiza and carbon flow to the soil. pp. 134-160. In: Allen M.J. (ed.), Mycorrhizal functioning. Chapman & Hall, New York.

Fitter, A.H. & Nichols, R. 1988. The use of benomyl to control infection by vesicular-arbuscular mycorrhizal fungi. New Phytologist 110: 201-206.

Fleming, T.R. & Harrington, D.P. 1981. A class of hypothesis tests for one and two sample censored survival data. Communications in Statistics A10(8): 763-794.

Fomin, S.V., Onipchenko, V.G. & Sennov, A.V. 1989. Feeding and digging activities of the pine vole (*Pitymys majori* Thos.) in alpine communities of the Northwestern Caucasus. Byull. Mosk. o-va ispyt. prir. otd. biol. 94(3): 6-13. (in Russian).

Forcella, F. 1984. A species-area curve for buried viable seeds. Australian Journal of Agricultural Research 35: 645-652.

Fowler, N.L. 1981. Competition and coexistence in a North Carolina grassland. II. The effects of the experimental removal of species. Journal of Ecology 69: 843-854.

Fowler, N.L. 1986. The role of competition in plant communites in arid and semiarid regions. Annual Review of Ecology and Systematics 17: 89-110.

Francis, R. & Read, D.J. 1994. The contributions of mycorrhizal fungi to the determination of plant community structure. Plant & Soil 159: 11-25.

Frank, D.A. & del Moral, R. 1986. Thirty-five years of secondary succession in a *Festuca viridula - Lupinus latifolius* dominated meadow at Sunrise, Mount Rainier National Park, Washington. Canadian Journal of Botany 64: 1232-1236.

Frankland, J.C., Dighton, J. & Boddy, L. 1990. Methods for studying fungi in soil and forest litter. Methods in Microbiology 22: 343-404.

Franz, G. 1975. Beitrag zur Verbreitung der mikroskopischen Bodenpilze in Nepal. Nova Hedwigia 26: 105-123.

Franz, G. 1981. Vergleich der mikroskopischen Bodenpilzflora des Grossglocknergebietes mit derjenigen des Himalaya. Österreichische Akademie der Wissenschaften 4: 295-300.

Franz, H. 1979. Ökologie der Hochgebirge. Ulmer, Stuttgart.

Franz, H.(ed) 1981: Bodenbiologische Untersuchungen in den Hohen Tauern 1974-1978.- Veröff. des Österreich. MaBHochgebiergsprogramms Hohe Tauern. Innsbruck, 4: 1-300.

Fredlund, G.G. & Tieszen, L.T. 1994. Modern phytolith assemblages from the North American Great Plains. J. of Biogeog. 21: 321-335.

Gadzhiev, V.D. 1979. Pecularities of demutation dymamics in the Caucasian high mountains. Problemy botaniki 14(1): 178-182 (in Russian).

Galushko, A.I. 1978-1980. Flora of the North Caucasus [Flora Severnogo Kavkaza]. Izd-vo Rostov Univ., Rostov-na-Donu, v. 1-3. (in Russian).

Gams, W. 1959. Die Bodenpilze im zentralalpinen Rohhumus. Dissertation zur Erlangung des philosophischen Doktorgrades. Leopold-Franzens-Universität Innsbruck. Innsbruck.

Gams, W. 1960. Studium zellulolytischer Bodenpilze mit Hilfe der Zellophanstreifen und mit Carboxymethyl-Zellulose. Sydowia 14: 295-307.

Gams, W. 1977. A key to the species of *Mortierella*. Persoonia 9: 381-391.

Gams, W. 1992. The analysis of communities of saprophytic microfungi with special reference to soil fungi. pp. 182-223. In: Winterhoff, W. (ed), Fungi in Vegetation Science. Kluwer, Dordrecht.

Gams, W., Hoekstra, E.S., Aptroot, A., van der Aa, H.A., Boekhout, T., de Cock, A.W.A.M.,, de Hoog, G.S., Samson, R.A., Smith, M.T., Stalpers, J.A., Tan, C.S. & Verkley, G.J.M. 1998. CBS course of mycology. Centraalbureau voor Schimmelcultures, Baarn, Delft.

Gamtsemlidze, Z.G. 1979. On winter buds of sunbnival plants in Kazbegi. Izvestiya Akad. Nauk GruzSSR, ser. biol. 5: 342-350 (in Russian).
Gardes, M. & Dahlberg, A. 1996. Mycorrhizal diversity in arctic and alpine tundra: an open question. New Phytologist 133: 147-157.
Gatsuk, L.E. Smirnova, O.V., Vorontsova, L.I., Zaugolnova, L.B. & Zhukova, L.A. 1980. Age stages of plants of various growth forms: a review. Journal of Ecology 68: 675-696.
Gauch, H.G. Jr. 1982. Multivariate analysis in community ecology. Cambridge University Press, Cambridge.
Gerasimov, I.P., Bravard, Y. & Dresch, J. (eds) 1980. Alps – Caucasus: modern problems of constructive geography of mountain countries. Nauka, Moscow (in Russian).
Gigon, A. 1971. Vergleich alpiner Rasen auf Silikat- und Karbonatboden. Veröff. des Geobot. Inst. der ETH, Stiftung Rübel. Zürich. 38: 1-159.
Gigon, A. 1987. A hierarchic approach in causal ecosystem analysis: the calcifuge-calcicole problem in alpine grasslands. pp. 228-244. In: Schulze, E.-D. & Zwölfer, H. (eds), Potentials and limitations of ecosystem analysis. Springer, Berlin.
Gigon, A. 1994. Positive Interaktionen bei Pflanzen in Trespen-Halbtrockenrasen Halbtrockenrasen bei Schaffhausen. Verhandlungen der Gesellschaft für Ökologie 23: 1-6.
Gigon, A. 1999. Positive Interaktionen in einem alpinen Blumenpolster. Ber. d. Reinh.-Tüxen-Ges. 11: 321-330.
Gigon, A. & Ryser, P. 1986. Positiv Interaktionen zwischen Pflanzenarten. I.Definition und Beispiele aus Gruenlandoekosystemen. Veröff. des Geobot. Inst. der ETH, Stiftung Rübel. Zürich. 87: 372-387.
Gilyarov, M.S. 1975. Investigations of large soil invertebrates (mesofauna). pp. 12-29. In: Gilyarov, M.S. (ed.) Metody pochvenno-zoologicheskikh issledovanii. Nauka, Moscow (in Russian).
Gilyarov, M.S. & Chernov, Y.I. 1975. Soil invertebrates in communities of the temporal zone. pp. 218-240. In: Rodin, L.E. & Smirnov, N.N. (eds), Resursy biosfery, 1. Nauka, Leningrad, (in Russian).
Gleason, H.A. 1922. On the relation between species and area. Ecology 3: 158-162.
Goldberg, D. & Novoplatsky, A. 1997. On the relative importance of competition in unproductive environments. Journal of Ecology 85: 409-419.
Golikov, K.A. & Onipchenko, V.G. 1999. Influence of some meteorological factors on the generative shoot's dynamics in an alpine lichen heath. pp. 64-72. In: Pavlov, V.N. (ed.), High-mountain ecosystems of the Teberda Reserve. Transactions of Teberda State Reserve [Trudy Teberdinskogo Gos. Zapovednika.]. Moscow, 15. (in Russian).
Golubev, V.N. 1969. On the method of flower curves of plant communities. Byull. Mosk. o-va ispyt. prir. otd. biol. 74(2): 90-97. (in Russian).
Gorchakovskii, P.L. 1975. Flora and vegetation of high mountain zone of the Ural. [Rastitelnyi mir visocogornogo Urala] Nauka, Moscow (In Russian).
Gorchakovskii, P.L. & Bulatova, I.K. 1979. Syasonal dynamics of mountain tundra plant communities in the Ural. Ekologiya (4):16-24.
Goryshina, T. K. 1979. Ekologiya Rastenii (Plant ecology). Vysshaya Shkola, Moscow. (in Russian).
Gracanin, Z. 1972. Die Boden der Alpen. S. 172-191. In: Ganssen, R. & Gracanin, Z. Bodengeographie mit besonderer Berucksichtigung der Boden Mitteleuropas. Koehler, Stuttgart.
Grabherr, G. 1987a. Produktion und Produktionsstrategien im Krummseggenrasen (Caricetum curvulae) der Silikatalpen und ihre Bedeutung fur die Bestandesstruktur. Veröff. Osterr. Mass-Programms Osterr. Akad. Wiss. 10: 233-241.

Grabherr, G. 1987b. High alpine flora and vegetation of the Tyrolean Alps (W. Austria). 14 Int. Botan. Congress. Excursion Guide (exc. N 18). Berlin.
Grabherr, G. 1989. On community structure in high alpine grasslands. Vegetatio 83: 223-227.
Grabherr, G. 1993a. Carici rupestris-Kobresietea bellardii. S. 373-381. In: Grabherr, G. & Mucina, L. (eds.) Die Pflanzengesellschaften Osterreichs. Gustav Fischer, Jena e.a., Teil 2.
Grabherr, G. 1993b. Caricetea curvulae. S. 343-372. In: Grabherr, G. & Mucina, L. (eds.) Die Pflanzengesellschaften Osterreichs. Gustav Fischer, Jena e.a., Teil 2.
Grabherr, G. 1993c. Loiseleurio-Vaccinietea. S. 447-467. In: Grabherr, G. & Mucina, L. (eds.) Die Pflanzengesellschaften Osterreichs. Gustav Fischer, Jena e.a., Teil 2.
Grabherr, G., Gottfried, M., Gruber, A. & Pauli, H. 1995. Patterns and current changes in alpine plant diversity. pp. 167-181. In: Chapin, F.S. & Korner, C. (eds.). Arctic and alpine biodiversity: patterns, causes and ecosystem, consequences. Springer, Berlin e.a.
Grabherr, G., Mahr, E. & Reisigl, H. 1978. Nettoprimarproduktion und reproduktion in einem Krummseggenrasen (Caricetum curvulae) der Otztaler Alpen, Tirol. Oecologia Plantarum 13: 227-252.
Grabherr, G., Mair, A. & Stimpfl, H. 1987. Wachstums und Reproduktionsstrategien von Hochgebirgspflanzen und ihre Bedeutung für die Begrünung von Schipisten und anderen hochalpinen Erosionsflächen. Verh. Ges. Ökol. 15: 183-188.
Grace, J. B. 1993. The effects of habitat productivity on competition intensity. Trends in Ecology and Evolution 8: 229-230.
Graglia, E., Jonasson, S., Michelsen, A. & Schmidt, I. K. 1997. Effects of shading, nutrient application and warming on leaf growth and shoot densities of dwarf shrubs in two arctic-alpine plant communities. Ecoscience 4: 191-198.
Grant, W.E., French, N.R. & Folse L.J. 1980. Effects of pocket gopher mounds on plant production in shortgrass prairie ecosystems. Southwest. Natur. 25: 215-224.
Gregory, P.H. 1973. Microbiology of the atmosphere. 2nd ed. Leonard Hill Books, Aylesbury, Bucks.
Grichuk, V.P. 1982. Flora and vegetation. pp. 337-373. In: Shantser, E.V. (ed.) Stratigrafiya SSSR: Chetvertichnaya Sistema. Nedra. Moscow (in Russian).
Grime, J.P. 1977. Evidence for the existence of three primary strategies in plants and its relevance to ecological and evolutionary theory. American Naturalist 111: 1169-1194.
Grime, J.P. 1979. Plant strategies and vegetation processes. John Willey & Sons, New York.
Grime, J.P. 1981. Plant strategies in shade. pp. 159-186. In: Smith H. (ed.) Plants and the daylight spectrum. Acad. Press, London.
Grime, J.P. 2001. Plant strategies, vegetation processes, and ecosystem properties. 2nd Edition. John Willey & Sons Ltd, Chichester e.a.
Grime, J.P., Hodgson, J.G. & Hunt, R. 1988. Comparative plant ecology: a functional approach to common British species. Hyman, London.
Grime, J.P. & Hunt, R. 1975. Relative growth-rate: its range and adaptive significance in a local flora. Journal of Ecology 63: 393-422.
Grishina, L.A., Cherkinskii, A.E., Zhakova, O.A. & Onipchenko, V.G. 1987. Radiocarbon age of mountain meadow soils in the Northwestern Caucasus. Doklady Akad. Nauk SSSR 296: 218-220 (in Russian).
Grishina, L.A., Onipchenko, V.G., Makarov, M.I., Petrova-Nikitina, A.D., Ignat'eva, O.V., Dobrolyubova, T.V., Makarova, N.A., Baikalova, A.S. & Fomin, S.V. 1985. Results of the complex study of alpine lichen heaths in the Teberda Reserve. pp. 25-43. In: Polivanova, N.N. (ed), Ecological investigations in the Northwestern Caucasus [Ekologicheskie issledovaniya na severo-zapadnom Kavkaze]. CNIL Glavokhoty RSFSR, Moscow (in Russian).

Grishina, L.A., Onipchenko, V.G., Makarov, M.I. et al. 1986. Composition and structure of alpine heath biogeocoenosis [Sostav i struktura biogeocenozov alpijskih pustoshei]. Izd-vo Moskovskogo Universiteta, Moscow (in Russian).
Grishina, L.A., Onipchenko, V.G., Makarov, M.I. & Vanyasin, V.A. 1993. Changes in properties of mountain-meadow alpine soils of the northwestern Caucasus under different ecological conditions. Eurasian Soil Sci 25: 1-12.
Gromov, I.M. & Vorob'ev, B.L. 1955. Pine vole (*Pitymys majori* Thos.) in high mountain area, the west part of the Great Caucasian Range. Trudy Zoologicheskogo inst. AN SSSR 17: 135-159 (in Russian).
Gross, K.L. 1990. A comparison of methods for estimating seed numbers in the soil. Journal of Ecology 78: 1079-1093.
Grossheim, A.A. 1936. Analysis of the Caucasian flora. Trudy Botan. inst. Azerb. filial AN SSSR 1: 1-260 (in Russian).
Grossheim, A.A. 1948. Plant cover of the Caucasus [Rastitel'nyi pokrov Kavkaza]. Izd-vo MOIP, Moscow (in Russian).
Grossheim, A.A. 1949. Keybook for the Caucasian plants [Opredelitel rasteniy Kavkaza]. Sovetskaya nauka, Moscow (in Russian).
Grubb, P. 1977. The maintenance of species richnes in plant communities: the importance of the regeneration niche. Biol. Rev. 52: 107-145.
Grubb, P.J., Ford, M.A. & Rochefort, L. 1997. The control of relative abundance of perennials in chalk grassland: is root competition or shoot competition more important? Phytocoenologia 27: 289-309.
Gulisashvili, V.Z. 1964. Prirodnye zony i estestvenno-istoricheskie oblasti Kavkaza. Nauka, Moscow (in Russian).
Gulisashvili, V.Z., Makhatadze, L.B. & Prilipko, L.I. 1975: Vegetation of the Caucasus. [Rastitel'nost' Kavkaza] Nauka, Moscow. (in Russian).
Gurevitch, J. & Collins, S.L. 1994. Experimental manipulation of natural plant communities. Trends in Ecology and Evolution 9: 94-98.
Gurevitch, J., Morrow, L.L., Wallace, A. & Walsh, I.S. 1992. A meta-analysis of field experiments on competition. American Naturalist 140: 539-572.
Gurevitch, J. & Unnasch, R.S. 1989. Experimental removal of a dominant species at two levels of soil fertility. Canadian Journal of Botany 67: 3470-3477.
Hardy, R., Holston, R., Jackson, E. & Burns, R. 1968. The acetylene-ethylene assay for N_2-fixation: laboratory and field evaluation. Plant Physiology 43: 1185-1207.
Harley, J.L. & Harley, E.L. 1987. A check-list of mycorrhiza in the British flora. New Phytologist 105(Suppl.): 1-102.
Haselwandter, K. & Read, D.J. 1980. Fungal associations of roots of dominant and subdominant plants in high-alpine vegetation systems with special reference to mycorrhiza. Oecologia 45: 57-62.
Hatt, M. 1991. Semenvorrat von zwei alpinen Böden. Berichte des Geobotanischen Instituts ETH, Stiftung Rübel, Zürich 57: 41-71.
Heal, O.W., Bailey, A.D. & Latter, P.M. 1967. Bacteria, fungi and protozoa in Signy Island soils compared with those from a temperate moorland. Philosophical Transactions of Royal Society of London. Ser. B. 252: 191-197.
Henry, G.H.R. & Svoboda, J. 1986. Dinitrogen fixation (acetylene reduction) in high Arctic sedge meadow communities. Arctic and Alpine Research 18: 124-127.
Hester, A. J. 1987. Succesional vegetation change: the effect of shading on *Calluna vulgaris* (L.) Hull. Transactions of the Botanical Society of Edinburgh 45: 121-126.
Hils, M.H. & Vankat, J.L. 1982. Species removals from a first-year old-field plant community. Ecology 63: 705-711.

Hirsch, G. & Braun, U. 1992. Communities of parasitic microfungi. pp. 225-250. In: W.Winterhoff (ed.), Fungi in vegetation science. Kluwer Academic Publishers, Dordrecht e.a.

Hobbie, S.E. 1992. Effects of plant species on nutrient cycling. Trends in Ecology and Evolution 7: 336-339.

Hodachek, E.A. 1985. Soil seed banks of tundras (Taimyr) and polar deserts (Novaja Zemla). Botanicheskii Zhurnal 70: 896-908 (in Russian).

Hodge, K.T., Krasnoff, S.B. & Humber, R.A. 1996. *Tolypocladium inflatum* is the anamorph of *Cordyceps subsessilis*. Mycologia 88: 715-719.

Hofer, H. 1981. Der Einfluss des Massenschilaufes auf alpine Sauerbodenrasen am Beispiel der Gurgler Heide (Otztal/Tirol) und Beobachtungen zur Phanologie des Curvuletums. Ber. Naturwiss.-med. Ver. Innsbruck 68: 31-56.

Hohenacker, F.R. 1833. Enumeratio plantarum in territorio Elisabethopolensi et in provincia Karabach sponte nascentium. Bull. Soc. Natur. Moscou (6): 210-261.

Hohenacker, F.R. 1838. Enumeratio plantarum, quas in itinere per provinciam Talysch collegit. Bull. Soc. Natur. Moscou (3): 231-330, (4): 337-414.

Horak, E. 1960. Die Pilzvegetation im Gletschervorfeld (2290-2350 m) des Rotmoosferns in den Otztaler Alpen. Nova Hedvigia 2: 487-507.

Huber, R. 1994. Changes in plant species richness in a calcareous grassland following changes in environmental conditions. Folia Geobotanica et Phytotaxonomica 29: 469-482.

Huntly, N.J. 1987. Influence of refuging consumers (pikas: *Ochotona princeps*) on subalpine meadow vegetation. Ecology 68: 274-283.

Iason, G. R. & Hester, A. J. 1993. The response of heather (*Calluna vulgaris*) to shade and nutrients - predictions of the carbon-nutrient balance hypothesis. Journal of Ecology 81: 75-80.

Ignatov, M.S. & Afonina, O.M. 1992. Check-list of mosses from the former USSR. Arctoa 1: 1-85.

Ignatova, E.A., Vana, J. & Vorob'eva, F.M. 1990. Bryoflora of the Teberda Reserve. Transactions of Teberda State Reserve [Trudy Teberdinskogo Gos. Zapovednika.]. Stavropol, 12: 1-40. (in Russian).

Igosheva, N.I. 1984. Seasonal dynamics of snowbed alpine meadows in the Polar Ural. pp. 128-135. In: Gorchakovskii, P.L. (ed), Rastitel'nye soobshchestva Urala i ikh antropogennaya degradatsiya. Sverdlovsk (in Russian).

Igosheva, N.I. 1986. On above- and belowground phytomass in kriohil meadows in the Polar Ural. pp. 113-117. In: Kamelin, R.V. (ed.) Plant cover in high mountains [Rastitel'nyi pokrov vysokogorii]. Nauka, Leningrad (in Russian).

Isachenko, T.I. & Lavrenko, E.M. 1980. Botanical and geographical regionalization. pp. 14-20. In: Gribova, S.A. et al. (eds), Vegetation of the European part of the USSR [Rastitel'nost' Evropeiskoi chasti SSSR]. Nauka, Leningrad (in Russian).

Jackson, R. B. & Caldwell, M. M. 1992. Shading and the capture of localized soil nutrients: nutrient contents, carbohydrates, and root uptake kinetics of a perennial tussock grass. Oecologia 91: 457-462.

Janssen, C.R. 1981. On the reconstruction of past vegetation by pollen analysis. Proc. Kon. Nederl. Akad. Wetensch. 84(2): 197-210.

Jasper, D.A., Abbott, L.K. & Robson, A.D. 1989. Soil disturbance reduces the infectivity of external hyphae of vesicular-arbuscular mycorrhizal fungi. New Phytologist 112: 93-99.

Jensch, F. 1843. List of plants and their seeds collected in the southwest Caucasus. Zapiski o-va sel'skogo khozyaistva Yuzhnoi Rossii. (4): 54-94.

Johnson, P.L. & Billings, W.D. 1962. The alpine vegetation of the Beartooth plateau in relation to cryopedogenic processes and patterns. Ecological Monographs 32: 105-135.

Johnsson, D.D. & Cline, A.J. 1965. Colorado mountain soils. Advances in Agronomy 17: 233-281.
Jonasson, S. 1992. Plant responce to fertilization and species removal in tundra related to community structure and clonality. Oikos 63: 420-429.
Jongman, R.H.G., ter Braak, C.J.F. & van Tongeren, O.F.R. (eds). 1995. Data analysis in community and landscape ecology. Cambridge Univ. Press, Cambridge.
Jackson, R.B. & Caldwell, M.M. 1993. The scale of nutrient heterogeneity around individual plants and its quantification with geostatistics. Ecology 74: 612-614.
Janečková, V., Fassatiová, O., Daniel, M. & Křivanec, K. 1977. Nálezy půdních mikromicetů ve Vysokém Himlaji (Nepál). Česká Mykologie 31: 206-213.
Julve, Ph. 1993. Synopsis phytosociologique de la France (communautes de plantes vasculaires). Lejeunia 140: 1-160.
Kal'vish, T.K. 1990. Morphophysiological features of the entomopathogenic fungus *Tolypocladium niveum* (O.Rostr.) Bissett (Deuteromycotina: Moniliales) new to the USSR flora. Mikologiya i Fitopatologiya 24: 210-215 (in Russian).
Karagatzides, J.D., Lewis, M.C. & Schulman, H.M. 1983. Nitrogen fixation in the high arctic tundra at Sarcpa Lake, Northwest Territories. Canadian Journal of Microbiology 29: 851-859.
Karner, P. & Mucina, L. 1993. *Mulgedio-Aconitetea*. S. 468-505. In: Grabherr, G. & Mucina, L. (eds.) Die Pflanzengesellschaften Osterreichs. Gustav Fischer, Jena e.a., Teil 2.
Keddy, P.A. 1989. Competition. Chapman & Hall, London.
Keddy, P. A. 1992. Assembly and response rules: two goals for predictive community ecology. Journal of Vegetation Science 3: 157-164.
Keddy, P.A., Twolan-Strutt, L. & Wisheu, I.C. 1994. Competitive effect and response rankings in 20 wetland plants: are they consistent across three environments? Journal of Ecology 82: 635-643.
Kelly, E.F., Chadwick, O.A. & Hilinski, T.E. 1998. The effect of plants on mineral weathering. Biogeochemistry 42: 21-53.
Keppen, F.P. 1885. Geographic distribution of coniferous trees in European Russia and the Caucasus. Appendix for the 2nd vol. Zapiski Imper. Akademii Nauk 4. Sankt Peterburg (in Russian).
Khapaev, S.A. 1976. Fluctuation of the upper treeline in Teberda Reserve. pp. 55-57. In: Agahanjanz O.E. et al. (eds), Vysokogornaya geoekologiya. Nauka. Moscow (in Russian).
Khramtsov, V.N. 1982. Productivity of *Bromopsis variegata* (Poaceae) in plant communities of Teberda Reserve. Botanicheskii Zhurnal 67: 951-959 (in Russian).
Kirtsideli, I.I. 1999a. Soil micromycetes from tundra of the Khibine Mountains (Kola Peninsula). Mikologiya i Fitopatologiya 33(6): 386-391.
Kirtsideli, I.I. 1999b. Soil micromycetes of arctic tundra on the Taimyr cost of Kara Sea. Mikologiya i Fitopatologiya 33(1): 19-24.
Kirtsideli, I.I. 2001a. Micromycetes of Solnechnaia Bay area (Bol'Shevik Island, Severnaia Zemlia Archipelago). Mikologiya i Fitopatologiya 35(6): 22-28.
Kirtsideli, I.I. 2001b. Soil micromycetes of mountain tundra ecosystems (Polar Ural and Putorana Plateau). Mikologiya i Fitopatologiya 35(4): 48-53.
Kirtsideli, I.I. & Tomilin, B.A. 1997. Soil micromycetes of Severnaya Zemlia Archipelago. Mikologiya i Fitopatologiya 31(6): 1-6.
Kirtsideli, I.I., Vorob'ev, N.I. & Tereshenkov, O.M. 1996. Micromycetes accompliances in different kinds of soil in area of typical tundra on the Northern part of lake Taimyr. Mikologiya i Fitopatologiya 30(2): 9-13 (in Russian).

Kiseleva, N.K. 1992. Botanical and phytolith analysis of zoogenic deposits in North Ossetia. pp. 24-83. In: Dinesman, L.G. (ed.) Istoricheskaya ekologiya dikikh I domashnikh kopytnykh. Istoriya pastbishchnykh ekosistem. Nauka, Moscow (in Russian).

Kjøller, A. & Struwe, S. 1982. Microfungi in ecosystems: fungal occurence and activity in litter and soil. Oikos 39: 389-422.

Klevezal, G.A. & Kleinberg, S.E. 1967. Determination of mammals' age by lamella structure of teeth and bone [Opredelenie vozrasta mlekopitayushchikh po sloistym strukturam zubov i kosti]. Nauka, Moscow (in Russian).

Knyasev, A.V., Savinetsky, A.B. & Gey, N.A. 1992. Holocene history of vegetation in North Osetia. pp. 84-108. In: Dinesman, L.G. (ed.) Istoricheskaya ekologiya dikikh I domashnikh kopytnykh. Istoriya pastbishchnykh ekosistem. Nauka, Moscow (in Russian).

Koch, K.H.E. 1848-1851. Beitrage zu einer Flora des Orientes. Schneider in Comm., Berlin, H. 1-6.

Kohia, M.S. 1987. A role of the Diplopoda in cellulose destruction and their energetic needs on the subalpine meadows of the Great Caucasus. pp. 358-359. In: Striganova B.R. (ed), Soil fauna and soil fertility. 9 Int. soil zool. Colloq., Moscow, Aug. 16-20, 1985. Moscow, (in Russian).

Komárkova, V. 1993. Vegetation type hierarchies and landform disturbance in arctic Alaska and alpine Colorado with emphasis on snow patches. Vegetatio 106: 155-181.

Kononov, V.N. 1957. Vegetation of the Teberda Reserve. Transactions of Teberda State Reserve [Trudy Teberdinskogo Gos. Zapovednika.]. Stavropol, 1: 85-112 (in Russian).

Konstantinova, N.A., Potemkin, A.D. & Shlyakov, A.N. 1992. Check-list of liverworts and anthocerotaes from the former USSR. Arctoa 1: 87-127.

Kontrimavichus, V.L. 1969. Helminthofauna of Mustellids and ways of its formation [Gelmintofauna kun'ikh i puti ee formirovaniya]. Nauka, Moscow (in Russian).

Körner, C. 1982. CO_2-exchange in the alpine sedge *Carex curvula* as influenced by canopy structure, light and temperature. Oecologia 53: 98-104.

Körner, C. 1995. Alpine plant diversity: a global survey and functional interpretations. pp. 45-62. In: Chapin, F.S. & Korner, C. (eds.). Arctic and alpine biodiversity: patterns, causes and ecosystem, consequences. Springer, Berlin e.a.

Körner, C. 1999. Alpine plant life: functional plant ecology of high mountain ecosystems. Springer, Berlin.

Körner, C., Diemer, M., Schappi, B., Niklaus, P. & Arnone, J.A. III 1997. The responses of alpine grassland to four seasons of CO_2 enrichment: a synthesis. Acta Oecologica 18: 165-175.

Körner, C., & Renhardt, U. 1987. Dry matter partitioning and root/leaf area ratios in herbaceous perennial plants. Oecologia 74: 411-418.

Kosenko, I.S. 1970. Flora of vascular plants of the Northwestern Caucasus and Predkavkaze (handbook) [Opredelitel' sosudistykh rastenii severo-zapadnogo Kavkaza i Predkavkaz'ya]. Kolos, Moscow. (in Russian).

Kotov, V.A. 1968. Kuban ibex, its ecology and economic importance. Transactions of Caucasian State Reserve [Trudy Kavkazskogo Gos. Zapovednika.] 10: 201-293.

Kotov, V.A. & Ryabov, L.S. 1963. Trade and valuable mammals of foothills and mountains areas of Krasnodarsk region. Transactions of Caucasian State Reserve [Trudy Kavkazskogo Gos. Zapovednika.]. Maikop, 7: 1-238 (in Russian).

Krahulec, F. & Moravec, J. 1995. Nardo-Callunetea. pp. 81-86. In: Moravec, J. a Kol. Red list of plant communities of the Czech Republic and their endangerment. 2nd ed. Severoceskou Prirodou, Litomeroce.

Kranz, J. 1990. Fungal diseases in multispecies plant communities. New Phytologist 116: 383-405.

Krasnoff, S.B., Gupta, S., StLéger, R.J., Renwick, J.A.A. & Roberts, D.W. 1991. Antifungal and insecticidal properties of the efrapeptins: metabolites of the fungus *Tolypocladium niveum*. Journal of Invertebrate Pathology 58: 180-188.

Krever, V., Zazanashvili, N., Jungius, H., Williams, L. & Petelin, D. (eds). 2001. Biodiversity of the Caucasus Ecoregion. WWF, Baku e.a.

Kudaktin, A.N. 1977. Wolf in the Western Caucasus. Okhota i okhotnich'e khozyaistvo (9): 18-19 (in Russian).

Kudo, G. 1991. Effects of snow-free period on the phenology of alpine plants inhabiting snow patches. Arctic and Alpine Research 23: 436-443.

Kudo, G. & Ito, K. 1992. Plant distribution in relation to the length of the growing season in a snow-bed in the Taisetsu Mountains, Northern Japan. Vegetatio 98: 165-174.

Kurapova, M.V. 1949. Fauna of the Teberda Reserve [Zhivotnyi mir Teberdinskogo zapovednika]. Moscow (in Russian).

Kuthubutheen, A.J. & Pugh, G.J.F. 1979. Effects of temperature and fungicides on *Chrysosporium pannorum* (Link) Hughes. Antonie van Leeuwenhoek 45: 65-79.

Kutschera, L., Lichtenegger, E. & Sobotik, M. 1992 Wurzelatlas mitteleuropaischer Grunlandpflanzen Band 2. Pteridophyta und Dicotyledoneae (Magnoliopsida). Teil 1. Gustav Fischer, Stuttgart.

Kuznetsov, N.I. 1901. Map of botanical-geographical provinces of the Caucasus. Trudy Botan. sada Imperat. Yur'evskogo Univ. 11: 1-5 (in Russian).

Kuznetsov, N.I. 1909. Principles of identifying phytogeographic provinces in the Caucasus. Zapiski Akademii nauk, ser.8, 24(1): 1-177.

Kvavadze, E.S. 1985 Earthworms (Lumbricidae) of the Caucasus [Dozhdevye chervi Kavkaza]. Metsniereba, Tbilisi (in Russian).

Larcher, W. 1995. Physiological Plant Ecology. 3rd edition. Berlin: Springer.

Larin, I.V., Agababyan, S.M., Rabotnov, T.A. et al. 1950-1956. Fodder crops of the hay meadows and pastures of the USSR [Kormovye rasteniya senokosov i pastbishch SSSR]. Selkhozgiz, Moscow-Leningrad, v.1-3 (in Russian).

Larina, N.I. & Tarasov, M.A. 1980. Quantitative estimation of feed and defecation of *Apodemus sylvaticus* L. and *Pitymys majori* Thos. in the Northwest Caucasus. Ekologiya zhivotnykh (Saratov) 6(8): 138-153 (in Russian).

Latter, P.M. & Heal, O.W. 1971. A preliminary study of the growth of fungi and bacteria from temperate and antarctic soils in relation to temperature. Soil Biology and Biochemistry 3: 365-379.

Lavrenko, E.M. 1940. Steppes of the USSR. pp. 1-206. In: Keller, B.A. et al. (eds), Vegetation of the USSR [Rastitel'nost' SSSR], v. 2. Izd-vo AN SSSR, Moscow, Leningrad (in Russian).

Lavrenko, E.M. 1964. Types of vertical belt structure in mountains of the USSR. pp. 189-195. In: Gerasimov, I.P. et al. (eds), Sovremennye problemy geografii. Nauka, Moscow (in Russian).

Law, R. A. 1983. Model for the dynamics of a plant population containing individuals classified by age and size. Ecology 64: 2: 224-230.

Lebedeva, L.P. 1986. Biomorphological features of subalpine plants in the Kirgiziya Range. pp. 141-146. In: Kamelin, R.V. (ed.) Plant cover in high mountains [Rastitel'nyi pokrov vysokogorii]. Nauka, Leningrad (in Russian).

Ledebour, C.F. 1841-1853. Flora Rossica. Schweizerbalt, Stuttgart, v. 1-4.

Lefkovich, L.P. 1965. The study of population growth in organisms grouped by stages. Biometrics 21: 1-18.

Legendre, P. & Legendre, L. 1998. Numerical ecology. 2nd edit. Elsevier, Amsterdam e.a.

Lehman, C.L. & Tilman, D. 1998. Competition in spatial habitats. pp. 185-203. In: Tilman, D. & Kareiva, P. (eds.). Spatial ecology: The role of space in population dynamics and interspecific interactions. Princeton Univ. Press, Princeton.
Leinsoo, T.A., Onipchenko, V.G. & Shtina, E.A. 1987. Soil and epiphytic algae. pp. 31-35. In: Rabotnov, T.A. (ed), Biogeotsenozy alpiiskikh pustoshei. Nauka, Moscow (in Russian).
Leinsoo, T.N. 1994. Soil microfungi in alpine zone, the Northwest Caucasus. PhD dissertation. Moscow University, Biology Faculty.
Leinsoo, T.N., Leinsoo, T.A. & Sogonov, M.V. 1999. Soil fungi of alpine ecosystems in Teberda Reserve, Northwestern Caucasus. pp. 130-145. In: Pavlov, V.N. (ed.), High-mountain ecosystems of the Teberda Reserve. Transactions of Teberda State Reserve [Trudy Teberdinskogo Gos. Zapovednika.]. Moscow, 15. (in Russian).
Leinsoo, T.N., Onipchenko, V.G., Leinsoo, T.A. & Velikanov, L.L. 1991. Biomass variation of soil micromycetes in the alpine communities in the Northwest Caucasus. Mikologiya i fitopatologiya 25: 206-212.
Leishman, M.R. 1999. How well do plant traits correlate with establishment ability? Evidence from a study of 16 calcareous grassland species. New Phytologist 141: 487-496.
Leps, J. & Stursa, T. 1989. Species-area curve, life history strategies and succession: a field test of relationships. Vegetatio 83: 249-257.
Lesica, P. & Antibus, R.K. 1986. Mycorrhizae of alpine fell-field communities on soils derived from crystalline and calcareous parent materials. Canadian Journal of Botany 64: 1691-1697.
Lesica, P. & Steele, B.N. 1996. A method for monitoring long-term population trends: an example using rare arctic-alpine plants. Ecological Applications 6: 879-887.
Leskov, A.I. 1932. Upper limits of the forests in the West Caucasian Mountains. Botanicheskii zhurnal 17: 227-260 (in Russian).
Leslie, P.H. 1945. On the use of matrices in certain population mathematics. Biometrika 33: 183-212.
Lewis, E.G. 1942. On the generation and growth of population. Sankhya: The Indian Journal of Statistics 6: 93-96.
Li, X.-L., George, E. & Marchner, H. 1991. Extension of the phosphorus depletion zone in VA-mycorrhizal white clover in a calcareous soil. Plant & Soil 136: 41-48.
Lipskii, V.I. 1891. Study of the North Caucasus. Zapiski Kievskogo obshchestva estestvoispytatelei 11(2): 23-61 (in Russian).
Logvinenko, O.A. & Onipchenko, V.G. 1999. Seed production of alpine plants. pp. 51-63. In: Pavlov, V.N. (ed.), High-mountain ecosystems of the Teberda Reserve. Transactions of Teberda State Reserve [Trudy Teberdinskogo Gos. Zapovednika.]. Moscow, 15. (in Russian).
Lund, J.W.G. 1945. Observations of soil algae I. New Phytologist 44: 196-219.
Lynov, Y.S. 1982. Phenorhythmotypic composition of middle and high mountain plant communities in the Chatkal Range. Botanicheskii Zhurnal 67: 355-359 (in Russian).
Lynov, Y.S. 1986. Ecological and phenological features of plants and plant communities in the middle and high mountain zones in the Western Tien Shan. Byull. Mosk. o-va ispyt. prir. otd. biol. 91(2): 153-158 (in Russian).
MacArthur, R.H. 1972. Geographical Ecology: Patterns in the Distribution of Species. Harper & Row, New York.
MacArthur, R.H. & Wilson, E.O. 1967. The theory of island biogeography. Princeton Univ. Press, Princeton.
Magomedov, M.-R.D., Akhmedov, A.G., Yarovenko, Y.A. 2001. Daghestan ibex (Dagestanskii tur). Nauka, Moscow (in Russian).

Makarov, M.I., Malysheva, T.I., Haumaier, L., Alt, H.G. & Zech, W. 1997. The forms of phosphorus in humic and fulvic acids of a toposequence of alpine soils in the northern Caucasus. Geoderma 80: 61-73.

Makhova, Yu.V. & Patyk-Kara, N.G. 1961. History of vegetation of the high-mountains of the Greater Caucasus in the Holocene. pp. 125-130. In: Markov, K.K. (ed.), Paleogeographia chetvertichnogo perioda SSSR. Moscow (in Russian).

Maksimova, T.A. 1985. Mycorrhizas of mountain tundra plants in Khakassia. pp. 16-26. In: Selivanov, I.A. (ed), Mikoriza i drugie formy konsortivnykh otnoshenii v prirode. Perm'skii pedagog. institut, Perm' (in Russian).

Maksimova, T.A. 1986. Mycorrhizae of alpine plants in Kuznetskii Alatau, Khakassia. Byull. VNII Sel'skokhoz. Mikrobiol. 3: 54-57 (in Russian).

Maleev, V.P. 1939. On distribution of kolchic floristic elements on the north slope of the Western Caucasus. Izvestiya gosud. geografich. obshchestva 71: 844-855 (in Russian).

Maleev, V.P. 1941. Tertiary relicts in the flora of the West Caucasus and main stages of Quaternary history of its flora and vegetation. Materialy po istorii flory i rastitel'nosti SSSR 1: 61-144.

Maleev, V.P. 1947. Mediterranean forest area. pp. 72-80. In: Lavrenko, E.M. (ed), Geobotanicheskoe raionirovanie SSSR. Izd-vo AN SSSR, Moscow & Leningrad.

Manders, P.T. 1987. A transition matrix model of the population dynamics of the Clanwilliam Cedar *(Widdringtania cedarbergensis)* in natural stands subject to fire. Forest Ecol. and Manag. 20: 171-186.

Maranon, T. & Bartolome, J.W. 1993. Reciprocal transplants of herbaceous communities between *Quercus agrifolia* woodland and adjacent grassland. Journal of Ecology 81: 673-683.

Marchand, P.J. & Roach, D.A. 1980. Reproductive strategies of pioneering alpine species: seed production, dispersal, and germination. Arctic and Alpine Research 12: 137-146.

Mark, A.F. 1970. Floral initiation and development in New Zealand alpine plants. New Zealand Journal of Botany 8: 67-75.

Margalitadze, N.A. & Kimeridze, K. R. 1985. History of the Holocene vegetation of the upper Svanetia. pp. 240-260. In: Nakhutsishvili, G.S. (ed), Flora i rastitelnost' Svanetii. Mezniereba. Tbilisi (in Russian).

Mark, A.F. & Bliss, L.C. 1970. The high-alpine vegetation of Central Otago, New Zealand reserves in narrow-leaved snow tussock (*Chionochloa rigida*). New Zealand Journal of Botany 8: 381-451.

Matvejeva, E.P. 1980. Biologische Gesamtproduktivitat von oberund unterirdischer Phytomasse von Wiesen der Sowjetunion (SU) (Flachlandteil). Phytocoenologia 7: 237-251.

May, D.E., Webber, P.J. & May, T.A. 1982. Success of transplanted alpine plants on Niwot Ridge, Colorado. Univ. Colo. Inst. Arct. and Alp. Res. Occas. Pap. 37: 73-81.

Myers, N., Mittermeier, R.A., Mittermeier, C.G., da Fonseca, G.A.B. & Kent, J. 2000. Biodiversity hotspots for conservation priorities. Nature 403: 853-858.

McDonald, A.W., Bakker, J.P. & Vegelin, K. 1996. Seed bank classification and its importance for the restoration of species-rich flood-meadows. Journal of Vegetation Science 7: 157-164.

McGraw, J.B. 1985. Experimental ecology of *Dryas octopetala* ecotypes: relative response to competitors. New Phytologist 100: 233-241.

McKendrick, J.D., Batzli, G.O., Everett, K.R. & Swansons, J. 1980. Some effects of mammalian herbivores and fertilization on tundra soils and vegetation. Arctic and Alpine Research 12: 565-578.

McLellan, A.J., Fitter, A.H. & Law, R. 1995. On decaying roots, mycorrhizal colonization and the design of removal experiments. Journal of Ecology 83: 225-230.
McMahon, G.A. 1982. Successions of ecosystems: preliminary comparative analysis. pp. 21-38. In: Biosphernyi zapovednik. Trudy 2 sovetsko-amerikanskogo simposiuma, Florida 10-15 March 1980. Leningrad (in Russian).
Medvedev, J.S. 1882. Essay of transcaucasian forests. Sbornik kavkazskogo o-va sel'skogo khozyaistva. Tiflis (in Russian).
Medvedev, J.S. 1907. On vegetation areas in the Caucasus.Vestnik Tiflisskogo botan. sada 8: 5-32 (in Russian).
Mentler, A. & Brauer, C. 1986. Vergleich zweier Pflanzengesellschaften in den hohen Tauern bezuglich der Standortfaktoren und des Futterwertes. Veröff. Osterr. Mass-Programms Osterr. Akad. Wiss. 9: 201-206.
Meunier, J.D. & Colin, F. (eds) 2001. Phytoliths: applications in earth sciences and human history. A.A.Balkema, Lisse.
Meurk, C.D. 1978. Alpine phytomass and primary production in Central Otago, New Zealand. New Zealand Journal of Ecology 1: 27-50.
Meyer, C.A. 1831. Verzeichniss der Pflanzen, welche während der auf Allerhöchsten Befehl in den Jaren 1829 und 1830 unternommenen Reise im Caucasus und in den Provinzen am westlichen Ufer des Caspischen Meeres gefunden und eingesammelt sind. St.-Petersburg.
Meyer, E. 1981. Abundanz und Biomasse von Invertebraten in zentralalpinen Boden (Hohe Tauern, Österreich). In: Franz, H. (Hersg.), Bodenbiologische Untersuchungen in den Hohen Tauern 1974-1978. Veröff. Österr. MaB-Hochgebirgsprogramms Hohe Tauern. Innsbruck:Universitatsverlag Wagner, 4: 153-178.
Mikeladze, R.M. 1960. About the origin of the snowbed communities of the Caucasus. pp. 33-41. In: Proceedings of scientific works of young scientists commemorating 1500 years of Tbilisi and 40 years of Komsomol. Tbilisi. (in Russian).
Miles, J. 1979. Vegetation dynamics. Chapman and Hall, London.
Minckley, T. & Whitlock, C. 2000. Spatial variation of modern pollen in Oregon and southern Washington, USA. Rev. of Palaeobot & Palynol. 112: 97-123.
Mirchink, T.G. 1963. Distribution of fungi in the soils of the Pamir. Biologicheskie Nauki (1): 199-204 (in Russian).
Mirchink, T.G. 1971. Soil fungal flora of the Pamir. Vestnik Moskovskogo Universiteta. Ser. Biologiia i Pochvovedenie (2): 73-79 (in Russian).
Mirchink, T.G. 1988. Soil Mycology (Pochvennaia mikologiia). Izd-vo Moskovskogo Universiteta, Moscow.
Mirkin, B.M. 1983. On types of ecological and coenotic strategies in plants. Zhurnal Obshchei Biologii 44: 603-613 (in Russian).
Mirkin, B.M. 1994. Which plant communities do exist? Journal of Vegetation Science 5: 283-284.
Mirkin, B.M., Solomeshch, A.I., Ishbirdin, A.R. & Alimbekova, L.M. 1989. List and diagnostic criteria of the higher syntaxa of USSR's vegetation. Moscow (in Russian).
Mistral, M., Buck, O., Meier-Behrmann, D.C., Burnett, D.A., Barnfield, T.E., Scott, A.J., Anderson, B.J. & Wilson, J.B. 2000. Direct measurements of spatial autocorrelation at the community level in four plant communities. Journal of Vegetation Science 11: 911-916.
Muller, M. 1987. Bodenbildung auf Silikatunterlage in der Alpinen Stufe des Oberengadins (Zentralalpen, Schweiz). Catena 14: 419-437.
Moen, J. 1989. Diffuse competition - diffuse concept. Oikos 54: 260-263.
Moen, J. 1993. Positive versus negative interactions in a high alpine block field: Germination of *Oxyria digyna* seeds in a *Ranunculus nivalis* community. Arctic and Alpine Research 25: 201-206.

Moen, J. & Oksanen, L. 1998. Long-term exclusion of folivorous mammals in two arctic-alpine plant communities: a test of the hypothesis of exploitation ecosystems. Oikos 82: 333-346.

Monk, C.D. 1966. An ecological significance of evergreenness. Ecology 47:504-505.

Moore, P. 1980. The advantages of being evergreen. Nature 285: 535-536.

Moore, P. 1984. Why be an evergreen? Nature 312: 703.

Morin, H. & Payette, S. 1988. Buried seed populations in the montane, subalpine, and alpine belts of Mont Jaques-Cartier, Quebec. Canadian Journal of Botany 66: 101-107.

Montemartini Corte, A. & Gestro, C. 1992. Fungi from mud of Antarctic ponds. Scienza e Cultura. Proceedings of the 2nd meeting on "Antarctic Biology" (Padova, 26-28 February 1992), 33-42.

Mosca, A.M. 1957. Ricerche sulla microflore del terreno di una valletta nivale nel Parco Nationale del Gran Paradiso. Allionia 3: 83-107.

Mosca, A.M. 1960. Sulla microflora del terreno di un pascolo alpino in val di Lanzo (Alpie Graie). Allionia 6: 17-34.

Mulholland, S.C. 1989. Phytolith shape frequencies in North Dakota grasses: a comparison to general patterns. Journal of Archaeological Sciences 16: 489-511.

Muller, E. & Magnuson, J.A. 1987. On the origin and ecology of Alpine plant parasitic fungi. Arctic and Alpine Mycology: Proc. 2nd Int.Symp. (Fetan, Aug.26-Sept.2, 1984). New York, London, 3-16.

Nakhutsrishvili, G.S. 1975. On ecology of high mountain plants in winter time. Izvestiya Akad. Nauk GruzSSR, ser. biol. 1: 260-264 (in Russian).

Nakhutsrishvili, G.S. 1981. The specific features of structure and development of high mountains plants. pp. 249-264. In: Serebryakova T.I. (ed.) Life forms, structure, spectra and evolution [Zhiznennye formy, struktura, spektry i evolyutsiya]. Nauka, Leningrad (in Russian).

Nakhutsrishvili, G.S., Chkhikvadze, A.K. & Khezuriani, L.D. 1980. Production of high mountain herbaceuos communities in the Central Caucasus [Produktivnost' vysokogornykh travyanykh soobshchestv Tsentral'nogo Kavkaza]. Metsniereba, Tbilisi (in Russian).

Nakhutsrishvili, G.S. & Gamtsemlidze, Z.G. 1984. Plant life under extreme high mountain conditions: an example from the Central Caucasus [Zhizn' rastenii v extremalnyh usloviyakh vysokogorii: na primere Tsentralnogo Kavkasa]. Nauka, Leningrad (in Russian).

Nakhutsrishvili, G.S., Cernusca, A., Decker, P. & Chezuriani, L. 1987. Photosynthesis and respiration in *Caricetum tristis* - community in the alpine belt. pp. 66-77. In: Rabotnov, T.A. (ed), Ekologicheskie issledovaniya vysokogornykh lugov Kazbegi. Metsniereba, Tbilisi (in Russian).

Naniev, V.I. 1986. Ermine in Osset. pp. 61-63. In: Chopikashvili, L.V. et al. (eds), Fauna i ekologiya zhivotnykh Tsentral'nogo Kavkaza. Ordzhonikidze (in Russian).

Narinyan, S.G. 1959a. Preliminary results of ecological study of alpine snowbeds in Aragats. Izvestiya Akad. Nauk ArmSSR (biol. nauki) 12: 3-15 (in Russian).

Narinyan, S.G. 1959b. On some phenological features of alpine snowbed communities in Mt Aragaz. Izvestiya Akad. Nauk ArmSSR (biol. nauki) 12(11): 47-69.

Nasimovich, A.A. 1949. New data on chamois biology in the West Caucasus. Transactions of Caucasian State Reserve [Trudy Kavkazskogo Gos. Zapovednika.]. Moscow, 3: 51-64 (in Russian).

Nelson, W. 1982. Applied Life Data Analysis. John Wiley & Sons, New York.

Nestroy, O. 1984. Aspekte der Bodenentwicklung unter Almen der Ostalpen. Wien. geogr. Schr. 59-60: 67-72.

Netujilin, I.A., Petrova, A.D., Dobroliubova, T.V. & Onipchenko, V.G. 1990. The structure of microarthropod communities in alpine meadows of the Northwestern Caucasus and its changes due to human trampling. Proc. 14th General Meet. Europ. Grassland Feder. Lahti, Finland, 740-742.

Nozadze, L.M. 1968. Mycotrophy of some plants in high mountain vegetation in Kazbek region in connection with a vertical zonation. Uchenye zapiski Permskogo gosud. pedagog. instituta 64: 313-317 (in Russian).

Nye, P.H. 1981. Changes in pH across the rhizosphere inducted by roots. Plant and Soil 61: 7-26.

Oberdorfer, E. 1978. Suddeutsche Pflanzengesellschaften 2 Aufl. Teil 2. Gustav Fischer, Jena.

Oberdorfer, E. 1992. Suddeutsche Pflanzengesellschaften 2 Aufl. Teil 4 Gustav Fischer, Jena.

Ognev, S.I. 1950. Mammals of the USSR and surrounding countries [Zveri SSSR i prilezhashchikh stran]. V. 7 – Rodents. Akademiya Nauk SSSR, Moscow & Leningrad (in Russian).

Ogureeva, G.N. 1999. Type of belt structure in mountains. pp. 32-62. In: Ogureeva, G.N. (ed), Zony i tipy poyasnosti rastitel'nosti Rossii i sopredel'nykh territorii. Integratsiya, Moscow.

Oksanen, L., Fretwell, S.D., Arrud, J. & Niemala, P. 1981. Exploitation ecosystems in gradients of primary productivity. American Naturalist 118: 240-261.

Oksanen. L. & Oksanen, T. 1989. Natural grazing as a factor shaping out barren landscapes. Journal of Arid Environment 17: 219-233.

Oksanen, L. & Ranta, E. 1992. Plant strategies along mountain vegetation gradients: a test of two theories. Journal of Vegetation Science 3: 175-186.

Olshwang, V.N. & Fileva, O.N. 1982. Investigation of invertebrates in the tundras of the Northern Urals. p. 9. In: Postnikov, S.N. (ed), Problemy ekologii zhivotnykh. Sverdlovsk (in Russian).

Olsvig-Whittaker, L. 1988. Relating small-scale vegetation patterns to the environment. pp. 87-84. In: During, H.J. et al. (eds.) Diversity and pattern in plant communities. SPB Acad. Publ., Hague.

Onipchenko, V.G. 1983. Seasonal dynamics of a phytocoenosis of the alpine heathland in the Northern Caucasus. Byull. Mosk. o-va ispyt. prir. otd. biol. 88(5): 106-114 (in Russian).

Onipchenko, V.G. 1984. Experimental investigation of alpine lichen heath's structure. pp. 78-81. In: Doklady Mosk. o-va ispyt. prir. Zoologia i botanika. Nauka, Moscow (in Russian).

Onipchenko, V.G. 1985. The structure, phytomass and production of alpine lichen heaths. Byull. Mosk. o-va ispyt. prir. otd. biol. 90(1): 59-66 (in Russian).

Onipchenko, V.G. 1987a. Lichens and higher plants. pp. 19-30. In: Rabotnov, T.A. (ed), Biogeotsenozy alpiiskikh pustoshei. Nauka, Moscow (in Russian).

Onipchenko, V.G. 1987b: Mechanisms of ecological niches separation in terrestrial plants. Zhurnal obshchei biologii 48(5): 687-695 (in Russian).

Onipchenko, V.G. 1990. Phytomass of the alpine communities in the Northwestern Caucasus. Byull. Mosk. o-va ispyt. prir. otd. biol. 95(6): 52-62 (in Russian).

Onipchenko, V.G. 1994a. Study area and general description of the investigated communities. Veröff. des Geobot. Inst. der ETH, Stiftung Rübel. Zürich 115: 6-22.

Onipchenko, V.G. 1994b. The spatial structure of the alpine lichen heaths (ALH): hypothesis and experiments. Veröff. des Geobot. Inst. der ETH, Stiftung Rübel. Zürich 115: 100-111.

Onipchenko, V.G. 1994c. The structure and dynamics of alpine plant communities in the Teberda Reserve, the Northwestern Caucasus. Oecologia Montana 3: 40-50.

Onipchenko, V.G. 1996. Mycosymbiotrophism and vegetative mobility of alpine plants in the Northwestern Caucasus. Vestnik Moskovskogo Universiteta, ser. 16 Biology (1): 43-46 (in Russian).

Onipchenko, V.G. 2002. Alpine vegetation of the Teberda Reserve, the Northwestern Caucasus. Veröff. des Geobot. Inst. der ETH, Stiftung Rübel. Zürich 130: 1-168.

Onipchenko, V.G., Blinkova, O.V., Zakharov, A.A., Blinnikov, M.S. & Abzeeva, M.A. 2001. Dynamics of two contrasted alpine communities in the Caucasus: results of 14-years of permanent plot observations. Oecologia Montana 10: 1-8.

Onipchenko, V. G., Blinnikov, M.S. & Semenova, G.V. 1994. Shading experiments in alpine grasslands. Veröff. des Geobot. Inst. der ETH, Stiftung Rübel. Zürich 115: 89-99.

Onipchenko, V.G., Blinnikov M.S. & Aksenova, A.A. 2001a. Experimental evaluation of shading effects in seasonal dynamics of four alpine communities in northwestern Caucasus, Russia. Arctic, Antarctic, and Alpine Research 33: 330-339.

Onipchenko, V.G. & Golikov, K.A. 1996. Microscale revegetation of alpine lichen heath after wild boar digging: fifteen years of observation on permanent plots. Oecologia Montana 5: 35-39.

Onipchenko, V.G. & Ignatova, E.A. 1996. New species of vascular plants and bryophytes in the Teberda Reserve. Byull. Mosk. o-va ispyt. prir. otd. biol. 101(5): 92-98. (in Russian).

Onipchenko, V.G. & Komarov, A.S. 1997. Population dynamics and life history features of three alpine plant species in the Northwest Causasus. Zhurnal obshchei biologii 58: 64-75 (in Russian).

Onipchenko, V.G., Makarov, M.I. & van der Maarel, E. 2000. Change soil properties by alpine plant species: a monocultural experiment. Proceedings 41 IAVS Symposium, Opulus Press, p. 18-20.

Onipchenko, V.G., Makarov, M.I. & van der Maarel, E. 2001b. Influence of alpine plants on soil nutrient concentrations in a monoculture experiment. Folia Geobotanica 36: 225-241.

Onipchenko, V.G. & Pokarzhevskaya, G.A. 1994. "Mass-effect" in alpine communities of the Northwestern Caucasus. Veröff. des Geobot. Inst. der ETH, Stiftung Rübel. Zürich 115: 61-68.

Onipchenko, V.G. & Rabotnova, M.V. 1994. Natural "gaps" in alpine meadows and plant population strategies. Veröff. des Geobot. Inst. der ETH, Stiftung Rübel. Zürich 115: 83-88.

Onipchenko, V.G. & Semenova, G.V. 1995. Comparative analysis of the floristic richness of alpine communities in the Caucasus and the Central Alps. Journal of Vegetation Science 6: 299-304.

Onipchenko, V.G., Guzhova, G.A., Semenova, G.V. & Rabotnova, M.V. 1991. Population strategies of alpine plants of the Northwestern Caucasus. pp. 165-180. In: Shilov, I.A. (ed), Population ecology [Ekologiya populyatsii]. Nauka, Moscow (in Russian).

Onipchenko, V.G., Semenova, G.V. & van der Maarel, E. 1998a. Population strategies in severe environments: alpine plants in the Northwestern Caucasus. Journal of Vegetation Science 9: 27-40.

Onipchenko, V.G. & Ustinova, Y.A. 1993. Seasonal dynamics of alpine plant communities in the Teberda Reserve, the Northwestern Caucasus. Byull. Mosk. o-va ispyt. prir. otd. biol. 98(6): 71-79. (in Russian).

Onipchenko, V.G., Vertelina, O.S. & Makarov, M.I. 1998b. Spatial heterogeneity of alpine communities and soil properties. Pochvovedenie (6): 689-695 (in Russian).

Onipchenko, V.G.& Zhakova, O.E. 1997. The structure of large soil invertebrate communities (mesofauna) in the alpine ecosystems of the Teberda Reserve, the Northwestern Caucasus. Oecologia Montana 6: 35-38.

Onipchenko, V.G. & Zobel, M. 2000. Mycorrhiza, vegetative mobility and responses to disturbance of alpine plants in the Northwestern Caucasus. Folia Geobotanica 35: 1-11.

Ovchinnikov, P.N. 1941. *Sibbaldia tetrandra* Bge. and problem of cryophilic vegetation origin in Middle Asia. Sovetskaya botanika 1-2: 145-152 (in Russian).

Owen, W.R. 1995. Growth and reproduction in an alpine cushion plant: *Astragalus kentrophyta* var. *implexus*. Great Basin Naturalist 55: 117-123.
Partridge, T.R., & Wilson, J.B. 1988. The use of field transplants in determining environmental tolerance in salt marshes of Otago, New Zealand. New Zealand Journal of Botany 28: 183-192.
Paulech, C., Romaczewska-Salata, J. & Salata, B. 1991. Fytopatogenne mikromycety cel'ade Erysiphaceae rozsirene v oblasti Vysokich Tatier. Zbornik prac o Tatranskom Narodnom Parku 31: 70-71.
Pavlov, N.V. 1948. Botanical geography of the USSR [Botanicheskaya geografiya SSSR]. Izd-vo AN KazSSR, Alma-Ata (in Russian).
Pavlov, V.N. 1986. Ecological and biological features of subnival plants in the North-west Pamir. pp. 204-209. In: Kamelin, R.V. (ed.) Plant cover in high mountains [Rastitel'nyi pokrov vysokogorii]. Nauka, Leningrad (in Russian).
Pavlova, I.V. & Onipchenko, V.G. 1992. Holocene dynamics of alpine vegetation in the Northwest Caucasus. pp. 109-129. In: Dinesman, L.G. (ed.) Istoricheskaya ekologiya dikikh I domashnikh kopytnykh. Istoriya pastbishchnykh ekosistem. Nauka, Moscow (in Russian).
Perelman, A.I. 1975. Geochemistry of landscapes. [Geokhimiya landshaftov]. Vysschaya shkola, Moscow. (in Russian).
Petrov, G. 1880. Upper part of the Kuban river – Karachai. Pamyatnaya knizhka Kubanskoi oblasti na 1880 (Ekaterinodar): 13-17 (in Russian).
Petrova, G.K. Phenological types of plants of the cryophilic meadow. 1990. Izvestiya Akad. Nauk TadzhSSR, otd. biol. (2): 3-7 (in Russian).
Petersen, H. & Luxton,M. 1982.A comparative analysis of soil fauna populations and their role in decomposition processes. Oikos 39: 288-388.
Pfandenhauer, J. & Maas, D. 1987. Samenpotential im Niedermoorboden des Alpenvorlandes bei Grunlandnutzung unterschiedlicher Intensitat. Flora 179(2): 85-97.
Pianka, E.R. 1970. On r- and K-selection. American Naturalist 104: 592-597.
Pianka, E.R. 1983. Evolutionary Ecology. 3nd ed. Harper and Row, New York.
Pinder, J.E.III. 1975. Effects of species removal on an old-field plant community. Ecology 56: 747-751.
Piperno, D. 1988. Phytolith analysis: an archeological and geological perspective. Academic Press. San Diego.
Pokarzhevskaya, G.A. 1995. Morphological analysis of alpine communities of the North-Western Caucasus. Folia Geobotanica et Phytotaxonomica 30: 197-210.
Pokarzhevskaya, G.A. 1998a. Influence of the dominant grass Festuca varia Haeke on the spatial patterns of alpine grasslands in the Northwestern Caucasus, Russia. Arctic and Alpine Research 30: 11-18.
Pokarzhevskaya, G.A. 1998b. Patterns of variation in snowbed vegetation of the northwestern Caucasus. Patterns and process in Alpine communities of the Northwestern Caucasus, Russia. Acta Universitatis Upsaliensis, Uppsala 400: 1-22.
Pokarzhevskaya, G.A. & Onipchenko, V.G. 1995. Biomorphological analisys of alpine communities in the Northwestern Caucasus. Byull. Mosk. o-va ispyt. prir. otd. biol. 100(2): 50-58 (in Russian).
Pokarzhevskiy, A.D. 1985. Geochemical ecology of terrestrial animals [Geokhimicheskaya ekologiya nazemnykh zhivotnykh]. Nauka, Moscow (in Russian).
Politova, N.Y., Luk'yanova, L.M. & Nikonov, V.V. 1989. Store and structure of plant organic matter in mountain tundras of the Far North. pp. 64-65. In.: Actual problems of biology and rational use of natural resources in Karelia. Conf. abstr. Petrozavodsk (in Russian).
Polivanova, N.N. 1989. Teberda Reserve [Teberdinskii Zapovednik]. Cherkessk.

Polivanova, N.N. 1990. Teberda State Reserve. pp. 18-34. In: Sokolov, V.E. & Syroechkovskii, E.E. (eds.), Zapovedniki Kavkaza. Mysl', Moscow (In Russian).
Polozova, T.G. 1981. Life forms of vascular plants in different subzones of the Taimyr tundra. pp. 265 - 280. In: Serebryakova, T.I. (ed.) Life forms: structure, spectra, and evolution [Zhiznennye formy: srtruktura, spektry i evolutsiya]. Nauka, Moscow (in Russian).
Polushina, N.A. 1975. Burrowing activity of mammals on the Carpatian mountains. pp. 119-121. In: Zlotin, R.I., Isakov, Y.A. & Khodashova, K.S. (eds), Rol' zhivotnykh v funktsionirovanii ekosistem. Mater. soveshch. Nauka, Moscow.
Polynov, V.V. 1956. Selected Proceedings. [Izbrannye trudy]. Izdatel'stvo AN SSSR, Moskva. (in Russian).
Popkova, I.F. 1967. Chamois on the south slopes of the Grate Caucasian Range (ecology, morphology and economic importance). Transactions of Teberda State Reserve [Trudy Teberdinskogo Gos. Zapovednika.]. Moscow, 7: 160-211 (in Russian).
Popov, K.P. 1983. Wintergreen plants in the flora of the North Caucasus. pp. 71-90. In: Semenov-Tyan-Shanskii, O.I. (ed), Seasonal and between year's dinamics in National Reserves of Russian Federation [Sezonnaya i raznogodichnaya dinamika v zapovednikakh RSFSR]. CNIL Glavokhoty RSFSR, Moscow (in Russian).
Posch, A. 1977: Bodenkundliche Untersuchungen im Bereich der Glocknerstrasse in den Hohen Tauern (2300-2600 m MH). In: Cernusca, A. (ed.) Alpine Grasheide Hohe Tauern. Ergebnisse der Okosystemstudie. Innsbruck, 1: 111-121.
Pott, R. 1995. Die Pflanzengesellschaften Deutschlands. 2, uberarb. Auflage. Eugen Ulmer, Stuttgart.
Prikhod'ko, S.L. 1991. Mikromitsety dikorastushchikh travyanistykh rastenii basseina reki Ala-Archa (severnyi sklon Kirgizskogo khrebta). Ilim, Bishkek (in Russian).
Prock, S. & Körner, C. 1996. A cross-continental comparison of phenology, leaf dynamics and dry matter allocation in arctic and temperate zone herbaceous plants from contrasting altitudes. Ecological Bulletin 45: 93-103.
Pumpel, B. 1977. Bestandesstruktur, Phytomassevorrat und Produktion verschiedener Pflanzengesellschaften im Glocknergebiet. S. 83-101. In: Cernuska, A. (Schr.) Alpine Grasheide Hohe Tauern. Ergebnisse der Okosystemstudie Veröff. Osterr. MaB-Hochgebirgsprogramms Hohe Tauern. Universitatsverlag Wagner, Innsbruck, Bd. 1.
Rabotnov, T.A. 1950. The life cycle of perennial herbaceous plants in meadow coenoses. Trudy Botanicheskogo instituta Akademii Nauk SSSR (ser. 3) 6: 7-204 (in Russian).
Rabotnov, T.A. 1958. On the methodology of studies of the germinable seed contents in soils of meadows. Botanicheskii Zhurnal 43: 1572-1581 (in Russian).
Rabotnov, T.A. 1964. Determination of age composition of species populations in a community. Polevaya Geobotanika 3: 132-208 (in Russian).
Rabotnov, T.A. 1972. Studies of fluctuation (inter-year dynamics) of phytocoenoses. Polevaya geobotanica 4: 95-136 (in Russian).
Rabotnov, T.A. 1973. Influence of mineral fertilizers on meadow plants and plant communities [Vliyanie mineral'nykh udobrenii na lugovye rasteniya i lugovye fitotsenosy]. Nauka, Moscow (in Russian).
Rabotnov, T.A. 1975a. On the cyclic variability of phytocoenosysis. Ekologia (5): 5-8. (in Russian).
Rabotnov, T.A. 1975b. The study of coenotic populations for elucidation of "strategy of life" of plant species. Byull. Mosk. o-va ispyt. prir. otd. biol. 80(2): 5-17 (in Russian).
Rabotnov, T.A. 1982. Viable seeds in soils of natural biogeocoenoses of the USSR. pp. 35-59. In: Drosdov, N.N. (ed), Teoreticheskie i prikladnye aspekty biogeografii. Nauka, Moscow (in Russian).

Rabotnov, T.A. 1983. Phytocoenology [Phytocenologiya]. Izdat. Moscow Univ., Moscow (in Russian).
Rabotnov, T.A. 1985. On plant strategy types. Ekologiya (3): 3-12 (in Russian).
Rabotnov, T.A. 1986. Viable seeds in the soils of the natural biogeocoenoses. Byull. Mosk. o-va ispyt. prir. otd. biol. 91(3): 3-18 (in Russian).
Rabotnov, T.A. (ed). 1987. Biogeotsenozy alpiiskikh pustoshei. Nauka, Moscow (in Russian).
Rabotnov, T.A. 1993. Experience of using ecological scales to study plants' patience. Russian Journal of Ecology 25: 9-13.
Rabotnova M.V., Onipchenko, V.G.& Ustinova, J.A. 1992. Influence of artificial shading on alpine lichen heaths. Vestnik Moskovskogo Universiteta. Biologia (1): 57-65 (In Russian).
Radde, G.I. 1901. Main features of plant world in the Caucasus. Kollektsii Kavkazskogo muzeya. Botanika (Tiflis). 2: 1-101 (in Russian).
Rall, G. 1965. Soil fungi from the alpine zone of the Medicine Bow Mountains, Wyoming. Mycologia 57: 872-881.
Randall, R.E., Andrew, R. & West, R.G. 1986. Pollen catchment in relation to local vegetation. New Phytologist 104: 271-310.
Raunkiaer, C. 1934. Life forms of plants and statistical plant geography. Clarendon Press, Oxford.
Ramenskii, L.G. 1938. Introduction to complex soil and geobotanical investigation of lands [Vvedenie v kompleksnoe pochvenno-geobotanicheskoe izuchenie zemel']. Selhozgiz, Moscow (in Russian).
Read, D. J. & Birch, C.P.D. 1988. The effects of implications of disturbance of mycorrhizal mycelial systems. Proc. Roy. Soc. Edinburgh 94B: 13-24.
Read, D.J. & Haselwandter, K. 1981. Observations on the mycorrhizal status of some alpine plant communities. New Phytologist 81: 341-352.
Reisigl, H. & Keller, R. 1987. Alpenpflanzen im Lebensraum: Alpine Rasen, Schutt- und Felsvegetation. Gustav Fischer, Stuttgart, New York.
Retzer, J.L., 1956. Alpine soils of the Rocky Mountains. Journal of Soil Science 7: 22-32.
Retzer, J.L., 1974. Alpine soils. pp. 771-802. In: Ives, J.D. & Barry, R.G. (eds.) Arctic and alpine environments. Methuen and Co., London.
Reynolds, D.G. 1984. Population dynamics of three annual species of alpine plants in the Rocky Mountains. Oecologia 62: 250-255.
Reynolds, H.L. Hungate, B.A. Chapin, F.S. & D'Antonio, C.M. 1997. Soil heterogenety and plant competition in an annual grassland. Ecology 78: 2076-2090.
Rhodes, L.H. & Gerdemann, J.W. 1975. Phosphate uptake zones of mycorrhizal and nonmycorrhizal onions. New Phytologist 75: 555-561.
Robbins, W.L. 1957. Physiological aspects of ageing in plants. Amer. J. of Botany 44: 289-294.
Roberts, H.A. 1981. Seed banks in soil. Advances in Applied Biology 6: 1-55.
Romanovsky, Y. E. 1985. Food limitation and life-history strategies in cladoceran crustaceans. Arch. Hydrobiol. Beih. Ergebn. Limnol. 21: 363-372.
Romanovsky, Y. E. 1989a. Competition for a fluctuating resource: evolutionary and ecological consequences. Zhurnal Obshchei Biologii 50: 304-315 (in Russian).
Romanovsky, Y. E. 1989b. The current state of the life-history strategy concept. Biologicheskie Nauki (11): 18-31 (in Russian).
Romanovsky, Yu. E. 1992. Competition, productivity, and species diversity in natural communities. pp. 139-152. In: Yurtsev, B.A. (ed), Biologicheskoe raznoobrazie: podkhody k izucheniyu i sokhraneniyu (Biological diversity: approaches to investigation and conservation). Sankt-Peterburg (in Russian).

Roxburgh, S.H., Wilson, J.B. & Mark, A.F. 1988. Succession after disturbance of a New Zealand high-alpine cushionfield. Arctic and Alpine Research 20: 230-236.

Runge, F. 1998. Bibliography of phytolith research. University of Paderborn, Paderborn.

Ryser, P. 1990. Influence of gaps and neighbouring plants on seedling establishment in limestone grassland. Veröff. des Geobot. Inst. der ETH, Stiftung Rübel. Zürich 104: 1-71.

Sandra, V., Popescu, A. & Peicea, I. 1988. Les associations de la classe Salicetea herbaceae Br.-Bl. 47 des Carpates roumaines. Rev. roum. biol. S`er. biol.veg. 33: 93-102.

Sarukhan, J. & Gadgil, M. 1974. Studies in plant demography: *Ranunculus repens* L., *Ranunculus bulbosus* L., *Ranunculus acris* L. III. A mathematical model incorporating multiple models of reproduction. Journal of Ecology 62: 921-936

Savina, L. P. 1982. Taiga forests of North Asia in the Holocene. Nauka. Novosibirsk (in Russian).

Savinetsky, A.B. 1992. Centennial changes of the pasture ecosystems in North Osetia. pp. 165-171. In: Dinesman, L.G. (ed.) Istoricheskaya ekologiya dikikh I domashnikh kopytnykh. Istoriya pastbishchnykh ekosistem. Nauka, Moscow (in Russian).

Schappi, B. & Korner, C. 1996. Growth responses of an alpine grassland to elevated CO_2. Oecologia 105: 43-52.

Schappi, B. & Korner, C. 1997. In situ effects of elevated CO_2 on the carbon and nitrogen status of alpine plants. Functional Ecology 11: 290-299.

Schatz, H. 1981. Abundanz, Biomasse und Respirationsrate der Arthropoden-Mesofauna im Hochgebirge (Obergurgl, Tiroler Zentralalpen). Pedobiologia 22: 52-70.

Schibler, W. 1937. Flora von Davos. Verzeichnis der Gefasspflanzen der Landschaft Davos und der angrenzenden Gebiete. Bischofberger & Co, Chur.

Schimpf, D.J. & Bayn, R.L. 1983. Reproductive attributes of some Rocky Mountain subalpine herbs in successional context. Great Basin Naturalist 43: 573-578.

Schröter, C. 1926. Pflanzenleben der Alpen. 2 Aufl. Raustein, Zürich.

Sedel'nikov, V.P. 1979. Flora and vegetation of high mountain area of the Kuznetskii Alatau. [Flora i rastitel'nost' vysokogorii Kuznetskogo Alatau] Nauka, Novosibirsk. (in Russian).

Sedel'nikov, V.P. 1985. Production of high mountain plant communities in Altai-Sayan Region. Geographiya i prirodnye resursy 1: 87-91 (in Russian).

Sedelnikov, V.P. 1988. High mountain vegetation of the Altai and Sajan mountains [Vysokogornaya rastitel'nost' Altae-Sayanskoi gornoi oblasti] Nauka, Novosibirsk. (In Russian).

Seel, W.E. & Press, M.C. 1994. Influence of the host on three subarctic annual facultative root hemiparasites II. Gas exchange characteristics and resource use-efficiency. New Phytologist 127: 37-44.

Seliskar, D.M. 1985. Effect of reciprocal transplanting between extremes of plant zones on morphometric plasticity of five plant species in an Oregon salt marsh. Canadian Journal of Botany 63: 2254-2262.

Selivanov, I.A. 1981. Mycosymbiotrophy as a form of consortic relationships in vegetation of the Soviet Union [Mikosimbiotrofizm kak forma konsortivnykh svyazei v rastitel'nom pokrove Sovetskogo Soyuza. Nauka, Moscow (in Russian).

Semenova, G.V. & Onipchenko, V.G. 1989. The composition and dynamics of coenotic populations of dominants of alpine communities in the Teberda National Reserve. pp. 105-118. In: Sokolov, V. E. (ed) Populatsionnye issledovaniya rasteniy v zapovednikach. Nauka, Moscow (in Russian).

Semenova, G.V. & Onipchenko, V.G. 1990. Soil seed banks in the alpine communities in the Teberda Reserve, the Northwestern Caucasus. Byull. Mosk. o-va ispyt. prir. otd. biol. 95(5): 77-87 (in Russian).

Semenova, G.V.& Onipchenko, V.G. 1991. Field investigation of the soil diasporae bank in alpine communities. Byull. Mosk. o-va ispyt. prir. otd. biol. 96(4): 117-122 (in Russian).
Semenova, G.V. & Onipchenko, V.G. 1994. Soil seed banks. Veröff. des Geobot. Inst. der ETH, Stiftung Rübel. Zürich 115: 69-82.
Semenova, G.V. & Onipchenko, V.G. 1996. Soil seed bank of an alpine lichen heath in the Northwestern Caucasus: species richness. Oecologia Montana 5: 83-86.
Sennov, A.V. & Onipchenko, V.G. 1994. Experimental research of alpine communities with use of reciprocal transplantations. Veröff. des Geobot. Inst. der ETH, Stiftung Rübel. Zürich 115: 41-60.
Serebryakov, A.K. 1957. Soils of the Teberda Reserve. Transactions of Teberda State Reserve [Trudy Teberdinskogo Gos. Zapovednika.]. Stavropol, 1: 51-84 (in Russian).
Serebryakov, I.G. 1962. Ecological morphology of plants [Ecologicheskaya morphologiya rastenii]. Nauka, Moscow (in Russian).
Serebryakova, T.I. 1971. Morphogenesis and evolution of the grass life forms [Morfogenez Pobegov i Evolutsia Zhiznennych Form Zlakov]. Nauka, Moscow (in Russian).
Serebryakova, T.I. 1977. On main "architectural" models and possibilities of their change. Byull. Mosk. o-va ispyt. prir. otd. biol. 82(5): 112-128 (in Russian).
Serebryakova, T.I. 1981. Life forms and models of shoot formation in aboveground-creeping herbs. pp. 78-97. In: Serebryakova, T.I. (ed.) Life forms: structure, spectra, and evolution [Zhiznennye formy: srtruktura, spektry i evolutsiya]. Nauka, Moscow (in Russian).
Serebryanny, L.R., Gey, N.A. & Dzhaparidze, R.N. 1980. Vegetation of the central Caucasus in the Holocene. Byulleten komissii po izucheniyu chetvertichnogo perioda 50: 123-137 (in Russian).
Sergeev, M.G., Kopaneva, L.M., Rubtsov, I.A. et al. 1995. Siberian grasshopper (*Aeropus sibiricus* L.) [Sibirskaya kobylka]. Nauka, Novosibirsk (in Russian).
Sewell, G.W.F. 1959. Ecology of *Mucor ramannianus* Moller. Nature 183: 1344-1345.
Shalnev, V.A., Serebryakov, A.K. & Chikalin, A.N. 1977. Mountain meadow soils of the Mt. Malaya Khatipara. Transactions of Teberda State Reserve [Trudy Teberdinskogo Gos. Zapovednika.]. Stavropol, 9: 88-104. (in Russian).
Shannon, C.E. & Weaver, W. 1964. The mathematical theory of communication. University of Illinois Press, Urbana, IL.
Shatvorjan, P.V. 1981. Influence of different fertilizers on production of alpine meadows with *Carum caucasicum*. Armen. agricul. inst. Erevan 4: 105-111 (In Russian).
Shaver, G.R. & Billings, W.D. 1975. Root production and root turnover in a wet tundra ecosystem, Barrow, Alaska. Ecology 56: 401-409.
Shavkunova, V.F. 1987. Mycosymbiotrophy of plants in mountain meadows. pp. 54-56. In: Selivanov, I.A. (ed), Mykoriza i drugie formy konsortivnykh svyazei v prirode. Permskii Pedagogicheskii Institut, Perm' (in Russian).
Shcherbakova, A.A. 1979. History of the bonany in Russia (until 60s XIX century) [Istoriya botaniki v Rossii]. Nauka, Novosibirsk.
Shcherbakova, E.M. 1973. Pleistocene glaciation of the Great Caucasus [Drevnee oledenenie Bol'shogo Kavkaza]. Moscow Univ. Press, Moscow (In Russian).
Shennikov, A. P. 1942. Natural factors of plant distribution from experimental perspective. Zhurnal Obshchei Biologii 3: 331-361 (in Russian).
Shevtsova, A., Ojala, A., Neuvonen, S., Vieno, M. & Haukioja, E. 1995. Growth and reproduction of dwarf shrubs in a subarctic plant community: annual variation and above-ground interactions with neighbours. Journal of Ecology 83: 263-275.
Shiffers, E.V. 1953. Vegetation and natural forage areas of the North Caucasus [Rastitel'nost' Severnogo Kavkaza i ego kormovye ugod'ya]. Izd-va Akad.Nauk SSSR, Moscow.

Shiosaka H., & Shibata, O. 1993. Morphological changes in *Polygonum cuspidatum* Sieb. et Zucc. reciprocally transplanted among different altitudes. Japanese Journal of Ecology 43: 31-37.

Shlothauer, S.D. 1990. Vegetation of the suboceanic high mountains. [Rastitel'nost' suokeanicheskikh vysokogorii]. Nauka, Moscow (in Russian).

Shmida, A A. & Ellner, S. 1984. Coexistence of plant species with similar niches. Vegetatio 58: 29-55.

Shtina, E.A. 1984. Methods of soil algae investigation. pp. 58-72. In: Mishustin, E.N. (ed), Mikroorganizmy kak komponent biogeotsenoza. Nauka, Moscow (in Russian).

Shtina, E.A. & Gollerbakh, M.M. 1976. Ecology of soil algae [Ekologiay pochvennykh vodoroslei]. Nauka, Moscow (in Russian)..

Shtina, E.A., Onipchenko, V.G. & Rasran, L.M. 1995. Soil algae of alpine communities in Teberda Reserve (the NW Caucasus). Algologia 5: 17-28 (in Russian).

Silander, J.A. & Antonovicz, J. 1982. Analysis of interspecific interactions in a coastal plant community: A perturbation approach. Nature 298: 557-560.

Simpson, R.L., Leck, M.A. & Parker, V.T. 1989. Seed banks: general concepts and methodological issues. pp. 3-8. In: Leck, M.A., Parker, V.T. & Simpson, R.L. (eds), Ecology of soil seed banks. Academic Press, New York.

Sizov., I.E. Komarov, A.S. & Onipchenko, V.G. 1999a. Life span estimations of three alpine non clonal plants (*Anemone speciosa, Campanula tridentata, Carum caucasicum*). pp. 64-72. In: Pavlov, V.N. (ed), High-mountain ecosystems of the Teberda Reserve. Transactions of Teberda State Reserve [Trudy Teberdinskogo Gos. Zapovednika.]. Moscow, 15. (in Russian).

Sizov, I.E. Onipchenko, V.G. & Komarov, A.S. 1999b. Life span indirect evaluation of three alpine perennial plants. Oecologia Montana 8: 21-26.

Smith, B., Mark, A. F. & Wilson, J. B. 1995. A functional analysis of New Zealand alpine vegetation: variation in canopy roughness and functional diversity in response to an experimental wind barrier. Functional Ecology 9: 904-912.

Smith, J.M.B. & Klinger, L.F. 1985. Aboveground:belowground phytomass ratios in Venezuelan paramo vegetation and their significance. Arctic and Alpine Research 17: 189-198.

Smith, S.E. & Read, D.J. 1996. Mycorrhizal symbiosis. Academic Press, London.

Smith, S.E., Robson, A.D. & Abbott, L.K. 1992. The involvment of mycorrhizas in assessment of genetically dependent efficiency of nutrient uptake and use. Plant & Soil 146: 169-179.

Smith, T. M. & Huston, M. A. 1989. A theory of the spatial and temporal dynamics of plant communities. Vegetatio 83: 49-69.

Soares, G.G., Pinnock, D.E. & Samson, R.A. 1979. *Tolypocladium*, a new fungal pathogen of mosquito larvae with promise for use in microbial control. Proc. Calif. Mosquito Vector Control Assoc. 47: 51-54.

Sochava, V.B. 1946. Some botanical and geographical results of the expedition of Akademy of Sciences USSR to the Northwest Caucasus in 1945. Sovetskaya botanika 3: 182-187.

Sochava, V.B. 1949. On origin of *Fagus*-forests in the Caucasus. Izvestiya AN SSSR, ser.biol. 2: 224-236 (in Russian).

Sohlberg, E.H. & Bliss, L.C. 1984. Microscale pattern of vascular plant distribution in two high arctic plant communities. Can. J. Bot. 62: 2033-2042.

Sokolov, V.E. & Tembotov, A.K. 1989. Mammals: Insectivorous [Mlekopitayushchie: Nasekomoyadnye]. Nauka, Moscow (in Russian).

Sokolov, V.E. & Tembotov, A.K. 1993. Mammals: Ungulata [Mlekopitayushchie: Kopytnye]. Nauka, Moscow (in Russian).

Sokolova, N.S. 1982. About the history of vegetation of the Greater Caucasus. Vestnik MGU seriya 5 Geographiya 5: 40-44 (in Russian).
Solbrig, O.T. 1980. Demography and natural selection. pp. 1-20. In: Solbrig, O.T. (ed), Demography and evolution in plant populations. Blackwell, Oxford.
Sommier, C.P.S. & Levier, E. 1900. Enumeratio plantarum anno 1890 in Caucaso lectarum. Acta Hort. Petrop. 16: 1-586.
Spassov, N.& Barichnikov, G. 1986. Premiere decouverte de l'hermine sur le versant meridional du Grand Caucase (Georgie). J. Mammalia 50: 123-125.
Stanton, M.L., Rejmanek, M. & Galen, C. 1994. Changes in vegetation and soil fertility along a predictable snowmelt gradient in the Mosquito Range, Colorado, USA. Arctic and Alpine Research 26: 364-374.
Stanyukovich, K.B. 1955. Main types of vertical belt structure in the mountains of USSR. Izvestiya Vsesoyuznogo geograficheskogo obshchestva 87(3): 232-244 (in Russian).
Stanyukovich, K.B. 1960. High mountain vegetation of the USSR. [Rastitelnost visokogorii SSSR]. Acad.sci. Tadjik SSR, Stalinobad (in Russian).
Stanyukovich, K.B. 1973. Mountain vegetation of the USSR [Rastitel'nost' gor SSSR]. Donish, Dushanbe.
Stebaev, I.V. & Pshenitcina, L.B. 1984. Opportunities of study of feeding specialization by grasshoppers (Orthoptera, Acridoidea) on the basis of excrement analysis. Entomologicheskoe obozrenie 63: 434-446 (in Russian).
Steiner, G.M. 1993. Scheuchzerio-Caricetea fuscae. S. 131-165. In: Grabherr, G & Mucina, L. (eds.) Die Pflanzengesellschaften Osterreichs. Gustav Fischer, Jena. Teil 2.
Stepanov, A.L. & Onipchenko, V.G. 1989. Assessment of respiration rate and nitrogen-fixing and denitrifying activity in high-meadow alpine soils of Northwestern Caucasus. Vestnik Moskovskogo Universiteta. Pochvovedenie 44(2): 55-56.
Stolyarov, M.V. 1975. Role of Orthoptera insects as consuments in ecosystems of the high mountain area, Lagodekh Reserve. Izvestiya AN GruzSSR, sr. biol. 1(1): 51-55 (in Russian).
Stoyke, G. & Currah, R.S. 1991. Endophytic fungi from the mycorrhizae of alpine ericoid plants. Canadian Journal of Botany 69: 347-352.
Striganova, B.R. 1980. Nutrition of soil saprophages [Pitanie pochvennykh saprofagov]. Nauka, Moscow. (in Russian).
Striganova, B.R. & Loginova, N.G. 1984. The role of Diplopoda in biological turnover of alpine meadows ecosystems of the Small Caucasus. Zhurnal Obshchei biologii 45: 196-202 (in Russian).
Strom, L. 1995. Root exudation of organic acids: importance to nutrient availability and the calcifuge and calcicole behaviour of plants. Oikos 80: 459-466.
Stuefer, J.F., De Kroon, H. & During, H.J. 1996. Exploitation of environmental heterogeneity by spatial division of labour in a clonal plant. Functional Ecology 10: 328-334.
Stuefer, J.F. & Hutchings, M.J. 1994. Environmental heterogeneity and clonal growth: a study of the capacity for reciprocal translocation in *Glechoma hederacea* L. Oecologia 100: 302-308.
Sukhova, D.V., Onipchenko, V.G. & Sukhov, S.V. 1997. Does "island effect" influence the species richness of a plant community? An alpine grassland as an example. Byull. Mosk. o-va ispyt. prir. otd. biol. 102(2): 48-53 (in Russian).
Svensson, B.M. & Callaghan, T. V. 1988. Small-scale vegetation pattern related to the growth of *Lycopodium annotinum* and variations in its micro-environment. Vegetatio 76: 167-177.
Tanasevich, A.V. 1990. Spiders of Linyphiidae of the Caucasus (Arachnida, Aranei). pp. 5-114. In: Striganova, B.R. (ed.), Fauna i ekologiya nazemnykh bezpozvonochnykh Kavkaza. Nauka, Moscow (in Russian).

Tarasov, M.A. 1973. Some data on ungulate population dynamics in the Teberda State Reserve. pp. 147-148. In: Sokolov, V.E. (ed), Redkie vidy mlekopitayushchikh fauny SSSR i ikh okhrana. Nauka, Moscow (in Russian).
Tembotov, A.K. 1960. Ermine from the Central Caucasus. Byull. Mosk. o-va ispyt. prir. otd. biol. 65(5): 105-106 (in Russian).
Tembotov, A.K. 1977. Ermine. The North-Western Caucasus and Pre-Caucasus. pp. 126-127. In: Nasimovich, A.A. (ed), Kolonok, gornostai, vydra. Nauka, Moscow (in Russian).
ter Braak, C.J.F. 1987. CANOCO – A FORTRAN program for canonical community ordination by (particular) (detrended) (canonical) correspondence analysis, principal component analysis and redundancy analysis (ver. 2.1). TNO Institute of Applied Computer Sci., Wageningen.
Theodose, T.A. & Bowman, W.D. 1997. The influence of interspecific competition on the distribution of an alpine graminoid: evidence for the importance of plant competition in the extreme environment. Oikos 79: 101-114.
Thomas, B.D. & Bowman, W.D. 1998. Influence of N2-fixing Trifolium on plant species composition and biomass production in alpine tundra. Oecologia 115: 26-31.
Thompson, K. 1992. The functional ecology of seed banks. pp. 231-258. In: Fenner, M. (ed), Seeds: the Ecology of Regeneration in Plant Communities. CAB International, Wallingford.
Thompson, K. 1993. Persistence in soil. pp. 199-202. In: Hendry, G.A.F. & Grime, J.P. (eds) Methods in Comparative Plant Ecology. Chapman & Hall, London.
Thompson, K., Band, S.R. & Hodgson, J.G. 1993. Seed size and shape predict persistence in soil. Functional Ecology 7: 236-241.
Thompson, K. & Grime, J.P. 1979. Seasonal variation in the seed banks of herbaceous species in ten contrasting habitats. Journal of Ecology 67: 893-921.
Thompson, K., Hillier, S.H., Grime, J.P., Bossard, C.C. & Band, S.R. 1996: A functional analysis of a limestone grassland community. Journal of Vegetation Science 7: 371-380.
Tilman, D. 1982. Resource competition and community structure. Princeton Univ. Press, Princeton NJ.
Tilman, D. 1988. Plant strategies and the dynamics and structure of plant communities. Princeton Univ.Press, Princeton, New Jersey.
Tilman, D. 1990. Constraints and tradeoffs: toward a predictive theory of competition and succession. Oikos 58: 3-15.
Tilman, D. 1994. Competition and biodiversity in spatially structured habitats. Ecology 75: 2-16.
Tilman, D. & Wedin, D. 1991. Dynamics of nitrogen competition between successional grasses. Ecology 72: 1038-1049.
Tkachenko, V.I. 1962. Ermine in the North-Western Caucasus. Transactions of Teberda State Reserve [Trudy Teberdinskogo Gos. Zapovednika.]. Stavropol, 4: 214-218 (in Russian).
Tosca, C. & Labrone, L. 1981. Le cycle de l'azote dans les milieux supra-forestiers des Pyrenees Centrales: contribution a l'evaluation des gain. Annales de Biologie 20: 44-55.
Traaen, A.E. 1914. Untersuchungen über Bodenpilze aus Norwegen. Nyt Magazin for Naturvidenskaberne 52: 19-121.
Tsareva, V.T. 1978. Distribution and shoot formation of *Luzula spicata* (L.) DC. in the Khibiny Mountains. pp. 39-56. In: Luk'yanova, L.M. (ed), Botanicheskie issledovaniya za polyarnym krugom. Apatity (in Russian).
Tsepkova, N.L. 1986. An ancient agriculture and grazing ecosystems in the Central Caucasus. Trans. High Mountain Geophys. Inst.[Trudy vysokogornogo geophizicheskogo instituta] 64: 47-65 (in Russian).

Tsuyuzaki, S. 1991. Species turnover and diversity during early stages of vegetation recovery on the volcano Usu, northern Japan. Journal of Vegetation Science 2: 301-306.
Tubaki, K. 1961. On some fungi isolated from the Antarctic materials. Biological results of the Japanese Antarctic research expedition. Special publications from the Seto Marine Biological Laboratory 14: 1-9.
Tumadzhanov, I.I., 1940. Beech and beech forest in the mountain Dagestan. Trudy Tbilis. botan. instituta 7: 57-80 (in Russian).
Tumadzhanov, I.I., 1959. On the history of plant landscape development in the west part of the Skalistyi Range. Trudy Tbilis. botan. instituta 20: 363-410 (in Russian).
Tumadzhanov, I.I. 1960a. On typology of subalpine elfin woods in the Teberda Reserve. Transactions of Teberda State Reserve [Trudy Teberdinskogo Gos. Zapovednika.]. Stavropol, 2: 63-105 (in Russian).
Tumadzhanov, I.I. 1960b. Elfin woods in the Teberda basin in connection with landscape changes in subalpine belt on the north slope of the Great Caucasus. Problemy botaniki 5: 148-162 (in Russian).
Tumadzhanov, I.I. 1963. Opyt drobnogo geobotanicheskogo raionirovaniya severnogo sklona Bol'shogo Kavkaza (na primere Karachaya). Izd-vo AN GruzSSR, Tbilisi (in Russian).
Tumadzhanov, I.I. 1980. Caucasian meadows and elfinwoods. pp. 198-202. In: Gribova, S.A. et al. (eds), Vegetation of the European part of the USSR [Rastitel'nost' Evropeiskoi chasti SSSR]. Nauka, Leningrad (in Russian).
Tushinskii, G.K. 1957. Geomorphology of the Teberda Reserve. Transactions of Teberda State Reserve [Trudy Teberdinskogo Gos. Zapovednika.]. Stavropol, 1: 3-49 (in Russian).
Tushinskii, G.K. 1962. Rythms in glaciation and snowness in Teberda State Reserve. Transactions of Teberda State Reserve [Trudy Teberdinskogo Gos. Zapovednika.]. Stavropol, 4: 57-71 (In Russian).
Twiss, P.C., Suess,C.E. & Smith, R.M. 1969. Morphological classification of grass phytoliths. Soil Sci. Soc. Amer. Proc. 33: 109-115.
Tyurin, I.V. 1937. On the biological accumulation of silica in soils. Problemy sovetskogo pochvovedeniya 4: 3-23 (in Russian).
Uranov, A.A. 1975. Age spectrum of phytocoenopopulations as a function of time and energy wave processes. Biologicheskie Nauki (2): 7-34 (in Russian).
Uranov, A.A. 1977. The problems of investigation of structure of phytocenosys and coenopopulations of species. pp. 8-20. In: Uranov, A.A., Zaugolnova, L.B, Smirnova O.V. & Serebriakova, T.I. (eds), Tsenopopulyatsii rastenii. Razvitie i vzaimootnosheniya. Nauka, Moscow (in Russian).
Uranov, A.A., Zaugolnova, L.B, Smirnova, O.V. & Serebriakova, T.I. (eds) 1977. Coenopopulations of Plants. Development and Interrelations. [Tsenopopulyatsii rastenii] Nauka, Moscow (in Russian).
Urbanska, K.M. 1984. Plant reproductive strategies. pp.211-228. In: Grant, F.W. (ed), Plant biosystematics. Academic Press, Canada.
Urbanska, K.M. 1986. High altitude revegetation research in Switzerland – problems and perspectives. Veröff. des Geobot. Inst. der ETH, Stiftung Rübel. Zürich 87: 155-167.
Urbanska, K.M. 1995. Ecological restoration above the timberline and its demographic assessment. pp. 15-36. In: Urbanska, K.M., & Grodzinska, K. (eds.). Restoration Ecology in Europe. Geobotanical Institute SFIT, Zürich.
Urbanska, K. & Schütz, M. 1986. Reproduction by seed in alpine plants and revegetation research above timberline. Botanica Helvetica 96: 43-60.
Utyakov, P.A. 1960. Some observation under snow avalanches and their influence on the vegetation in the Teberda Reserve. Transactions of Teberda State Reserve [Trudy Teberdinskogo Gos. Zapovednika.]. Stavropol, 2: 43-60 (in Russian).

van der Maarel, E. 1988. Floristic diversity and guild structure in the grasslands of Öland's Stora Alvar. Acta Phytogeogr. Suec. 76: 53-65.
van der Maarel, E., Janssen, J.G.M. & Louppen, J.M.W. 1978. TABORD, a program for structuring phytosociological tables. Vegetatio 38: 143-156.
van Oorschot, C.A.N. 1980. A revision of *Chrysosporium* and allied genera. Studies in Mycology 20: 1-89.
Väre, H., Vestberg, M. & Ohtonen, R. 1997. Shifts in mycorrhiza and microbial activity along an oroarctic altitudinal gradient in Northern Fennoscandia. Arctic and Alpine Research 29: 93-104.
Vertelina, O.V., Onipchenko, V.G. & Makarov, M.I. 1996. Primary minerals and processes of weathering in high-mountain soils of the Teberda Reserve. Moscow University Soil Sci. Bull. [Vestnik Movskovskogo Universiteta. Pochvovedenie] 51: 1-8. (in Russian).
Vladychenskii, A.S. & Grishina, L.A. 1987. Soils of the Teberda Reserve. pp.65-87. In: Grishina L.A. (eds), Dynamics, soil structure and processes. [Dinamika, struktura pochv i sovremennye pochvennye protsessy]. ZNIL Glavokhoty RSFSR,. Moscow (in Russian).
Voeykov, A.I. 1871. Beitrage zur Kentniss der Wald- und Regenzonen des Kaukasus. Zeitschrift der österreichischen Gesellschaft für Meteorologie 4(14): 241-246.
Volkov, A.V. 1999. Effect of vegetation and relief on properties of high-mountain soils. pp. 14-41. In: Pavlov, V.N. (ed.), High-mountain ecosystems of the Teberda Reserve. Transactions of Teberda State Reserve [Trudy Teberdinskogo Gos. Zapovednika.]. Moscow, 15. (in Russian).
Volkova, E.V. & Onipchenko, V.G. 1999. Experimental transplantation among the alpine communities: 9-year's results. pp. 86-105. In: Pavlov, V.N. (ed.), High-mountain ecosystems of the Teberda Reserve. Transactions of Teberda State Reserve [Trudy Teberdinskogo Gos. Zapovednika.]. Moscow, 15. (in Russian).
Volkova, E.V., Onipchenko, V.G. & Blinnikov, M.S. 1995. Subrecent phytolith assemblages of alpine communities in the Teberda Nature Reserve, the north-western Caucasus, Russia. Oecologia Montana 4: 1-8.
Volkova, E.V., Onipchenko, V.G. & Burtin, A.Y. Competition in alpine plant communities: results of 7-year transplantation experiment (Floristic richness and shoot biomass). Byull. Mosk. o-va ispyt. prir. otd. biol. 104(2): 21-28 (in Russian).
Vorob'eva, F.M. 1977. Vegetation of the alpine zone in the Teberda Reserve. Transactions of Teberda State Reserve [Trudy Teberdinskogo Gos. Zapovednika]. Stavropol', 9: 37-87 (in Russian).
Vorob'eva, F.M. & Kononov, V.N. 1991. Flora of the Teberda Reserve (vascular plants). Transactions of Teberda State Reserve [Trudy Teberdinskogo Gos. Zapovednika.]. Stavropol, 13. (in Russian).
Vorob'eva, F.M. & Onipchenko, V.G. 2001. Vascular plants of Teberda Reserve. Flora i fauna zapovednikov. Moscow 99: 1-100 (in Russian).
Vorob'eva, F.M., Onishchenko, V.V., Onipchenko, V.G., Malyshev, A.A. & Khramtsova, N.F. 1986. Green cover of the Teberda. pp. 50-72. In: Polivanova, N.N. (ed), Zapovednaya Teberda. Stavropol'skoe knizhnoe izd-vo, Stavropol'.
Voronina, I.N., Onipchenko, V.G. & Ignat'yeva, O.V. 1986. Components of the biological cycle in alpine lichen barrens of the Nortwestern Caucasus. Sov. Soil Sci. 18(3): 20-29.
Voskanyan, V.E. 1977. On flora and vegetation of upper part of alpine zone, Mt.Kaputdzhukh. Biolog. Zhurnal Armenii 30(12): 33-36 (in Russian).
Voskanyan, V.E. & Ziroyan, A.N. 1979. Production and phytomass of main plant communities on the south slope in Mt Aragaz. Problemy botaniki 14(1): 144-147 (in Russian).

Wallnoffer, S. 1993. Vaccinio-Piceetea. S. 283-337. In: Mucina, L. et al. (eds.) Die Pflanzengesellschaften Osterreichs. Gustav Fischer, Jena e.a., Teil 3.
Walter, H. & Breckle, S.-W. 1994. Ökologie der Erde. 2 Aufl. Bd. 3. Gustav Fischer, Stuttgart, Jena.
Walter, H., Harnickell, E. & Mueller-Dombois, D. 1975. Klimadiagramm-Karten. Gustav Fischer, Stuttgart.
Warcup, J.H. 1957. Studies on the occurrence and activity of fungi in a wheat-field soil. Trans. Brit. Mycol. Soc. 40: 237-262.
Watt, A.S. 1947. Pattern and process in the plant community. Journal of Ecology 35: 1-22.
Wedin, D. & Tilman, D. 1993. Competition among grasses along a nitrogen gradient: initial conditions and mechanisms of competition. Ecol. Monogr. 63: 199-229.
Weiser, J. 1987. Mosquito-killing activity of strains of *Tolypocladium cylindrosporium* and *T.niveum*. Česká Mykologie 41: 219-224.
Weiser, J. & Matha, V. 1988. The insecticidal activities of cyclosporines on mosquito larvae. Journal of Invertebrate Pathology 51: 92-93.
Weiser, J. & Pillai, J.S. 1981. *Tolypocladium cylindrosporum* (Deuteromycetes, Moniliaceae), a new pathogen of mosquito larvae. Entomophaga 26: 357-361.
White, J. (ed) 1985. The Population Structure of Vegetation. Junk, Dordtrecht e.a.
Whittaker, R.H. & Goodman, D. 1979. Classifying species according to their demographic strategy. I. Population fluctuations and environmental heterogeneity. American Naturalist 113: 185-200.
Widden, P. & Parkinson, D. 1979. Populations of fungi in a high arctic ecosystems. Canadian Journal of Botany 57: 2408-2417.
Wielgolaski, F.E. 1980. Tundra plant structure and production in relation to the environment. Int. Journal of Biometeorology 24: 23-30.
Wielgolaski, F. E. & Benedict, F. 1986. Transplantation of mountain vegetation. pp. 292-297. In: Proceedings of the international symposium on mountain vegetation. Botanical Society of China.
Wielgolaski, F. E., and Benedict, F. 1993. Transplantation of mountain plant communities. p. 70. In Gilmanov, T.et al. (eds), Abstracts, Global change and Arctic terrestrial ecosystems. Oppdal, Norway.
Wilding, L.P. & Drees, L.R. 1971. Biogenic opal in Ohio soils. Soil Sci. Soc. Amer. Proc. 35: 1004-1010.
Wilkinson, D.M. 1998. The evolutionary ecology of mycorrhizal networks communities: a test of the hypothesis of exploitation ecosystems. Oikos 82: 407-410.
Willard, B.E. & Marr J.W. 1971. Recovery of alpine tundra under protection after damage by human activities in the Rocky Mountains of Colorado. Biol. Conservation 3: 181-190.
Wilson, J.B. 1990. Mechanisms of species coexistence: twelve explanations for Hutchinson's 'Paradox of the plankton': evidence from New Zealand plant communities. New Zealand Journal of Ecology 13: 17-42.
Wilson, S. D. & Tilman, D. 1995. Competitive responses of eight old-field plant species in four environments. Ecology 76: 169-171.
Wirth, V. 1980. Flechtenflora. Ulmer, Stuttgart.
Yang Fu-tun, Wang, O., & Shi, S. 1987. The allocation of the biomass and energy in Kobresia humilis meadow, Haibei district, Oinghai province. Acta Phytoecol. et geobot. sin. 11: 106-112.
Yoshinari, T., Hynes, R. & Knowles, R. 1977. Acetylene inhibition of nitrous oxide reduction and measurement of denitrification and nitrogen fixation in soil. Soil Biology and Biochemistry 9: 177-183.

Zabawski, J. 1982. Soil microfungi of peats in the Hornsund region (West Spitsbergen). Acta Universitatis Wratislaviensis 525: 269-279.

Zakharov, A.A., Abzeeva, M.A. & Onipchenko, V.G. 2001. Natural dynamics of an alpine snowbed community. Byull. Mosk. o-va ispyt. prir. otd. biol. 106(5): 74-82 (in Russian).

Zar, J.H. 1999. Biostatistical Analysis. 4-th edit. Prentice-Hall, Upper Saddle River.

Zaugolnova, L.B., Sugorkina, N.S. & Shcherbakova, E.G. 1991. Life forms and population behaviour of perennial herbaceous plants. pp. 5-22. In: Shilov, I.A. (ed.) Population ecology [Ekologiya populatsii] Nauka, Moscow (in Russian).

Zaugolnova, L.B., Zhukova, A.A., Komarov, A.S. & Smirnova, O.V. 1988. Coenotic populations of plants (essays on population biology) [Tsenopopulyatsii rastenii. Ocherki populyatsionnoi biologii]. Nauka, Moscow (in Russian).

Zechmeister, H. 1993. *Montio-Cardaminetea*. S. 213-240. In: Grabherr, G. & Mucina, L. (eds.) Die Pflanzengesellschaften Osterreichs. Gustav Fischer, Jena e.a., Teil 2.

Zenyakin, S.A. & Onipchenko, V.G. 1997. Burrowing activity of the caucasian mole (*Talpa caucasica* Satunin) on an alpine meadow in the Teberda Reserve, the Northwestern Caucasus. Byull. Mosk. o-va ispyt. prir. otd. biol. 102: 52-53. (in Russian).

Zhang, Y. & Welker, J. M. 1996. Tibetan alpine tundra responses to simulated changes in climate: aboveground biomass and community responses. Arctic and Alpine Research 28: 203-209.

Zharkov, I.V. 1959. On relationship between chamois and domestic animals. Transactions of Caucasian State Reserve [Trudy Kavkazskogo Gos. Zapovednika.] 5: 3-38 (in Russian).

Zhmylev, P.Y., Alekseev, Y.E. & Karpukhina, E.A. 1993. Main terms and concepts of modern plant biomorphology [Osnovnye terminy i ponyatiya sovremennoi biomorfologii rastenii]. Izd-vo Moskovskogo universiteta, Moscow (in Russian).

Zhukova, L.A. 1986. Multiplicity of ontogeny's ways of meadow plants. pp. 104-114. In: Serebryakova, T.I. (ed), Zhiznennye formy v ekologii i systematike rastenii. Prometei, Moscow (in Russian).

Zhukova, L.A. & Komarov, A.S. 1990. Multiplicity of ontogeny's ways, and the dynamics of plant coenopopulations. Zhurnal obshchei biologii 51: 450-461 (in Russian).

Zimina, R.P. & Yasnyi, E.V. 1977. New findings of rare mammals in the Bolshoi Caucasus. pp. 63-64. In: Pokrovskii, A.V. (ed), Ekologiya, methody izucheniya i organizatsiya okhrany mlekopitayushchikh gornykh oblastei. Sverdlovsk (in Russian).

Zingg, Th. 1961. Beitrag zum Klima von Weissfluhjoch. Winterberg. Eidg. Inst. Schnee- und Lawinenforsch. 24: 102-127.

Ziroyan, A.N. 1981. On the method for aboveground production determination in Aragats grasslands. Biologicheskii Zhurnal Armenii 34: 390-395 (in Russian).

Ziroyan, A.N. 1983. Seasonal dynamisc of semidesert, meadow-steppe and alpine plants in Mt Aragats. Biologicheskii Zhurnal Armenii 36: 212-218 (in Russian).

Ziroyan, A.N. 1988. Phytocoenotic properties and productivity of main vegetation types of Armenia. Autoreferat doct. disser. Erevan. (in Russian).

Zlotin, R.I. 1975. Life in high mountains [Zhizn' v vysokogor'yakh]. Mysl', Moscow (in Russian).

Zlotin, R.I. 1984. Functional role of herbivores in terrestrial ecosystems and its geographical specificity. pp. 22-31. In: Abaturov, B.D. & Tishkov, A.A. (eds), Rastitel'noyadnye zhivotnye v biogeotsenozakh sushi. Mater. Vsesoyuz. sov. Moscow (in Russian).

Zobel, K., Zobel, M. & Rosen, E. 1994. An experimental test of diversity maintenance mechanisms, by a species removal experiment in a species-rich wooded meadow. Folia Geobotanica et Phytotaxonomica 29: 449-457.

Zvyagintsev, D.S. (ed) 1991. Methods of soil microbiology and biochemistry [Metody pochvennoi mikrobiologii i biokhimii]. 2nd ed. Izd-vo Moskovskogo Universiteta, Moscow (in Russian).

Zvyagintsev, D.S., Aseeva, I.V., Bab'eva, I.P. & Mirchink, T.G. 1980. Methods in Soil Microbiology and Biochemistry biochemistry [Metody pochvennoi mikrobiologii i biokhimii]. Izd-vo Moskovskogo Universiteta, Moscow (in Russian).

Species index

Plants and fungi

Animals

Geobotany

1. J.B. Hall and M.D. Swaine (eds.): *Distribution and Ecology of Vascular Plants in a Tropical Rain Forest.* Forest Vegetation in Ghana. 1981 ISBN 90-6193-681-0
2. W. Holzner and M. Numata (eds.): *Biology and Ecology of Weeds.* 1982 ISBN 90-6193-682-9
3. N.J.M. Gremmen: *The Vegetation of the Subantarctic Islands Marion and Prince Edward.* 1982 ISBN 90-6193-683-7
4. R.C. Buckley (ed.): *Ant-Plant Interactions in Australia.* 1982 ISBN 90-6193-684-5
5. W. Holzner, M.J.A. Werger and I. Ikusima (eds.): *Man's Impact on Vegetation.* 1983 ISBN 90-6193-685-3
6. P. Denny (ed.): *The Ecology and Management of African Wetland Vegetation.* 1985 ISBN 90-6193-509-1
7. C. Gómez-Campo (ed.): *Plant Conservation in the Mediterranean Area.* 1985 ISBN 90-6193-523-7
8. J.B. Faliński: *Ecological Studies in Białowieza Forest.* 1986 ISBN 90-6193-534-2
9. G.A. Ellenbroek: *Ecology and Productivity of an African Wetland System.* The Kafue Flats, Zambia. 1987 ISBN 90-6193-638-1
10. J. van Andel, J.P. Bakker and R.W. Snaydon (eds.): *Disturbance in Grasslands.* Causes, Effects and Processes. 1987 ISBN 90-6193-640-3
11. A.H.L. Huiskes, C.W.P.M. Blom and J. Rozema (eds.): *Vegetation Between Land and Sea.* Structure and Processes. 1987 ISBN 90-6193-649-7
12. G. Orshan (ed.): *Plant Pheno-morphological Studies in Mediterranean Type Ecosystems.* 1988 ISBN 90-6193-656-X
13. B. Dell, J.J. Havel and N. Malajczuk (eds.): *The Jarrah Forest.* A Complex Mediterranean Ecosystem. 1988 ISBN 90-6193-658-6
14. J.P. Bakker: *Nature Management by Grazing and Cutting.* 1989 ISBN 0-7923-0068-8
15. J. Osbornová, M. Kovářová, J. Lepš and K. Prach (eds.): *Succession in Abandoned Fields.* Studies in Central Bohemia, Czechoslovakia. 1990 ISBN 0-7923-0401-2
16. B. Gopal (ed.): *Ecology and Management of Aquatic Vegetation in the Indian Subcontinent.* 1990 ISBN 0-7923-0666-X
17. B.A. Roberts and J. Proctor (eds.): *The Ecology of Areas with Serpentinized Rocks.* A World View. 1991. ISBN 0-7923-0922-7
18. J.T.A. Verhoeven (ed.): *Fens and Bogs in the Netherlands.* Vegetation, History, Nutrient Dynamics and Conservation. 1992 ISBN 0-7923-1387-9
19. Woo-seok Kong and D. Watts: *The Plant Geography of Korea.* With an Emphasis on the Alpine Zones. 1993 ISBN 0-7923-2068-9

20. R. Aerts and G.W. Heil (eds.): *Heathlands.* Patterns and Processes in a Changing Environment. 1993 ISBN 0-7923-2094-8

21. W. van Vierssen, M. Hootsmans and J. Vermaat (eds.): *Lake Veluwe, a Macrophyte-dominated System under Eutrophication Stress.* 1994 ISBN 0-7923-2320-3

22. Y. Laumonier: *The Vegetation and Physiography of Sumatra.* 1997 ISBN 0-7923-3761-1

23. C.M. Finlayson and I. von Oertzen (eds.): *Landscape and Vegetation Ecology of the Kakadu Region, Northern Australia.* 1996 ISBN 0-7923-3770-0

24. R. Peters: *Beech Forests.* 1997 ISBN 0-7923-4485-5

25. S.A. Ghazanfar and M. Fisher (eds.): *Vegetation of the Arabian Peninsula.* 1998 ISBN 0-7923-5015-4

26. P.D. Gunin, E.A. Vostokova, N.I. Dorofeyuk, P.E. Tarasov and C.C. Black (eds.): *Vegetation Dynamics of Mongolia.* 1999 ISBN 0-7923-5582-2

27. Y. Gutterman: *Regeneration of Plants in Arid Ecosystems Resulting from Patch Disturbance.* 2001 ISBN 0-7923-6725-1

28. J. Kolbek, M. Šrůtek and E.O. Box (eds.): *Forest Vegetation of Northeast Asia.* 2003 ISBN 1-4020-1370-1

29. V.G. Onipchenko (ed.): *Alpine Ecosystems in the Northwest Caucasus.* 2004 ISBN 1-4020-2382-0

KLUWER ACADEMIC PUBLISHERS – DORDRECHT / BOSTON / LONDON

GPSR Compliance

The European Union's (EU) General Product Safety Regulation (GPSR) is a set of rules that requires consumer products to be safe and our obligations to ensure this.

If you have any concerns about our products, you can contact us on ProductSafety@springernature.com

In case Publisher is established outside the EU, the EU authorized representative is:

Springer Nature Customer Service Center GmbH
Europaplatz 3
69115 Heidelberg, Germany

Batch number: 10762579

Printed by Printforce, the Netherlands